Software and Programming Tools in Pharmaceutical Research

Edited by

Dilpreet Singh
Department of Pharmaceutics
ISF College of Pharmacy
Moga, India

&

Prashant Tiwari
Department of Pharmaceutical Sciences
Dayanand Sagar University
Bengaluru, India

Software and Programming Tools in

Pharmaceutical Research

Editors: Dilpreet Singh and Prashant Tiwari

ISBN (Online): 978-981-5223-01-9

ISBN (Print): 978-981-5223-02-6

ISBN (Paperback): 978-981-5223-03-3

need for a court order if at any point you breach any terms of this License Agreement. In no event will any delay or failure by Bentham Science Publishers in enforcing your compliance with this License Agreement constitute a waiver of any of its rights.

3. You acknowledge that you have read this License Agreement, and agree to be bound by its terms and conditions. To the extent that any other terms and conditions presented on any website of Bentham Science Publishers conflict with, or are inconsistent with, the terms and conditions set out in this License Agreement, you acknowledge that the terms and conditions set out in this License Agreement shall prevail.

Bentham Science Publishers Pte. Ltd.
80 Robinson Road #02-00
Singapore 068898
Singapore
Email: subscriptions@benthamscience.net

BENTHAM SCIENCE

CONTENTS

PREFACE i

LIST OF CONTRIBUTORS ii

CHAPTER 1 INTRODUCTION TO COMPUTER-BASED SIMULATIONS AND METHODOLOGIES IN PHARMACEUTICAL RESEARCH 1
Samaresh Pal Roy
1. INTRODUCTION 2
    1.1. Types of Computer-Based Simulations in Pharmaceutical Research 2
        1.1.1. Challenges and Limitations of Computer-Based Simulations 3
        1.1.2. Advances in Computer-Based Simulations 3
2. MOLECULAR MODELLING: PRINCIPLES AND APPLICATIONS IN DRUG DISCOVERY 4
    2.1. Principles of Molecular Modelling 4
    2.2. Applications of Molecular Modelling in Drug Discovery 4
    2.3. Molecular Modeling Techniques 5
3. COMPUTER-AIDED DRUG DESIGN: CONCEPTS AND TECHNIQUES 6
    3.1. Principles of Computer-Aided Drug Design 7
        3.1.1. Virtual Screening 7
    3.2. Applications of Computer-Aided Drug Design in Drug Discovery 7
        3.2.1. Computer-Aided Drug Design Techniques 8
        3.2.2. Molecular Docking: Predicting Protein-ligand Interactions 11
        3.2.3. Quantitative Structure-Activity Relationship (QSAR) Modeling 13
        3.2.4. Virtual Screening: Accelerating Drug Discovery Through Computational Techniques 14
        3.2.5. Standardization of Methods for Data Collection and Analysis 16
CONCLUSION 16
ACKNOWLEDGEMENT 17
REFERENCES 17

CHAPTER 2 TOOLS FOR THE CALCULATION OF DISSOLUTION EXPERIMENTS AND THEIR PREDICTIVE PROPERTIES 25
Ram Babu S., Sakshi T. and Amardeep K.
1. INTRODUCTION 25
2. THEORIES OF DISSOLUTION 27
    2.1. The Diffusion Layer Model 27
    2.2. The Interfacial Barrier Model 27
    2.3. Danckwert's Model 28
3. IVIVC (*IN VITRO - IN VIVO* CORRELATION) STUDIES 29
4. EXPERIMENTAL TECHNIQUES FOR DISSOLUTION 30
5. FUNDAMENTALS OF DISSOLUTION TESTING 32
    5.1. Signifance of Dissolution Testing 33
    5.2. Product Stability 33
    5.3. Comparability Assessment 33
    5.4. Noyes-Whitney Rule 34
    5.5. Nernst and Brunner Film Theory 34
6. MATHEMATICAL AND STATISTICAL TOOLS FOR *IN VITRO* DISSOLUTION METHODS 34
    6.1. Kinetics in Dosage Form 34
    6.2. Empirical and Semi-empirical Mathematical Modeling 35
        6.2.1. Zero order Kinetics 35

6.2.2. *First-order Kinetics* 35

6.2.3. *Higuchi Model* 36

6.2.4. *Hixson- Crowell Model* 37

6.2.5. *Korsmeyer-Peppas Model* 37

6.2.6. *Baker Lonsdale Model* 38

6.2.7. *Weibull Model* 38

6.2.8. *Hopfenberg Model* 38

7. PREDICTION OF *IN VITRO* DISSOLUTION STUDIES USING ADVANCED

MEASUREMENTS TECHNIQUES 39

CONCLUSION 40

CONSENT FOR PUBLICATON 41

CONFLICT OF INTEREST 41

ACKNOWLEDGEMENT 41

REFERENCES 41

CHAPTER 3 THE ROLE OF PRINCIPAL COMPONENT ANALYSIS IN

PHARMACEUTICAL RESEARCH: CURRENT ADVANCES 45

Diksha Sharma, Anjali Sharma, Punam Gaba, Neelam Sharma, Rahul Kumar

Sharma and *Shailesh Sharma*

1. INTRODUCTION 45

1.1. Definition of PCA 46

1.1.1. Definition 46

1.1.2. Goals 46

1.2. History of PCA 47

2. TERMINOLOGY IN THE PCA ALGORITHM 48

2.1. Dimensionality 48

2.2. Correlation 48

2.3. Orthogonal 48

2.4. Eigenvectors 48

2.5. Covariance Matrix 48

2.6. The PCA Algorithm's Steps 49

2.6.1. Getting the Dataset 49

2.6.2. Structure of Representing Data 49

2.6.3. Standardizing the data 49

2.6.4. Covariance of Z 49

2.6.5. Eigenvalues and Eigenvectors 49

2.6.6. Sorting the Eigenve+-ctors 49

2.6.7. Calculating the New Features 50

2.6.8. Unimportant Features from the New Data 50

2.7. PCA for Feature Engineering 50

2.7.1. Dimensionality Reduction 50

2.7.2. Anomaly Detection 50

2.7.3. Noise Reduction 50

2.7.4. Decorrelation 51

2.8. Role of Principal Component Analysis in Pharmaceutical Research 51

2.8.1. Covariance 52

2.8.2. Eigen vectors and eigen values 53

2.9. PCA in Drug Excipients Interaction Studies 54

2.10. Role of PCA in Various Pharmaceutical Fields 55

2.10.1. Adaptations 55

2.10.2. Functional PCA 55

       2.10.3. Simplified PCA 56
   2.11. Symbolic Data Principal Component Analysis 57
      2.11.1. Advantages of PCA 57
      2.11.2. Disadvantages of PCA 58
   2.12. Software's used to Perform Principal Component Analysis 58
3. APPLICATIONS OF PCA IN PHARMACEUTCAL RESEARCH 59
   3.1. Neuroscience 59
   3.2. Role of PCA in Drug Discovery 60
   3.3. Image Recognition 60
      3.3.1. Advantages 60
   3.4. QSAR Studies 61
4. MEDICAL DATA IN PCA REPOSITORIES 61
   4.1. As Sensory Assessment Tool for Fermented Food Products 61
   4.2. In Nanomaterials 61
   4.3. Bimolecular Molecule Dynamics 62
   4.4. ECG Signal Determination 62
CONCLUSION 62
REFERENCES 63

CHAPTER 4 QUALITY BY DESIGN IN PHARMACEUTICAL DEVELOPMENT: CURRENT ADVANCES AND FUTURE PROSPECTS 68
Popat Mohite, Amol Gholap, Sagar Pardeshi, Abhijeet Puri and *Tanavirsing Rajput*
1. INTRODUCTION 69
   1.1. Conventional Approach vs Design Approach 69
   1.2. QbD Paradigm and Regulatory Authorities 70
   1.3. Contribution of Ishikawa Diagram 71
   1.4. Impact of FEMA on Quality Improvement 71
   1.5. RRMA into QbD and Process Failures 72
   1.6. Role of CQA, KPI and CPP within QbD 72
2. ICH Q8 PHARMACEUTICAL DEVELOPMENT 73
   2.1. Overview of ICH Q8 (R2) Pharmaceutical Development 73
   2.2. Quality by Testing Approach vs QbD Approach 73
   2.3. QbD Elements in Pharmaceutical Development 74
   2.4. Formulation Development 74
3. QUALITY BY DESIGN (QBD) TOOLS: APPLICATION IN PRODUCT DEVELOPMENT 75
   3.1. Quality Product Quality Profile (QTPP) 76
   3.2. Critical Material Attributes (CMA's), Critical Material Parameters (CPPs), and Critical Quality Attributes (CQA's) 76
   3.3. Risk Assessment 78
   3.4. Design of Experiment (DOE) and Design Space 78
   3.5. Control Strategy 79
      3.5.1. Process Analytical Technology (PAT) 79
4. QUALITY-BY-DESIGN: CURRENT TRENDS IN PHARMACEUTICAL DEVELOPMENT 80
5. STRATEGIES FOR EVALUATING RISK IN PHARMACEUTICAL PRODUCTION PROCEDURES 81
   5.1. Risk Identification 81
   5.2. Risk Analysis 81
   5.3. Risk Evaluation 81
6. QBD IN PHARMACEUTICAL DEVELOPMENT 83

6.1. QbD Approach in Process Control 83
6.2. QbD in the Development of Analytical Methods and Pharmaceutical Manufacturing ... 83
    6.2.1. QbD Approach in Chromatographic Techniques 84
    6.2.2. Strategy for HPLC Method Development 85
    6.2.3. Strategy for HPTLC Method Optimization and Development 86
    6.2.4. Method Development/Optimization Strategy for U.V. 87
6.3. Quality-based Design for Novel Drug Delivery Systems 88
    6.3.1. Polymeric nanoparticles 88
    6.3.2. Polymeric micelles 89
    6.3.3. Liposomes 89
    6.3.4. Microemulsions and nanoemulsions 90
    6.3.5. Solid lipid nanoparticles 90
6.4. QbD Approach in the Extraction of Phytochemicals & Polyherbal Formulation 91
6.5. QbD Approach in Green Synthesis 94
6.6. QbD on the Processing of Biotherapeutics 97
6.7. QbD Approach in Immunoassays 98
7. REGULATORY AND INDUSTRY VIEW ON QBD 99
CONCLUSION 100
REFERENCES 101

CHAPTER 5 VIRTUAL TOOLS AND SCREENING DESIGNS FOR DRUG DISCOVERY AND NEW DRUG DEVELOPMENT 108
Sonal Dubey
1. INTRODUCTION 108
2. CONCEPT OF DRUG DESIGN 110
2.1. Quantitative Structure-activity Relationship (QSAR) 110
    2.1.1. Topological Approach 111
    2.1.2. Physicochemical Approach 111
    2.1.3. Quantum Chemical Approach 111
    2.1.4. Molecular Mechanics Approach 111
    2.1.5. Hybrid Approach 111
2.2. 2D-QSAR 112
2.3. 3D-QSAR 112
3. MOLECULAR MODELLING 116
3.1. Protein Modelling 116
3.2. Lead Modelling 117
    3.2.1. Lead Optimization 117
    3.2.2. Scaffold Hopping 117
    3.2.3. Protein Engineering 117
    3.2.4. Virtual Screening 117
    3.2.5. Toxicity Prediction 117
3.3. Software for Molecular Modelling 117
    3.3.1. Schrödinger 118
    3.3.2. MOE 118
4. MOLECULAR DOCKING 119
4.1. Rigid Docking 119
4.2. Flexible Docking 119
5. MOLECULAR DYNAMIC SIMULATION (MDS) 120
6. VIRTUAL SCREENING 121
6.1. Types of Virtual Screening 121
    6.1.1. Ligand-Based Screening 122

   6.1.2. Structure-Based Screening 126
  CONCLUSION 127
  REFERENCES 127

CHAPTER 6 PREDICTING DRUG PROPERTIES: COMPUTATIONAL STRATEGIES FOR SOLUBILITY AND PERMEABILITY RATES 135
Anshita Gupta Soni, Renjil Joshi, Deependra Soni, Chanchal Deep Kaur, Swarnlata Saraf and *Pankaj Kumar Singh*
  1. INTRODUCTION 136
  2. COMPUTATIONAL MODEL FOR PREDICTING PERMEABILITY AND SOLUBILITY OF DRUG 137
   2.1. Computational Model for Predicting Permeability of Drug 137
   2.2. Parallel Artificial Membrane Permeability Assay (PAMPA) 138
   2.3. Immobilized Artificial Membrane (IAM) Method 139
   2.4. Immobilized Liposome Chromatography (ILC) Technique 140
   2.5. Caco-2 Model for Predicting Permeability 140
  3. SOLUBILITY PREDICTION MODEL 140
   3.1. Data Sources 141
   3.2. Descriptors 141
   3.3. Model Development 142
   3.4. Feature Selection 142
   3.5. Validation 142
   3.6. Applicability Domain 142
   3.7. Data Quality 142
   3.8. External Validation 142
   3.9. Computer-aided Drug Discovery using Ligands (LB-CADD) 143
   3.10. Quantum Mechanics 143
   3.11. Quantum Mechanical Methods 144
    3.11.1. ADF COSMO-RS Program 144
    3.11.2. MFPCP Method 144
    3.11.3. QSAR Method for Solubility Prediction of Drug 144
    3.11.4. QSPR Technique for Solubility Prediction of Drug 145
   3.12. Critical Factors Affecting Solubility and Permeability 145
    3.12.1. Factors Affecting the Solubility of the Drug 145
    3.12.2. Factors Impacting Drug Permeability 146
    3.12.3. Relationship Between the Drug's Solubility and Permeability 146
    3.12.4. Solubility and Permeability Impacts on Drug Bioavailability 146
  CONCLUSION 147
  ACKNOWLEDGEMENT 147
  REFERENCES 147

CHAPTER 7 PHARMACOKINETIC AND PHARMACODYNAMIC MODELING (PK/PD) IN PHARMACEUTICAL RESEARCH: CURRENT RESEARCH AND ADVANCES 153
Richa Sood and *Anita A.*
  1. INTRODUCTION 153
  2. PHARMACOKINETIC MODELS 155
   2.1. Methods of Pharmacokinetic Study on Experimental Data 156
    2.1.1. Non-compartmental Analysis (NCA) 156
    2.1.2. Compartmental Modeling 156
    2.1.3. Population Pharmacokinetic Modeling 157
  3. METHODOLOGY OF PK/PD MODELING SIMULATION 158
   3.1. Non-compartmental and Compartmental Pharmacokinetic Analysis 158

        3.1.1. Population Approach 158
   3.2. Approahes for popPK Data 159
        3.2.1. Two-Stage Approach 159
        3.2.2. Nonlinear Mixed-Effects Method (NLME) 159
   3.3. Bayesian Method 160
        3.3.1. Prior Distribution 160
        3.3.2. Likelihood Function 160
        3.3.3. Bayesian Inference 160
        3.3.4. Posterior Distribution 161
        3.3.5. Parameter Estimation 161
        3.3.6. Model Validation 161
        3.3.7. Predictions and Decision Making 161
4. APPLICATIONS OF MODELING AND SIMULATION IN DRUG DEVELOPMENT 161
   4.1. Preclinical Development 162
   4.2. Clinical Development 163
   4.3. Lifecycle Assessment 164
5. REGULATORY ASPECTS 164
FUTURE PROSPECTIVES AND CONCLUSION 165
REFERENCES 166

CHAPTER 8 EXPERIMENTAL TOOLS AS AN "ALTERNATIVE TO ANIMAL RESEARCH" IN PHARMACOLOGY 170
Kunjbihari Sulakhiya, Rishi Paliwal, Anglina Kisku, Madhavi Sahu, Shivam Aditya, Pranay Soni and *Saurabh Maru*
1. INTRODUCTION 171
2. BRIEF HISTORY OF ANIMAL RESEARCH IN PHARMACOLOGY 172
   2.1. Overview of Alternative Experimental Tools and Techniques 173
        2.1.1. Cell Cultures 173
        2.1.2. Computer Modeling and Simulation 173
        2.1.3. Microfluidic Sevices 173
        2.1.4. Human Tissue Samples 173
        2.1.5. In vitro Assays 173
        2.1.6. Epidemiological Research 173
        2.1.7. Non-invasive Imaging Techniques 173
        2.1.8. High-throughput Screening 173
   2.2. Need for Alternative Methods to Animal Research in Pharmacology 174
        2.2.1. Benefits of Using Alternative Methods 175
   2.3. Ethical Concerns and Criticisms of Animal Research 176
   2.4. *In vitro* Methods for Alternative to Animal Research 177
        2.4.1. Types of In vitro Models 178
        2.4.2. Advantages and Disadvantages of In vitro Methods 181
        2.4.3. In vitro Models Applied in Pharmacological Research 182
        2.4.4. Comparison with Animal Studies 183
3. *IN SILICO* METHODS (COMPUTER MODELING AND SIMULATION) 183
   3.1. Animal Eelfare 184
   3.2. Cost and Time 184
   3.3. Molecular Modeling and Simulations 185
        3.3.1. Quantitative Structure-Activity Relationship (QSAR) 185
        3.3.2. Virtual Screening 186
   3.4. Applications of Computer Modeling and Simulation in Pharmacological Research 186
        3.4.1. Drug Development 186

3.4.2. Drug Repurposing 187
3.4.3. Molecular Docking 187
3.4.4. In Silico Imaging in Clinical Trials 188
3.5. Advantages and Limitations of *In Silico* Methods 188
3.6. Comparison with Animal Studies 190
4. *IN VIVO* NON-ANIMAL METHODS (HUMAN-BASED METHODS) 190
4.1. Significance of Micro-Dosing 193
4.2. Advantages of Micro-dosing 193
4.3. Human-Based Methods Employed in Pharmacological Research [8] 194
4.3.1. Microbiological Systems 194
4.3.2. Tissue/Organ Culture Preparation 194
4.3.3. Human Dopaminergic Neurons 194
4.3.4. Plant Analysis 194
4.3.5. Stem Cells in Toxicological Research 194
5. EMERGING EXPERIMENTAL TOOLS 194
5.1. Microfluidic Devices 195
5.2. Organ-on-a-chip Technology 196
5.2.1. Dynamic Mechanical Stress 197
5.2.2. Fluid Shear 197
5.2.3. Concentration Dradients 197
6. REGULATORY PERSPECTIVES 197
6.1. Current Regulatory Guidelines 198
6.2. Challenges to Regulatory Acceptance and Implementation 198
6.3. Recommendations for Overcoming these Challenges 198
6.4. Future Directions for Regulatory Framework and Validation 199
CONCLUSION 199
ACKNOWLEDGEMENT 200
REFERENCES 200

**CHAPTER 9 NEWER SCREENING SOFTWARE FOR COMPUTER AIDED HERBAL
DRUG INTERACTIONS AND ITS DEVELOPMENT** 207
Sunil Kumar Kadiri and *Prashant Tiwari*
1. INTRODUCTION 208
2. HDI ESTIMATIONS 209
2.1. Current Approaches 209
2.2. Limitations of Current Strategies 209
2.3. Databases and Web Services 210
2.3.1. Drugbank 210
2.3.2. Supertarget 211
2.3.3. Database of Therapeutic Targets 211
2.4. TDR Methods 211
2.5. MATADOR 211
2.6. PDTD 211
2.7. Integrity 211
2.8. FAERS 212
2.9. ZINC 212
2.10. SIDER 212
2.11. ChemBank 212
2.12. CanSAR 212
2.13. The IUPHAR/BPS Pharmacology Manual 212
2.14. DCDB 213

2.15. DINIES 213
3. DIGITAL MODELING AND SIMULATION 213
    3.1. HDI Data Accessed Without Cost 213
        3.1.1. Chi Mei's Indexing Service (CMSS) 213
        3.1.2. SUPP.AI 214
        3.1.3. Information Stored in the PHYDGI Database 214
    3.2. Commercially-available HDI Databases 214
        3.2.1. Database of Drug Interactions at UW (DIDB) 214
        3.2.2. A One-Stop Source for Natural Health Products (NMCD) 215
        3.2.3. Herbal Drug Interactions from Stockley's (SHMI) 215
    3.3. HDI Screening: Cutting-edge Intelligent In silico Methods 216
        3.3.1. In silico Forecasting HDIs 216
        3.3.2. Lexicomp Drug Interactions (LDI) 218
        3.3.3. Literature Excerpt 218
        3.3.4. Frequency, Liability and Sustainability of Content Updates. 218
        3.3.5. Interfaces for Users 219
        3.3.6. AI's Role in Creating HDI Databases 219
        3.3.7. HDI Database Development Going Forward 220
  CONCLUDING REMARKS 220
  ACKNOWLEDGEMENTS 221
  REFERENCES 221

**CHAPTER 10 DELIBERATIONS AND CONSIDERATIONS OF MESODYN SIMULATIONS
IN PHARMACEUTICALS** 227
Manisha Yadav, Dhriti Mahajan, Om Silakari and *Bharti Sapra*
1. INTRODUCTION 227
2. MESOSCALE SIMULATIONS IN PHARMACEUTICS 229
    2.1. Theoretical Background of Mesodyn 230
        2.1.1. Dynamical Considerations 230
        2.1.2. Numerical Considerations 232
        2.1.3. Order parameters 232
3. PROPERTIES OF MESODYN 233
    3.1. Aggregation and Coagulation 233
    3.2. Phase Morphology 233
    3.3. Effect of Confinement on Miscibility 234
    3.4. Effect of Shear on Morphology 234
    3.5. Compositional Order Parameters 235
    3.6. Free Energy and Entropy Evolution 235
    3.7. Density Histograms 236
**4. APPLICATIONS OF MESODYN SIMULATIONS IN FORMULATION
DEVELOPMENT** 236
    4.1. Precipitation Membrane Formation 240
    4.2. Nanoscale Drug Delivery Systems 241
    4.3. Development of thin Films from Block Copolymers 241
    4.4. Solubility of Menthol by Platycodin D 241
    4.5. Prediction of Chitosan and Poly(e-caprolactone) Binary Systems, Miscibility and Phase
Separation Behaviour 242
    4.6. Nanotube Self-forming Process of Amphiphilic Copolymer 242
    4.7. Utilizing a Macrocyclic Terbium Compound with Hetero-ligands to Create an Effective
Luminous Soft Media in a Lyomesophase 242
    4.8. Design and Development of Polymersome Chimaera 243

4.9. Mesoscopic Simulation for the Phase Separation Behavior of the Pluronic Aqueous Mixture 243

4.10. Quercetin's Solubility and Release Characteristics 244

CONCLUSION 244

ACKNOWLEDGEMENT 245

REFERENCES 245

CHAPTER 11 COMPUTATIONAL TOOLS TO PREDICT DRUG RELEASE KINETICS IN SOLID ORAL DOSAGE FORMS 249

Devendra S. Shirode, Vaibhav R. Vaidya and *Shilpa P. Chaudhari*

1. INTRODUCTION 249

2. CALCULATIONS AND VALIDATION OF DISSOLUTION MODELS USING MS EXCEL 250

2.1. Zero-order Drug Release Model 251

2.2. First Order Drug Release Model 252

2.3. Higuchi Model 253

2.4. Hixson Crowell Model 254

2.5. Korsmeyer-Peppas Model 254

2.6. Similarity Factor Calculations Using Excel 255

    2.6.1. Using Excel for f2 Calculation 256

2.7. Difference Factor (f1) 256

    2.7.1. Interpretation 257

    2.7.2. Using Excel for f1 Calculation 257

3. INTEGRATED TOOLS FOR DISSOLUTION MODELING OF VARIOUS DOSAGE FORMS 258

3.1. DD Solver: Modeling and Comparison of Drug Dissolution Profiles 258

3.2. Key Features and Capabilities 259

    3.2.1. Mechanistic Modeling 259

    3.2.2. Curve Fitting and Prediction 259

    3.2.3. Dissimilarity and Similarity Assessment 259

    3.2.4. Visualization Tools 260

    3.2.5. Predictive Insights 260

    3.2.6. User-Friendly Interface 260

3.3. Procedure for Using DD Solver for Drug Dissolution Profile Analysis and Comparison 260

    3.3.1. Data Collection and Input Preparation 260

    3.3.2. Model Selection and Curve Fitting 260

    3.3.3. Simulation and Prediction 261

    3.3.4. Dissolution Profile Comparison 261

    3.3.5. Visualization and Reporting 261

    3.3.6. Interpretation and Decision-Making 261

3.4. Applications of DD Solver 263

    3.4.1. Generic Product Development 264

    3.4.2. Formulation Optimization 264

    3.4.3. Comparative Studies 264

    3.4.4. Risk Assessment 264

4. PCPDISSO (V3I) 264

4.1. Hypothetical Release Models for PCP Disso 266

    4.1.1. Zero-Order Release Model 266

    4.1.2. First-order Release Model 266

    4.1.3. Higuchi Release Model 266

    4.1.4. Weibull Release Model 266

4.1.5. Fractional Release Model 267

5. KINETDS 3.0 SOFTWARE 267

6. OTHER DISSOLUTION WORKSTATIONS/TOOLS 272

CONCLUSION 273

ACKNOWLEDGEMENTS 273

REFERENCES 273

**CHAPTER 12 WARP AND WOOF OF DRUG DESIGNING AND DEVELOPMENT: AN IN-
SILICO APPROACH** 280

Monika Chauhan, Vikas Gupta, Anchal Arora, Gunpreet Kaur, Parveen Bansal and
Ravinder Sharma

1. INTRODUCTION 281

    1.1. Drug Design 281

2. STRUCTURE-BASED DRUG DESIGN 281

3. LIGAND-BASED DRUG DESIGN 283

    3.1. Quantitative Structure-activity Relationship 283

    3.2. Pharmacophore Modelling 284

4. VIRTUAL SCREENING 284

    4.1. Molecular Docking 285

    4.2. ADMET Predictions 286

CONCLUSION 289

REFERENCES 290

**CHAPTER 13 DATA INTERPRETATION AND MANAGEMENT TOOLS FOR
APPLICATION IN PHARMACEUTICAL RESEARCH** 295

*Arvinder Kaur, Avichal Kumar, Kavya Manjunath, Deepa Bagur Paramesh, Shilpa
Murthy* and *Anjali Sinha*

1. INTRODUCTION 296

2. HISTORY 297

    2.1. Conventional Approach 297

    2.2. Computer and Software Resolutions 297

3. TYPES OF RESEARCH DATA 298

    3.1. Qualitative Data 298

    3.2. Quantitative Data 298

    3.3. Observational Data 298

    3.4. Experimental Data 299

    3.5. Survey Data 299

    3.6. Secondary Data 299

    3.7. Meta-Data 299

4. DATA ANALYSIS 299

    4.1. Intelligent Data Analysis 300

    4.2. Data Mining 302

    4.3. Data Abstracting 302

5. DATA REPRESENTATION 304

    5.1. Software Applications 304

        5.1.1. SPSS (Statistical Package for the Social Sciences) 304

        5.1.2. GraphPad Prism 305

    5.2. Implementation Application 306

        5.2.1. EHRs, or Electronic Health Records 307

        5.2.2. Systems for Dispensing Medications 307

        5.2.3. Information Systems for Pharmacies 308

        5.2.4. Apps for Mobile Devices 308

       5.2.5. Tele Pharmacy 308

   5.3. Examples of Implementation 308

       5.3.1. Electronic Recommending 308

       5.3.2. Automating Pharmacies 309

       5.3.3. The Clinical Choice was Emotionally Supportive Networks 309

       5.3.4. Management of Medical Treatment 309

       5.3.5. Purpose of Care Testing 309

CONCLUSION 309

REFERENCES 309

SUBJECT INDEX 313

PREFACE

The research and development of new pharmaceuticals are carried out with various pharmaceutical hurdles, ethical ramifications, and social duties, along with the approach seeming obscure. The current pharmaceutical research needs creativity, which is the prime lifeblood of every industry. Nowadays, only computer software in the field of pharmaceutical sciences makes it possible to comprehend complicated processes and manage resources, money, and labor effectively and efficiently. Computer software may relieve medical professionals of tedious multitasking processes and complicated prescreening and evaluation of resource materials. The use of computers in all stages of drug discovery, development, and marketing is addressed holistically and comprehensively in this unique contributed work. It explains the process of simulations such as added functions, data mining, predicting human response, quality by design and artificial intelligence to develop cost-effective drugs, multi-formulation approaches, and high-throughput screening which are applied at various phases in clinical development. The book gives readers a comprehensive overview and a systems viewpoint from which they can design strategies to fully utilize the use of computers in their pharmaceutical industry throughout all stages of the discovery and development process. Researchers working in informatics and ADMET, drug discovery, and technology development, as well as IT professionals and scientists in the pharmaceutical sector, should read this. The book's multifaceted, cross-functional approach offers a singular chance for a comprehensive investigation and evaluation of computer applications in pharmaceutics.

There are the following sections in the book: the role of computers in drug discovery, preclinical development, clinical development, drug delivery, systemic optimization, understanding diseases and repurposing, advanced computer-aided functions to optimize biopharmaceutical variables and future applications, and future development. To maintain a consistent structure and approach throughout the book, each chapter is thoroughly revised after being authored by one or more of the foremost authorities in the field. Figures are frequently used to explain intricate ideas and diverse processes. Each chapter includes references so that readers can keep digging into a particular subject. Finally, many of the chapters provide tables of software resources.

Dilpreet Singh
Department of Pharmaceutics
ISF College of Pharmacy
Moga, India

&

Prashant Tiwari
Department of Pharmaceutical Sciences
Dayanand Sagar University
Bengaluru, India

List of Contributors

Abhijeet Puri	AET's St. John Institute of Pharmacy and Research, Palghar-401404, Maharashtra, India
Amardeep K.	Himalayan Institute of Pharmacy, Kala Amb, Dist. Sirmour, Himachal Pradesh-173030, India
Amol Gholap	AET's St. John Institute of Pharmacy and Research, Palghar-401404, Maharashtra, India
Anjali Sharma	Amar Shaheed Baba Ajit Singh Jujhar Singh Memorial College of Pharmacy, Bela, (An Autonomous College) Ropar, India
Anshita Gupta Soni	Shri Rawatpura Sarkar Institute of Pharmacy, Kumhari, Durg, Chhattisgarh, India
Anita A.	College of Pharmaceutical Sciences, Dayananda Sagar University, Bengaluru, Karnataka-560078, India
Anglina Kisku	Neuro Pharmacology Research Laboratory (NPRL), Department of Pharmacy, Indira Gandhi National Tribal University, Amarkantak, Madhya Pradesh, India
Anchal Arora	All India Institute of Medical Sciences, Bathinda, India
Arvinder Kaur	Department of Pharmaceutics, KLE College of Pharmacy, Constituent Unit of KLE Academy of Higher Education and Research (Deemed to be University), Rajajinagar, Bengaluru-560010, Karnataka, India
Avichal Kumar	Department of Pharmaceutics, KLE College of Pharmacy, Constituent Unit of KLE Academy of Higher Education and Research (Deemed to be University), Rajajinagar, Bengaluru-560010, Karnataka, India
Anjali Sinha	Department of Pharmaceutics, KLE College of Pharmacy, Constituent Unit of KLE Academy of Higher Education and Research (Deemed to be University), Rajajinagar, Bengaluru-560010, Karnataka, India
Bharti Sapra	Department of Pharmaceutical Sciences & Drug Research, Punjabi University, Patiala, Punjab, India
Chanchal Deep Kaur	Rungta Institute of Pharmaceutical Sciences and Research, Raipur, Chhattisgarh, India
Deependra Soni	Faculty of Pharmacy, MATS University Campus, Aarang, Raipur, Chhattisgarh, India
Deepa Bagur Paramesh	Department of Pharmaceutics, KLE College of Pharmacy, Constituent Unit of KLE Academy of Higher Education and Research (Deemed to be University), Rajajinagar, Bengaluru-560010, Karnataka, India
Devendra S. Shirode	Department of Pharmacology, Dr. D. Y. Patil College of Pharmacy, Akurdi, Pune, Maharashtra, India
Dhriti Mahajan	Department of Pharmaceutical Sciences & Drug Research, Punjabi University, Patiala, Punjab, India
Diksha Sharma	Amar Shaheed Baba Ajit Singh Jujhar Singh Memorial College of Pharmacy, Bela, (An Autonomous College) Ropar, India
Gunpreet Kaur	University Center of Excellence in Research, BFUHS, Faridkot, India

Kavya Manjunath	Department of Pharmacology, KLE College of Pharmacy, Constituent Unit of KLE Academy of Higher Education and Research (Deemed to be University), Rajajinagar, Bengaluru-560010, Karnataka, India
Kunjbihari Sulakhiya	Neuro Pharmacology Research Laboratory (NPRL), Department of Pharmacy, Indira Gandhi National Tribal University, Amarkantak, Madhya Pradesh, India
Madhavi Sahu	Neuro Pharmacology Research Laboratory (NPRL), Department of Pharmacy, Indira Gandhi National Tribal University, Amarkantak, Madhya Pradesh, India
Manisha Yadav	Department of Pharmaceutical Sciences & Drug Research, Punjabi University, Patiala, Punjab, India
Monika Chauhan	School of Health Sciences and Technology, UPES, Dehradun, India
Neelam Sharma	Amar Shaheed Baba Ajit Singh Jujhar Singh Memorial College of Pharmacy, Bela, (An Autonomous College) Ropar, India
Om Silakari	Department of Pharmaceutical Sciences & Drug Research, Punjabi University, Patiala, Punjab, India
Parveen Bansal	University Center of Excellence in Research, BFUHS, Faridkot, India
Pankaj Kumar Singh	Department of Pharmaceutics, National Institute of Pharmaceutical Education and Research NIPER), Hyderabad, Telangana-500037, India
Popat Mohite	AET's St. John Institute of Pharmacy and Research, Palghar-401404, Maharashtra, India
Pranay Soni	Department of Pharmacy, Indira Gandhi National Tribal University, Amarkantak, Madhya Pradesh, India
Prashant Tiwari	Department of Pharmaceutical Sciences, Dayananda Sagar University, Bengaluru, India
Punam Gaba	Amar Shaheed Baba Ajit Singh Jujhar Singh Memorial College of Pharmacy, Bela, (An Autonomous College) Ropar, India
Ram Babu S.	Himalayan Institute of Pharmacy, Kala Amb, Dist. Sirmour, Himachal Pradesh-173030, India
Rahul Kumar Sharma	Amar Shaheed Baba Ajit Singh Jujhar Singh Memorial College of Pharmacy, Bela, (An Autonomous College) Ropar, India
Ravinder Sharma	University Institute of Pharmaceutical Sciences and Research, BFUHS, Faridkot, India
Renjil Joshi	Shri Rawatpura Sarkar Institute of Pharmacy, Kumhari, Durg, Chhattisgarh, India
Richa Sood	College of Pharmaceutical Sciences, Dayananda Sagar University, Bengaluru, Karnataka-560078, India
Rishi Paliwal	Nanomedicine and Bioengineering Research Laboratory (NBRL), Department of Pharmacy, Indira Gandhi National Tribal University, Amarkantak, Madhya Pradesh, India
Sagar Pardeshi	AET's St. John Institute of Pharmacy and Research, Palghar-401404, Maharashtra, India
Samaresh Pal Roy	Department of Pharmacology, Maliba Pharmacy College, Uka Tarsadia University, Bardoli-394350, Surat, Gujarat, India

Sakshi T.   Himalayan Institute of Pharmacy, Kala Amb, Dist. Sirmour, Himachal Pradesh-173030, India

Saurabh Maru   Department of Pharmacology, School of Pharmacy and Technology Management, SVKM's Narsee Monjee Institute of Management Studies, Maharashtra, India

Shailesh Sharma   Amar Shaheed Baba Ajit Singh Jujhar Singh Memorial College of Pharmacy, Bela, (An Autonomous College) Ropar, India

Shivam Aditya   Neuro Pharmacology Research Laboratory (NPRL), Department of Pharmacy, Indira Gandhi National Tribal University, Amarkantak, Madhya Pradesh, India

Shilpa P. Chaudhari   Department of Pharmacology, Dr. D. Y. Patil College of Pharmacy, Akurdi, Pune, Maharashtra, India

Shilpa Murthy   Department of Pharmaceutical Chemistry, KLE College of Pharmacy, Constituent Unit of KLE Academy of Higher Education and Research (Deemed to be University), Rajajinagar, Bengaluru-560010, Karnataka, India

Sonal Dubey   College of Pharmaceutical Sciences, Dayananda Sagar University, Kumaraswamy Layout, Bengaluru-560111, India

Sunil Kumar Kadiri   Department of Pharmacology, College of Pharmaceutical Sciences, Dayananda Sagar University, K.S Layout, Bengaluru-560111, Karnataka, India

Swarnlata Saraf   University Institute of Pharmacy, Pt.Ravishankar Shukla University, Raipur, Chhattisgarh, India

Tanavirsing Rajput   AET's St. John Institute of Pharmacy and Research, Palghar-401404, Maharashtra, India

Vaibhav R. Vaidya   Department of Pharmacology, Dr. D. Y. Patil College of Pharmacy, Akurdi, Pune, Maharashtra, India

Vikas Gupta   University Center of Excellence in Research, BFUHS, Faridkot, India

Introduction to Computer-Based Simulations and Methodologies in Pharmaceutical Research

Samaresh Pal Roy[1,*]

[1] *Department of Pharmacology, Maliba Pharmacy College, Uka Tarsadia University, Bardoli-394350, Surat, Gujarat, India*

Abstract: Pharmaceutical research is increasingly using computer-based simulations and approaches to hasten the identification and development of new drugs. These methods make use of computational tools and models to forecast molecular behavior, evaluate therapeutic efficacy, and improve drug design. Molecular modeling is a key application of computer-based simulations in pharmaceutical research. It allows researchers to build virtual models of molecules and simulate their behavior, which provides insights into their interactions and properties. Molecular docking is a computational method used in Computer-Aided Drug Design (CADD) to predict the binding mode and affinity of a small molecule ligand to a target protein receptor. Quantitative structure-activity relationship (QSAR) modeling is another pharmaceutical research tool. QSAR models predict molecular activity based on the chemical structure and other attributes using statistical methods. This method prioritizes and optimizes drug candidates for specific medicinal uses, speeding up drug discovery. Another effective use of computer-based simulations in pharmaceutical research is virtual screening. It entails lowering the time and expense associated with conventional experimental screening methods by employing computational tools to screen huge libraries of chemicals for prospective therapeutic candidates. While computer-based techniques and simulations have many advantages for pharmaceutical research, they also demand a lot of processing power and knowledge. Also, they are an addition to conventional experimental procedures rather than their replacement. As a result, they frequently work in tandem with experimental techniques to offer a more thorough understanding of drug behavior and efficacy. Overall, computer-based simulations and methodologies enable pharmaceutical researchers to gather and analyze data more efficiently, bringing new medications and therapies to market.

Keywords: Computer-based simulations, Computer-aided drug design (CADD), Drug behavior, Drug efficacy, Drug discovery, Molecular modeling, Molecular dynamics, Pharmaceutical research, Quantitative structure-activity relationship (QSAR) modeling, Virtual screening.

---
* **Corresponding author Samaresh Pal Roy:** Department of Pharmacology, Maliba Pharmacy College, Uka Tarsadia University, Bardoli-394350, Surat, Gujarat, India; Tel: +91 9377077710; E-mail: samaresh.roy@utu.ac.in

Dilpreet Singh and Prashant Tiwari (Eds.)

1. INTRODUCTION

Pharmaceutical research is a crucial aspect of healthcare that aims to discover, develop, and test new medications in order to improve human health. During this process, potential drug candidates are identified, their characteristics are optimized, and their effectiveness and safety are evaluated. The drug development process is expensive and time-consuming due to the low success rate of new medication candidates and the high average cost of bringing a new treatment to market [1, 2]. Computer-based simulations have become important tools in pharmaceutical research to speed up the identification of novel drugs and increase the likelihood that new medication candidates will be successful. To simulate the behaviour of molecules and their interactions with biological targets, these simulations make use of computer techniques and models. The molecular underpinnings of disorders like Parkinson's disease and Alzheimer's disease have been studied using computational simulations [3, 4]. Additionally, it aids in predicting the activity of prospective therapeutic targets including enzymes and receptors as well as their interactions with other molecules [5, 6]. Physicochemical parameters, including solubility, permeability, and stability, can be predicted using computer-based simulations in order to design and refine drug candidates [7, 8]. Additionally, it predicts aspects of drug candidates' pharmacokinetics and pharmacodynamics, such as absorption, distribution, metabolism, and excretion [9, 10]. Researchers can get a more thorough understanding of the molecular principles underlying drug action by combining computer-based simulations with experimental data, which eventually results in safer and more effective drugs [11].

1.1. Types of Computer-Based Simulations in Pharmaceutical Research

Pharmaceutical research extensively utilizes a range of computer-based simulation techniques, with molecular modeling being a prominent example. This technique entails constructing and analyzing three-dimensional (3D) molecular models to comprehend their dynamics and interactions with other molecules. The molecular modelling technique known as molecular docking, for instance, forecasts the binding mechanism and affinity of small molecule ligands to their target proteins [12]. Simulating the movements of molecules over time to examine their interactions and behaviour at the atomic level is known as molecular dynamics simulation. For instance, protein conformational changes and their interactions with other molecules can be studied using molecular dynamics simulations [13]. Using statistical techniques, quantitative structure-activity relationship (QSAR) modelling links the chemical structure of molecules with their biological activity. For instance, QSAR modelling can be used to infer new drug candidates' biological activity from their chemical structure [14]. Virtual screening involves the identification of potential therapeutic candidates from vast chemical libraries by evaluating the affinity and selectivity of each compound toward a specific protein. For instance, this method can be applied to discover inhibitors for viral proteins, which could then be utilized as a means to treat viral infections [15].

1.1.1. Challenges and Limitations of Computer-Based Simulations

While computer-based simulations offer numerous advantages for pharmaceutical research, they also come with significant drawbacks. The accuracy and reliability of computational models, which hinge on the quality of input data and the assumptions made during model creation, pose a considerable challenge [16]. Another concern is the need for high-performance computing resources, which can be costly and demand specialized expertise to carry out complex simulations. Integrating computational and experimental data can prove challenging due to variations in the data generated by each method [17]. Furthermore, comparing results across different studies can be difficult due to the lack of standardized data processing techniques and tools [18].

1.1.2. Advances in Computer-Based Simulations

Some of the difficulties and restrictions of these methodologies have been overcome by recent developments in computer-based simulations. For instance, the accuracy and dependability of computational models have increased as a result of the development of machine learning algorithms [19]. High-performance computer resources are now easier to obtain and more affordable thanks to cloud computing services [20]. Additionally, the introduction of novel techniques like cryo-electron microscopy has made the integration of computational and experimental data more practical [21].

To validate and refine computer models of protein-ligand interactions, cryo-electron microscopy can be employed to determine the three-dimensional structure of proteins at nearly atomic resolution. The future of pharmaceutical research is expected to continue relying significantly on computer-based simulations. The accuracy and reliability of these computer models are projected to greatly improve with the emergence of novel computational techniques such as deep learning [22]. The integration of computer-based simulations with other technologies like high-throughput screening and gene editing is also anticipated to facilitate the advancement of personalized medicine [23].

By providing insights into molecular behavior and interactions that are challenging to obtain solely through experimental methods, computer-based simulations have the potential to revolutionize the approach to pharmaceutical research. Future drug discovery and development efforts are anticipated to benefit much more from the continued development of computational approaches. To ensure these techniques' accuracy and dependability in drug development, it is crucial to overcome the difficulties and restrictions related to them.

2. MOLECULAR MODELLING: PRINCIPLES AND APPLICATIONS IN DRUG DISCOVERY

A computational method called molecular modelling is used to investigate the atomic-level dynamics, interactions, and structure of molecules [24]. This method has become a crucial component of drug discovery research because it enables scientists to see and examine the interactions between drug candidates and their target proteins as well as to create new substances with enhanced binding affinity and selectivity [25]. The development of safe and effective medicines depends on the ability to predict features of drug candidates such as pharmacokinetics and pharmacodynamics.

2.1. Principles of Molecular Modelling

The process of building and analysing three-dimensional (3D) representations of molecules is known as molecular modelling [26]. The features of potential therapeutic candidates can be predicted using these models, and novel compounds can be created with enhanced binding capabilities [27]. The fundamentals of molecular modeling encompass various aspects, including force field calculations. These calculations assess the stability and energy of molecular structures by considering atom-to-atom interactions like bond stretching, bond angle bending, and nonbonded interactions [28]. Simulating the movements of molecules over time to examine their interactions and behaviour at the atomic level is known as molecular dynamics simulation. This method sheds light on the kinetics, flexibility, and structural changes that occur in biomolecules [29]. Calculations based on quantum mechanics: Examining molecules' electronic structures to forecast their chemical properties, such as reactivity, stability, and electronic spectra. In comparison to conventional force fields, these computations can give a more precise representation of molecular interactions and characteristics [30]. Molecular docking: Investigating the conformational space of the protein-ligand complex to predict the binding mechanism and affinity of small molecule ligands to their target proteins. This process is frequently employed to find new medication candidates and to enhance their binding qualities [31].

2.2. Applications of Molecular Modelling in Drug Discovery

Drug discovery has benefited from the use of molecular modelling, which enables researchers to understand the molecular underpinnings of drug action and resistance and requires visualisation and analysis of interactions between drug candidates and their target proteins [32]. By examining the interactions between ligands and target proteins and making appropriate modifications to their chemical structures, it is possible to design and optimise therapeutic candidates with higher binding affinity and selectivity [33]. Computationalmodels are used to forecast the pharmacokinetics (drug absorption, distribution, metabolism, and excretion) and pharmacodynamics (drug-receptor interactions and effects) of drug candidates. This can help prioritise candidates for experimental testing and can also help with drug molecule design and optimisation [34]. Using structure-based and ligand-based virtual screening techniques, one may quickly and cheaply identify new therapeutic targets and check the activity of vast databases of compounds [35].

2.3. Molecular Modeling Techniques

There are different methodologies and procedures which are applied in molecular modelling techniques. The first model comprises building a 3D model of a protein based on its amino acid sequence and the structures of related proteins are known as homology modelling. When a target protein's experimental structure is unavailable, this strategy is especially helpful [36]. Simulations of molecular dynamics: As was already indicated, this method mimics the motion of molecules overtime to examine their interactions and behaviour at the atomic level, revealing information on the dynamics and function of proteins [37]. Designing new therapeutic candidates using the structure and characteristics of existing ligands is known as "ligand-based drug design." This strategy can be especially helpful when the target protein's structure is unknown or challenging to identify experimentally [38]. Designing new therapeutic candidates using the target protein's structure and characteristics is known as "structure-based drug design." This strategy enables the logical development of medications that can specifically interact with a target protein, potentially improving efficacy and reducing negative effects [39]. The accuracy and dependability of computational models: The quality of molecular models depends on the accuracy of force fields, quantum mechanics methods, and docking algorithms used, as well as the accessibility of high-quality experimental data to validate and refine the models [40]. While molecular modelling offers many benefits in drug discovery, there are also some limitations and challenges to take into account. High-performance computing resources are required for molecular modelling, especially for quantum mechanics calculations and molecular dynamics simulations, which could be a barrier for some research teams and institutions [41]. Combining data from computational modelling and experimental research can be difficult since molecular systems are not all represented with the same amount of information, resolution, and representation. For the field to evolve, methodologies and tools to analyze diverse data sources must be developed [42].

The area of molecular modelling is quickly developing new tools and methods securely for drug discovery and development. The repeatability and generalizability of findings may be constrained by the lack of integrated tools and procedures for data analysis, which can make it challenging to compare outcomes across studies and research groups [43].

Despite these difficulties, molecular modelling is still evolving and becoming a more crucial part of the drug discovery process. In the upcoming years, it is anticipated that improvements in computational hardware, software, and algorithms as well as the incorporation of multidisciplinary techniques and data sources will further increase the influence of molecular modelling on drug discovery and development.

3. COMPUTER-AIDED DRUG DESIGN: CONCEPTS AND TECHNIQUES

In the computational method of computer-aided drug design (CADD), drug candidates are designed and optimized using computer simulations and models [44]. To find prospective drug candidates and improve their properties, CADD integrates molecular modelling, virtual screening, and other computational tools [45]. To expedite drug discovery and identify novel drug candidates with improved potency, selectivity, and pharmacokinetic properties, Computer-Aided Drug Design (CADD) has emerged as a vital component of pharmaceutical research [46]. This collaborative effort involves experts in chemistry, biology, and computer science working together to develop superior medications that specifically target proteins or receptors while minimizing adverse effects [47]. In the computational approach known as computer-aided drug design (CADD), drug candidates are conceptualized and refined through the use of computer simulations and models [44]. CADD employs a combination of molecular modeling, virtual screening, and various other computational tools to identify potential drug candidates and enhance their properties [45]. In the pursuit of expediting the drug discovery process and uncovering new candidates with enhanced potency, selectivity, and pharmacokinetic attributes, Computer-Aided Drug Design (CADD) has risen as a pivotal aspect of drug discovery research [46]. CADD plays a crucial role in crafting enhanced medications that precisely target specific proteins or receptors while minimizing undesirable side effects. This involves collaborative efforts between experts in chemistry, biology, and computer science [47].

3.1. Principles of Computer-Aided Drug Design

Constructing and examining three-dimensional (3D) models of molecules to investigate their dynamics, structure, and interactions with other molecules are referred to as molecular modeling [48]. This approach often involves simulations rooted in quantum mechanics, molecular mechanics, and molecular dynamics to accurately capture the behavior of molecules and their interactions with target proteins [49].

3.1.1. Virtual Screening

Determining possible therapeutic candidates from huge chemical libraries by assessing the selectivity and affinity of the compounds for a target protein [50]. To forecast and rank molecules based on their potential biological activity, this procedure may incorporate the use of high-throughput docking algorithms, machine learning, and artificial intelligence approaches [51]. Designing new drug candidates using the target protein's structure and characteristics is known as "structure-based drug design" (SBD) [52]. This strategy creates compounds that can attach to a protein's active site using the 3D structure as a guide, regulating the protein's activity and producing therapeutic benefits [53]. Designing new therapeutic candidates using the structure and characteristics of existing ligands is known as "ligand-based drug design" [54]. This method seeks to increase the activity and selectivity of molecules by identifying or designing novel compounds with features that are comparable to those of known active molecules [55]. The key characteristics of a ligand are investigated that contribute to its biological activity through pharmacophore modelling [56]. A pharmacophore is a physical configuration of molecular elements required for ideal interactions with a particular target protein. Pharmacophore models can be applied to evaluate compound libraries for possible therapeutic candidates or to direct the design of novel compounds [57].

3.2. Applications of Computer-Aided Drug Design in Drug Discovery

The possible therapeutic possibilities are looked for by searching through extensive databases of chemicals [58]. Virtual screening, which evaluates millions of compounds in silico before choosing a smaller subset for experimental testing, can significantly cut the time and cost of identifying promising drug candidates as a result of the increasing availability of chemical libraries and the advancements in computational methods [59]. Enhanced binding affinity and selectivity can be attained through the design and optimization of therapeutic candidates [60]. Researchers can create compounds with heightened potency and selectivity by utilizing CADD methods to identify crucial interactions between a drug candidate and its target protein. This increases the probability of successful outcomes in clinical trials [61].

The drug candidates' pharmacokinetics and pharmacodynamics are predicted [62]. Researchers can forecast possible problems with therapeutic efficacy and safety by simulating the absorption, distribution, metabolism, excretion, and toxicity (ADMET) properties of drug candidates with CADD, allowing for early adjustments in the drug design process [63]. Identifying potential drug targets is a key application [64]. CADD isutilized to examine the structural and functional characteristics of proteins, which enables researchers to discover potential therapeutic targets and gain insights into the role that proteins play in disease development [65].

Molecular causes of medication resistance should be researched [66]. The effectiveness of treatment approaches can be increased by helping researchers build medications that can overcome or circumvent resistance mechanisms by analysing the structural alterations in target proteins and comprehending the molecular mechanisms underlying drug resistance [67].

3.2.1. Computer-Aided Drug Design Techniques

These techniques are used to predict the binding mechanism and affinity of small molecular ligands to their target proteins by molecular docking [68]. This method ranks the most advantageous for binding poses using a variety of scoring functions and algorithms, revealing details about the molecular interactions underlying ligand binding and activity [69]. Simulating the mobility of molecules over time to examine their interactions and behaviour at the atomic level is known as molecular dynamics simulation [70]. This method can help in the design of novel compounds with improved features by offering useful information regarding the conformational changes, flexibility, and stability of proteins and ligands [71]. *De novo* drug design is the process of applying computer approaches to create brand-new medication candidates [72]. In this method, new chemical scaffolds with desirable features are possibly discovered by creating novel molecular structures based on the binding site or pharmacophore models of the target protein [73]. Drug design using smaller pieces to create bigger molecules is known as fragment-based drug design [74]. In order to create therapeutic candidates with improved affinity and selectivity, this strategy entails finding and optimising tiny molecular fragments that bind to the target protein [75].

The fundamental Computer-Aided Drug Design (CADD) workflow is essential to the process of finding new drugs. Wet lab experiments, Structure-Based Drug Design (SBDD), and Ligand-Based Drug Design (LBDD) are only a few of the methods used in this methodology. The combination of these methods improves the effectiveness and success rate of the drug development process by enabling iterative rounds of ligand creation. Wet-lab experiments are the initial step in the CADD workflow, and they are shown as solid lines. In wet lab experiments, physical laboratory tasks including compound synthesis, testing, and biological activity analysis are all part of the process. These studies offer important information on the potency, pharmacokinetics, and interactions of the drugs with the target molecules. Future CADD approaches are built upon the data collected from wet-lab research.

Structure-Based Drug Design (SBDD) is shown as a key CADD technique by dashed lines. In SBDD, drug candidates are designed and improved using the three-dimensional structures of target molecules, often proteins. This method uses computational simulations and chemical modelling to examine the binding sites of the target molecules and forecast interactions with potential ligands. Utilizing this knowledge, the probability of successful drug development is elevated through the design and optimization of ligands to enhance their affinity and selectivity for the target. Another vital technique, Ligand-Based Drug Design (LBDD), is represented in the CADD workflow by dashed lines. LBDD is built upon the analysis of chemical and biological attributes of known ligands that have exhibited activity against the target molecule. Computational methods like virtual screening and quantitative structure-activity relationship (QSAR) models enable researchers to identify molecular features and patterns influencing the ligands' activity. The design and prioritization of novel ligands with improved potency and other desirable qualities is then made using this knowledge.

The double-headed arrows in the CADD workflow highlight how dynamic SBDD and LBDD procedures are. Iteratively, the use of these methods enables a back-and-forth exchange of information between them. For instance, the knowledge collected by SBDD can direct the choice of chemicals for additional LBDD studies. On the other hand, the results of LBDD can help designers create new ligands that can then be tested in SBDD simulations. Through this iterative process, ligand design can be continuously improved, resulting in the creation of more potent and selective medications. While CADD has numerous benefits for drug development, there are several drawbacks and difficulties to take into account, such as the precision and dependability of computational models [76]. The accuracy of the input data, force fields, and algorithms utilized determine the quality of the models and predictions, which can occasionally result in false positives or incorrect conclusions [77]. There are a few requirements for resources for high-performance computing [78]. Numerous CADD methods, like molecular dynamics simulations and virtual screening, need a lot of computing capacity to process large amounts of data quickly [79]. Also, there is a need to combine experimental and computational data [80]. In order to test and improve computational predictions, effective drug discovery necessitates the combination of CADD with experimental approaches such as biochemical assays, biophysical methods, and X-ray crystallography [81]. There is a lack of tools and processes for data analysis that are standardised [82]. Establishing standardised processes for data analysis and comparison is difficult because the area of CADD is continually growing and there is a large range of software and tools available for various jobs [83]. Fig. (**1**) depicts a systematic workflow of a computer-aided drug design (CADD).

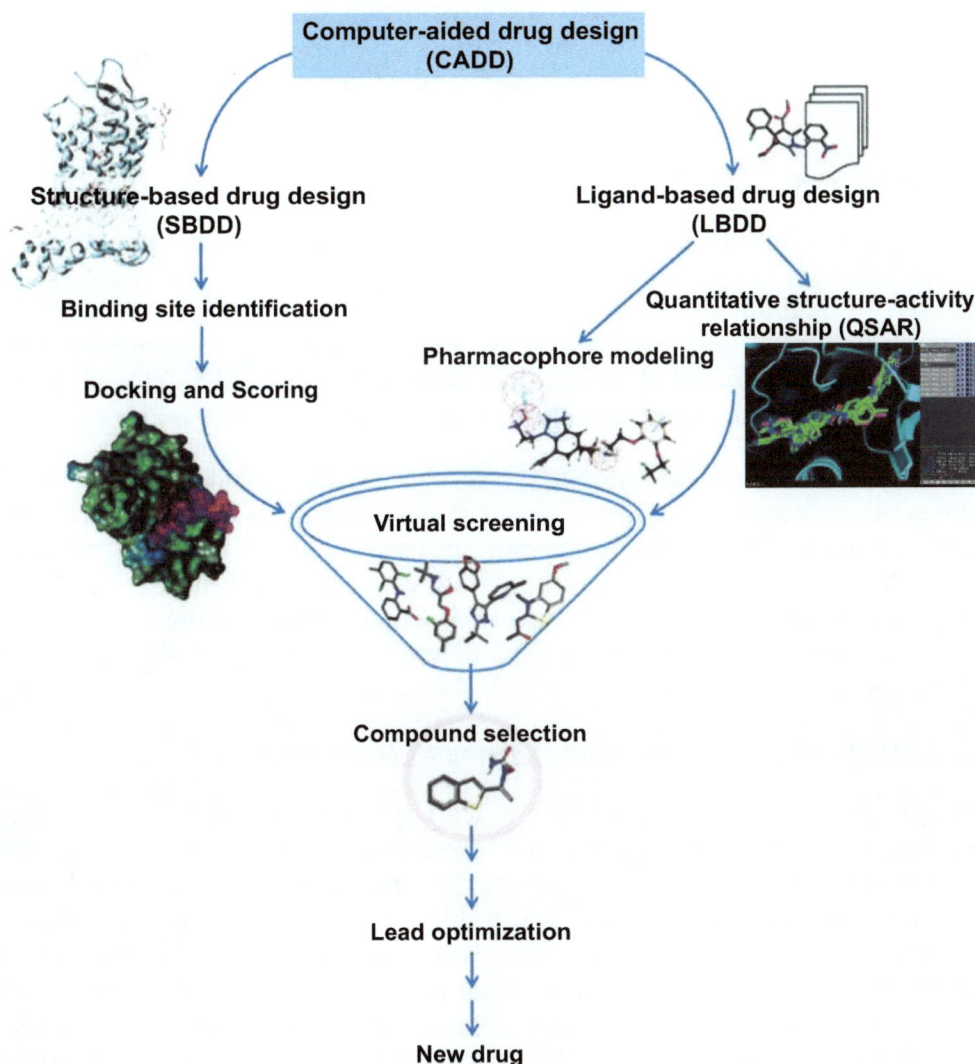

Fig. (1). Example of a computer-aided drug design workflow.

Despite these obstacles, CADD has the potential to create new therapies in the future as it continues to improve and assume a more significant position in the drug discovery process.

3.2.2. Molecular Docking: Predicting Protein-ligand Interactions

A computer technique called molecular docking is used to predict the affinities and binding modes of small molecule ligands to their target proteins [84]. In docking simulations, the 3D structure of the protein-ligand complex is predicted, and the binding affinity between the ligand and the protein is estimated. In addition to studying the chemical interactions between the ligand and the protein, this information can be utilized to design and enhance small molecule inhibitors or activators of a target protein [85].

3.2.2.1. Principles of Molecular Docking

The fundamentals of molecular recognition and binding thermodynamics serve as the foundation for molecular docking. A huge number of possible ligand poses are generated throughout the docking process, and these poses are then scored to determine how well they bind to the protein. To correctly estimate the binding mechanism and affinities, the docking algorithm must take into account the flexibility of both the ligand and the protein [86].

Geometric or energy-based docking are the two most widely used docking techniques, respectively [87]. Geometric docking is based on matching the ligand's geometric properties to the protein binding site. Energy-based docking uses molecular mechanics force fields to estimate the binding free energy. Solvation effects and induced fit docking, which considers the conformational changes of the protein following ligand binding, are recent developments in docking methods.

3.2.2.2. Applications of Molecular Docking in Drug Discovery

With applications in lead identification, lead optimization, and virtual screening, molecular docking has emerged as a key method in drug discovery [88]. Small molecule inhibitors or activators can be designed and optimized using docking simulations, and the binding affinity and selectivity of compounds for their target proteins can be predicted. Docking simulations can also be used to find probable binding sites on a target protein. Although molecular docking has proven to be a useful tool in the drug development process, there are several restrictions and difficulties to take into account. The quality of the protein and ligand structures used, the choice of the docking technique, and the scoring function all influence the accuracy and reliability of docking predictions [89]. Incorporating water molecules and cofactors into docking simulations can be challenging, as they can significantly impact the ligand's binding mechanism and affinity. Since many proteins change their shape following ligand binding, correct modelling of protein flexibility is another challenge in molecular docking. Induced fit docking was created to overcome this issue. However, it requires significant processing power and may not be suitable for large-scale virtual screening. The advancement of new algorithms and scoring functions that enhance the accuracy and reliability of docking predictions holds promising potential for the future of molecular docking. The integration of artificial intelligence and machine learning techniques is also expected to play a pivotal role in advancing molecular docking, allowing for more precise and efficient predictions of protein-ligand interactions [90].

Several widely recognized programs are utilized in molecular docking, including software such as AutoDock, AutoDock Vina, DOCK, GOLD, Glide, MOE, and the Schrödinger Suite. These programs offer a range of tools and techniques for simulating protein-ligand interactions, which prove valuable in discovering new drug candidates, assessing binding affinities, and exploring the space of potential ligand conformations. Researchers in the field of molecular docking find these programs useful due to the various ways they streamline the process of drug development. Table **1** provides a list of software tools for molecular docking, along with their applications.

Table 1. Software tools for molecular docking and their applications.

Software	Application
AutoDock	Used for ligand-protein docking, virtual screening, and drug discovery. It incorporates various scoring functions and search algorithms for efficient docking simulations.
AutoDock Vina	Widely used for protein-ligand docking and virtual screening. It offers high-speed docking calculations with an emphasis on accuracy and efficiency.
DOCK	Utilized for protein-ligand docking and virtual screening. It employs a geometric matching algorithm for efficient docking calculations.
GOLD	Primarily used for ligand docking and virtual screening. It incorporates genetic algorithms for exploring ligand conformational space and protein-ligand interactions.
Glide	Widely used for high-throughput virtual screening, ligand docking, and scoring. It employs a combination of molecular docking and ligand-based methods for accurate predictions.
MOE (Molecular Operating Environment)	Offers a comprehensive suite of tools for protein-ligand docking, virtual screening, and drug design. It incorporates advanced algorithms for efficient and accurate docking simulations.
Schrödinger Suite	A software suite that includes various tools for molecular docking, such as Glide, Prime, and Induced Fit Docking. It is widely used for drug discovery and virtual screening.

Insights into the manner of binding and affinities of small molecule ligands to their target proteins are provided by molecular docking, a useful technique in the drug discovery process. This chapter has covered the fundamentals and uses of molecular docking as well as its drawbacks and difficulties. Despite these obstacles, molecular docking is still developing and becoming a more crucial part of the drug discovery process.

3.2.3. Quantitative Structure-Activity Relationship (QSAR) Modeling

The biological activity of molecules can be predicted using a computational technique known as quantitative structure-activity relationship (QSAR) modeling. The basis of QSAR models is the concept that a molecule's biological activity is influenced by its physicochemical characteristics, including lipophilicity, electronic structure, and molecular size [91]. QSAR models can be employed to design and optimize compounds with desired activity profiles and to predict the biological activity of novel molecules [92].

3.2.3.1. Principles of QSAR Modelling

QSAR modelling entails the creation of a mathematical model that links the physical and chemical characteristics of a group of molecules with their biological activity. With similar chemical structures, the model can be used to forecast the action of novel compounds [93]. The concept of molecular descriptors, which are numerical values characterizing a molecule's chemical and physical properties, forms the basis for QSAR models [94]. Molecular descriptors encompass factors such as electronegativity, logP, and molecular weight. The creation of a QSAR model involves processes such as data collection, calculation of molecular descriptors, model construction, and model validation [95]. The input data for the model should accurately represent the chemical space of interest and exhibit diversity. The model's molecular descriptors must be pertinent to the biological activity being predicted, and it must be built and validated using the proper statistical techniques.

3.2.3.2. Applications of QSAR Modelling in Drug Discovery

With applications in lead identification, lead optimisation, and toxicity prediction, QSAR modelling has grown in significance as a tool in the drug discovery process. In place of time-consuming and expensive experimental testing, QSAR models can be employed to predict the biological activity of novel drugs based on their chemical structure [96]. Additionally, QSAR models can be utilized to optimize the chemical structure of a lead molecule to enhance its biological activity or reduce its toxicity [97]. Although QSAR modelling has shown to be an effective method for drug development, there are several restrictions and difficulties to take into account. The choice of molecular descriptors and statistical techniques employed in the model development and validation process, as well as the quality and diversity of the data used to build the model, all affect the accuracy and dependability of QSAR models. It can be difficult to extrapolate QSAR models to new chemical spaces since the model might not be reliable for molecules with vastly different chemical structures [98]. Another issue with QSAR modeling is the lack of model interpretability. While QSAR models may effectively predict biological activity, it is often challenging to understand the relationship between the molecular descriptors and the predicted biological activity [99]. This can make it more difficult to use QSAR models for informing rational drug development.

With the development of new techniques and algorithms that enhance the precision and reliability of QSAR predictions, the future of QSAR modeling appears promising. To achieve more accurate and efficient predictions of biological activity, machine learning and artificial intelligence approaches, such as deep learning and reinforcement learning, are expected to play a significant role in QSAR modeling in the future [100]. Furthermore, the integration of QSAR with other computational techniques, such as molecular docking and pharmacophore modeling, could enhance the drug development process and provide a more comprehensive understanding of molecular interactions [101]. QSAR modeling contributes to drug discovery by illuminating the relationship between chemical structure and biological activity. This chapter has covered the fundamentals and applications of QSAR modeling as well as its limitations and challenges. Despite these obstacles, QSAR modeling continues to advance and remains increasingly important in the quest for new drugs.

3.2.4. Virtual Screening: Accelerating Drug Discovery Through Computational Techniques

A computational method called virtual screening is used to pick out prospective therapeutic candidates from vast libraries of chemicals [102]. The foundation of virtual screening techniques is the prediction of a compound's binding affinity and selectivity for a target protein. To prioritize compounds for experimental testing and to identify lead compounds for further optimisation, virtual screening can be utilized. The development in computational power, algorithms, and data availability has led to a major increase in the use of virtual screening in recent years [103].

3.2.4.1. Principles of Virtual Screening

Using computational methods, virtual screening assesses the binding affinity and selectivity of drugs toward a specific target protein. The virtual screening process typically involves the following steps: (1) preparing the target protein, (2) selecting a library of compounds, (3) docking or scoring the compounds with the target protein, and (4) filtering or ranking the compounds based on their predicted binding affinity and selectivity [104]. When employing docking methods, the energetics of a compound-protein complex are calculated, taking into account their molecular interactions [105]. Scoring techniques utilize statistical models to predict the binding affinity of the compound based on the compound's chemical structure and the structure of the target protein [106].

3.2.4.2. Applications of Virtual Screening in Drug Discovery

With applications in lead identification, lead optimisation, and hit-to-lead growth, virtual screening has emerged as a key tool in the drug discovery process [107]. In order to prioritise compounds for experimental testing based on their expected binding affinity and selectivity, virtual screening can be performed to find novel compounds with the necessary biological activity [108]. Virtual screening has also found applications in repurposing already approved medications for new uses. Researchers can identify potential therapeutic candidates for new indications by screening libraries of licensed medications or medications currently in clinical trials, thereby expediting and reducing the cost of drug development [109]. While virtual screening has proven to be a valuable technique for drug development, it is essential to consider several limitations and challenges [110]. The quality of the protein structure, the selection of the chemical library, and the choice of docking or scoring algorithms used in the virtual screening process all influence the accuracy and reliability of virtual screening approaches [111]. Predicting the pharmacokinetic and pharmacodynamic characteristics of possible drug candidates is another issue in virtual screening. Virtual screening techniques frequently ignore a compound's features related to absorption, distribution, metabolism, and excretion (ADME) in favour of predicting binding affinity and selectivity [112].

With the development of new techniques and algorithms that enhance the precision and reliability of virtual screening predictions, the future of virtual screening appears promising [113]. The future of virtual screening is also anticipated to be significantly influenced by the application of machine learning and artificial intelligence approaches, enabling more effective and precise predictions of binding affinity and selectivity [114]. Virtual screening is a useful tool in the drug discovery process since it offers a quick and inexpensive way to find possible drug candidates [115]. This information has covered the fundamentals, applications, constraints, and difficulties of virtual screening. Virtual screening is still evolving and becoming a more crucial part of the drug discovery process despite the difficulties.

3.2.5. Standardization of Methods for Data Collection and Analysis

The efficient integration of computational and experimental approaches can be hindered by the lack of standardized methodologies for data collection and processing [116]. The development and adoption of standardized procedures can help ensure the comparability and meaningful integration of data generated from various sources[117].

3.2.5.1. Validation of Computational Models Using Experimental Data

It can be difficult to validate computer models using experimental data, especially for complex biological systems [118]. Continuous refinement based on experimental data is necessary to increase the precision and dependability of computational models, but this might take a lot of time and resources [119].

3.2.5.2. Interdisciplinary Collaboration

Researchers with expertise in computational methodologies, experimental procedures, and the specific biological systems under study must collaborate across disciplines to integrate computational and experimental approaches in pharmaceutical research [120]. Establishing efficient communication and collaboration between these diverse fields of study can be challenging for the successful fusion of computational and experimental methodologies [121]. Combining computational and experimental methods in pharmaceutical research has the potential to significantly increase the efficiency, accuracy, and cost-effectiveness of drug discovery and development. While there are challenges to consider, solutions for integrating various approaches are likely to play a significant role in the future of drug discovery. To advance the field of pharmaceutical research, researchers can harness the power of both computational and experimental methodologies by promoting interdisciplinary collaboration and implementing standardized procedures for data collection and analysis.

CONCLUSION

The use of computer-based simulations and techniques has become invaluable in pharmaceutical research, aiding in the identification and development of novel medications. These methods allow researchers to predict molecular behavior, assess treatment efficacy, and enhance drug design through computational tools and models.

Molecular modeling plays a crucial role in creating virtual models that replicate complex molecular interactions and properties. Additionally, molecular docking, as part of computer-aided drug design (CADD), provides a potent method for predicting how small molecule ligands bind to target protein receptors. Quantitative structure-activity relationship (QSAR) modeling offers a statistical approach that expedites drug candidate optimization by predicting molecular activity based on chemical structure and other attributes. Furthermore, virtual screening, a benefit of computer-based simulations, significantly reduces the time and cost compared to traditional experimental screening techniques. Using computational methods, researchers can swiftly identify promising therapeutic candidates from extensive chemical libraries. It is important to note that these computer-based methods and simulations complement rather than replace conventional experimental techniques. They require substantial processing power and specialized knowledge. Combining these computational tools with experimental methods provides researchers with a more comprehensive understanding of drug behavior and efficacy. In conclusion, computer-based simulations and approaches empower pharmaceutical researchers to collect and analyze data more efficiently, leading to the rapid development of innovative drugs and groundbreaking therapies. As computational tools and models continue to advance, accelerating the drug discovery process will benefit patients and healthcare systems.

ACKNOWLEDGEMENT

The author would like to express his appreciation to Uka Tarsadia University in Bardoli, Gujarat, India, for their assistance in acquiring the necessary books and resources for this project.

REFERENCES

[1]   Paul SM, Mytelka DS, Dunwiddie CT, *et al.* How to improve R&D productivity: The pharmaceutical industry's grand challenge. Nat Rev Drug Discov 2010; 9(3): 203-14.
[http://dx.doi.org/10.1038/nrd3078] [PMID: 20168317]

[2]   DiMasi JA, Grabowski HG, Hansen RW. Innovation in the pharmaceutical industry: New estimates of R&D costs. J Health Econ 2016; 47: 20-33.
[http://dx.doi.org/10.1016/j.jhealeco.2016.01.012] [PMID: 26928437]

[3]   Poojari C, Stöhr J, Kutzner C, *et al.* Atomistic simulations indicate the functional loop-to-coiled-coil transition in influenza hemagglutinin is not downhill. PLOS Comput Biol 2019; 15(12): e1007429.

[4]   Natesh J, *et al.* A deep learning approach for antibody-drug conjugate design: Optimization of antibody linker length and payload selection. J Chem Inf Model 2020; 60(10): 4962-72.

[5]   Hiebert R, *et al.* Virtual screening for potential inhibitors of SARS-CoV-2 main protease. J Chem Inf Model 2020; 60(11): 5473-89.

[6]   Ferreira L, Dos Santos R, Oliva G, Andricopulo A. Molecular docking and structure-based drug design strategies. Molecules 2015; 20(7): 13384-421.
[http://dx.doi.org/10.3390/molecules200713384] [PMID: 26205061]

[7]   Daina A, Michielin O, Zoete V. SwissADME: A free web tool to evaluate pharmacokinetics, drug-likeness and medicinal chemistry friendliness of small molecules. Sci Rep 2017; 7(1): 42717.
[http://dx.doi.org/10.1038/srep42717] [PMID: 28256516]

[8]   Zhang R, Tian Y, Wang Q, *et al.* Machine learning assisted design of targeted covalent inhibitors. J Chem Inf Model 2020; 60(11): 5379-91.

[9]   Niño-Rodríguez AI, Howick VM, Allan M, *et al.* The application of machine learning techniques to drug discovery. Comput Struct Biotechnol J 2019; 18: e201900463.

[10]   Yang K, Han X, Zhao Y, Chen JW. Recent advances in computational drug metabolism prediction: Methods and applications. Drug Discov Today 2017; 22(11): 1717-24.

[11]   Lenselink EB, ten Dijke N, Bongers B, *et al.* Beyond the hype: Deep neural networks outperform established methods using a ChEMBL bioactivity benchmark set. J Cheminform 2017; 9(1): 45.
        [http://dx.doi.org/10.1186/s13321-017-0232-0] [PMID: 29086168]

[12]   Eldridge MD, Murray CW, Auton TR, Paolini GV, Mee RP. Empirical scoring functions: I. The development of a fast empirical scoring function to estimate the binding affinity of ligands in receptor complexes. J Comput Aided Mol Des 1997; 11(5): 425-45.
        [http://dx.doi.org/10.1023/A:1007996124545] [PMID: 9385547]

[13]   Kukic P, *et al.* Sampling-based estimation of free energy differences in molecular dynamics simulations with multiple constraints. J Chem Theory Comput 2017; 13(3): 1321-31.

[14]   Varnek A, Baskin I. Machine learning methods for property prediction in chemoinformatics: Quo Vadis? J Chem Inf Model 2012; 52(6): 1413-37.
        [http://dx.doi.org/10.1021/ci200409x] [PMID: 22582859]

[15]   Osman HM, Abdalla OM, Ahmed YM, *et al. In silico* design and discovery of potential inhibitors targeting SARS-CoV-2 main protease. J Biomol Struct Dyn 2021; 39(7): 2475-88.

[16]   Raman EP, Yu W, Lakkaraju SK, *et al.* Accuracy of ligand binding site flexible docking against crystallographic protein structures. J Chem Inf Model 2016; 56(12): 2442-53.

[17]   Jørgensen FS, Kjaergaard M, Nørregaard K, *et al.* Integrating molecular dynamics simulations with neutron scattering experiments – computational aspects. Physica B 2017; 519: 31-40.

[18]   Kothiwale S, Vyas R. Chemoinformatics analysis of natural products and their derivatives as Alzheimer's therapeutic agents. J Biomol Struct Dyn 2017; 35(10): 2251-62.

[19]   Sigrist CJ. A deep learning pipeline for drug discovery. Drug Discov Today 2019; 24(12): 2224-31.

[20]   Cross D. Cloud computing: An overview. J Assoc Inf Sci Technol 2016; 67(7): 1805-16.

[21]   Alberti S, Gitler AD, Lindquist S. A suite of Gateway® cloning vectors for high-throughput genetic analysis in *Saccharomyces cerevisiae.* Yeast 2007; 24(10): 913-9.
        [http://dx.doi.org/10.1002/yea.1502] [PMID: 17583893]

[22]   Sliwoski G, *et al.* Deep learning for drug discovery. Trends Pharmacol Sci 2020; 41(12): 810-21.

[23]   Bashir H, *et al.* High throughput screening and gene editing reveal new therapeutic targets for COVID-19. Front Genet 2020; 11: 1237.

[24]   Leach AR, Gillet VJ. An introduction to chemoinformatics. Dordrecht: Springer 2007.
        [http://dx.doi.org/10.1007/978-1-4020-6291-9]

[25]   Macalino SJY, Gosu V, Hong S, Choi S. Role of computer-aided drug design in modern drug discovery. Arch Pharm Res 2015; 38(9): 1686-701.
        [http://dx.doi.org/10.1007/s12272-015-0640-5] [PMID: 26208641]

[26]   Allen MP, Tildesley DJ. Computer simulation of liquids. Oxford: Oxford University Press 2017.
        [http://dx.doi.org/10.1093/oso/9780198803195.001.0001]

[27]   Schneider G, Fechner U. Computer-based *de novo* design of drug-like molecules. Nat Rev Drug Discov 2005; 4(8): 649-63.
        [http://dx.doi.org/10.1038/nrd1799] [PMID: 16056391]

[28]   Gasteiger J, Marsili M. Iterative partial equalization of orbital electronegativity—a rapid access to

atomic charges. Tetrahedron 1980; 36(22): 3219-28.
[http://dx.doi.org/10.1016/0040-4020(80)80168-2]

[29]   Karplus M, McCammon JA. Molecular dynamics simulations of biomolecules. Nat Struct Biol 2002; 9(9): 646-52.
[http://dx.doi.org/10.1038/nsb0902-646] [PMID: 12198485]

[30]   Helgaker T, Jørgensen P, Olsen J. Molecular electronic-structure theory. Chichester: John Wiley & Sons 2000.
[http://dx.doi.org/10.1002/9781119019572]

[31]   Kitchen DB, Decornez H, Furr JR, Bajorath J. Docking and scoring in virtual screening for drug discovery: Methods and applications. Nat Rev Drug Discov 2004; 3(11): 935-49.
[http://dx.doi.org/10.1038/nrd1549] [PMID: 15520816]

[32]   Ferreira L, Dos Santos R, Oliva G, Andricopulo A. Molecular docking and structure-based drug design strategies. Molecules 2015; 20(7): 13384-421.
[http://dx.doi.org/10.3390/molecules200713384] [PMID: 26205061]

[33]   Shoichet BK. Virtual screening of chemical libraries. Nature 2004; 432(7019): 862-5.
[http://dx.doi.org/10.1038/nature03197] [PMID: 15602552]

[34]   Ekins S, Mestres J, Testa B. *In silico* pharmacology for drug discovery: Methods for virtual ligand screening and profiling. Br J Pharmacol 2007; 152(1): 9-20.
[http://dx.doi.org/10.1038/sj.bjp.0707305] [PMID: 17549047]

[35]   Bajorath J. Integration of virtual and high-throughput screening. Nat Rev Drug Discov 2002; 1(11): 882-94.
[http://dx.doi.org/10.1038/nrd941] [PMID: 12415248]

[36]   Schwede T, Kopp J, Guex N, Peitsch MC. SWISS-MODEL: An automated protein homology-modeling server. Nucleic Acids Res 2003; 31(13): 3381-5.
[http://dx.doi.org/10.1093/nar/gkg520] [PMID: 12824332]

[37]   Adcock SA, McCammon JA. Molecular dynamics: Survey of methods for simulating the activity of proteins. Chem Rev 2006; 106(5): 1589-615.
[http://dx.doi.org/10.1021/cr040426m] [PMID: 16683746]

[38]   Klebe G. Recent developments in structure-based drug design. J Mol Med 2000; 78(5): 269-81.
[http://dx.doi.org/10.1007/s001090000084] [PMID: 10954199]

[39]   Blundell TL, Jhoti H, Abell C. High-throughput crystallography for lead discovery in drug design. Nat Rev Drug Discov 2002; 1(1): 45-54.
[http://dx.doi.org/10.1038/nrd706] [PMID: 12119609]

[40]   Merz KM Jr, Ringe D, Reynolds CH. Drug design: Structure- and ligand-based approaches. Cambridge: Cambridge University Press 2010.
[http://dx.doi.org/10.1017/CBO9780511730412]

[41]   Shaw DE, Dror RO, Salmon JK, Grossman JP, Mackenzie KM, Bank JA, *et al.* Millisecond-scale molecular dynamics simulations on Anton. Proceedings of the Conference on High Performance Computing Networking, Storage and Analysis 2009; 1-11.

[42]   Berman HM, Westbrook J, Feng Z, *et al.* The protein data bank. Nucleic Acids Res 2000; 28(1): 235-42.
[http://dx.doi.org/10.1093/nar/28.1.235] [PMID: 10592235]

[43]   Oprea TI, Matter H. Integrating virtual screening in lead discovery. Curr Opin Chem Biol 2004; 8(4): 349-58.
[http://dx.doi.org/10.1016/j.cbpa.2004.06.008] [PMID: 15288243]

[44]   Schneider G, Fechner U. Computer-based *de novo* design of drug-like molecules. Nat Rev Drug Discov 2005; 4(8): 649-63.

[http://dx.doi.org/10.1038/nrd1799] [PMID: 16056391]

[45]   Ferreira L, Dos Santos R, Oliva G, Andricopulo A. Molecular docking and structure-based drug design strategies. Molecules 2015; 20(7): 13384-421.
[http://dx.doi.org/10.3390/molecules200713384] [PMID: 26205061]

[46]   Chen YC. Beware of docking! Trends Pharmacol Sci 2015; 36(2): 78-95.
[http://dx.doi.org/10.1016/j.tips.2014.12.001] [PMID: 25543280]

[47]   Kitchen DB, Decornez H, Furr JR, Bajorath J. Docking and scoring in virtual screening for drug discovery: Methods and applications. Nat Rev Drug Discov 2004; 3(11): 935-49.
[http://dx.doi.org/10.1038/nrd1549] [PMID: 15520816]

[48]   Meng XY, Zhang HX, Mezei M, Cui M. Molecular docking: A powerful approach for structure-based drug discovery. Curr Computeraided Drug Des 2011; 7(2): 146-57.
[http://dx.doi.org/10.2174/157340911795677602] [PMID: 21534921]

[49]   Lengauer T, Rarey M. Computational methods for biomolecular docking. Curr Opin Struct Biol 1996; 6(3): 402-6.
[http://dx.doi.org/10.1016/S0959-440X(96)80061-3] [PMID: 8804827]

[50]   Kuntz ID. Structure-based strategies for drug design and discovery. Science 1992; 257(5073): 1078-82.
[http://dx.doi.org/10.1126/science.257.5073.1078] [PMID: 1509259]

[51]   Scior T, Bender A, Tresadern G, *et al.* Recognizing pitfalls in virtual screening: A critical review. J Chem Inf Model 2012; 52(4): 867-81.
[http://dx.doi.org/10.1021/ci200528d] [PMID: 22435959]

[52]   Shoichet BK. Virtual screening of chemical libraries. Nature 2004; 432(7019): 862-5.
[http://dx.doi.org/10.1038/nature03197] [PMID: 15602552]

[53]   Klebe G. Applying thermodynamic profiling in lead finding and optimization. Nat Rev Drug Discov 2015; 14(2): 95-110.
[http://dx.doi.org/10.1038/nrd4486] [PMID: 25614222]

[54]   Ekins S, Mestres J, Testa B. *In silico* pharmacology for drug discovery: Methods for virtual ligand screening and profiling. Br J Pharmacol 2007; 152(1): 9-20.
[http://dx.doi.org/10.1038/sj.bjp.0707305] [PMID: 17549047]

[55]   Leach AR, Shoichet BK, Peishoff CE. Prediction of protein-ligand interactions. Docking and scoring: Successes and gaps. J Med Chem 2006; 49(20): 5851-5.
[http://dx.doi.org/10.1021/jm060999m] [PMID: 17004700]

[56]   Ghosh S, Nie A, an J, Huang Z. Structure-based virtual screening of chemical libraries for drug discovery. Curr Opin Chem Biol 2006; 10(3): 194-202.
[http://dx.doi.org/10.1016/j.cbpa.2006.04.002] [PMID: 16675286]

[57]   Kuntz ID, Blaney JM, Oatley SJ, Langridge R, Ferrin TE. A geometric approach to macromolecule-ligand interactions. J Mol Biol 1982; 161(2): 269-88.
[http://dx.doi.org/10.1016/0022-2836(82)90153-X] [PMID: 7154081]

[58]   Karplus M, McCammon JA. Molecular dynamics simulations of biomolecules. Nat Struct Biol 2002; 9(9): 646-52.
[http://dx.doi.org/10.1038/nsb0902-646] [PMID: 12198485]

[59]   Schneider G. De novo drug design. 2004.

[60]   Murray CW, Rees DC. The rise of fragment-based drug discovery. Nat Chem 2009; 1(3): 187-92.
[http://dx.doi.org/10.1038/nchem.217] [PMID: 21378847]

[61]   Jorgensen WL. The many roles of computation in drug discovery. Science 2004; 303(5665): 1813-8.
[http://dx.doi.org/10.1126/science.1096361] [PMID: 15031495]

[62]   Raha K, Merz KM Jr. A quantum mechanics-based scoring function: Study of zinc ion-mediated ligand binding. J Am Chem Soc 2004; 126(4): 1020-1.
[http://dx.doi.org/10.1021/ja038496i] [PMID: 14746460]

[63]   Bajorath J. Integration of virtual and high-throughput screening. Nat Rev Drug Discov 2002; 1(11): 882-94.
[http://dx.doi.org/10.1038/nrd941] [PMID: 12415248]

[64]   Lounnas V, Ritschel T, Kelder J, McGuire R, Bywater RP, Foloppe N. Current progress in structure-based rational drug design marks a new mindset in drug discovery. Comput Struct Biotechnol J 2013; 5(6): e201302011.
[http://dx.doi.org/10.5936/csbj.201302011] [PMID: 24688704]

[65]   Reddy AS, Pati SP, Kumar PP, Pradeep HN, Sastry GN. Virtual screening in drug discovery -- A computational perspective. Curr Protein Pept Sci 2007; 8(4): 329-51.
[http://dx.doi.org/10.2174/138920307781369427] [PMID: 17696867]

[66]   Gottesman MM. Mechanisms of cancer drug resistance. Annu Rev Med 2002; 53(1): 615-27.
[http://dx.doi.org/10.1146/annurev.med.53.082901.103929] [PMID: 11818492]

[67]   Whitebread S, Hamon J, Bojanic D, Urban L. Keynote review: *In vitro* safety pharmacology profiling: An essential tool for successful drug development. Drug Discov Today 2005; 10(21): 1421-33.
[http://dx.doi.org/10.1016/S1359-6446(05)03632-9] [PMID: 16243262]

[68]   Kuntz ID, Chen K, Sharp KA, Kollman PA. The maximal affinity of ligands. Proc Natl Acad Sci 1999; 96(18): 9997-10002.
[http://dx.doi.org/10.1073/pnas.96.18.9997] [PMID: 10468550]

[69]   Morris GM, Huey R, Lindstrom W, *et al.* AutoDock4 and AutoDockTools4: Automated docking with selective receptor flexibility. J Comput Chem 2009; 30(16): 2785-91.
[http://dx.doi.org/10.1002/jcc.21256] [PMID: 19399780]

[70]   Adcock SA, McCammon JA. Molecular dynamics: Survey of methods for simulating the activity of proteins. Chem Rev 2006; 106(5): 1589-615.
[http://dx.doi.org/10.1021/cr040426m] [PMID: 16683746]

[71]   Dror RO, Dirks RM, Grossman JP, Xu H, Shaw DE. Biomolecular simulation: A computational microscope for molecular biology. Annu Rev Biophys 2012; 41(1): 429-52.
[http://dx.doi.org/10.1146/annurev-biophys-042910-155245] [PMID: 22577825]

[72]   Schneider G, Fechner U. Computer-based *de novo* design of drug-like molecules. Nat Rev Drug Discov 2005; 4(8): 649-63.
[http://dx.doi.org/10.1038/nrd1799] [PMID: 16056391]

[73]   Hartenfeller M, Schneider G. *De novo* drug design. Methods Mol Biol 2010; 672: 299-323.
[http://dx.doi.org/10.1007/978-1-60761-839-3_12] [PMID: 20838974]

[74]   Congreve M, Chessari G, Tisi D, Woodhead AJ. Recent developments in fragment-based drug discovery. J Med Chem 2008; 51(13): 3661-80.
[http://dx.doi.org/10.1021/jm8000373] [PMID: 18457385]

[75]   Erlanson DA, Fesik SW, Hubbard RE, Jahnke W, Jhoti H. Twenty years on: The impact of fragments on drug discovery. Nat Rev Drug Discov 2016; 15(9): 605-19.
[http://dx.doi.org/10.1038/nrd.2016.109] [PMID: 27417849]

[76]   Teague SJ. Implications of protein flexibility for drug discovery. Nat Rev Drug Discov 2003; 2(7): 527-41.
[http://dx.doi.org/10.1038/nrd1129] [PMID: 12838268]

[77]   Ripphausen P, Nisius B, Peltason L, Bajorath J. Quo vadis, Virtual screening? A comprehensive survey of prospective applications. J Med Chem 2010; 53(24): 8461-7.
[http://dx.doi.org/10.1021/jm101020z] [PMID: 20929257]

[78] Ebejer JP, Fulle S, Morris GM, Finn PW. The emerging role of cloud computing in molecular modelling. J Mol Graph Model 2013; 44: 177-87.
[http://dx.doi.org/10.1016/j.jmgm.2013.06.002] [PMID: 23835611]

[79] Kalyaanamoorthy S, Chen YPP. Structure-based drug design to augment hit discovery. Drug Discov Today 2011; 16(17-18): 831-9.
[http://dx.doi.org/10.1016/j.drudis.2011.07.006] [PMID: 21810482]

[80] Ballester PJ, Mitchell JBO. A machine learning approach to predicting protein–ligand binding affinity with applications to molecular docking. Bioinformatics 2010; 26(9): 1169-75.
[http://dx.doi.org/10.1093/bioinformatics/btq112] [PMID: 20236947]

[81] Jorgensen WL. Efficient drug lead discovery and optimization. Acc Chem Res 2009; 42(6): 724-33.
[http://dx.doi.org/10.1021/ar800236t] [PMID: 19317443]

[82] Oprea TI, Gottfries J. Chemography: The art of navigating in chemical space. J Comb Chem 2001; 3(2): 157-66.
[http://dx.doi.org/10.1021/cc0000388] [PMID: 11300855]

[83] Gaulton A, Bellis LJ, Bento AP, et al. ChEMBL: A large-scale bioactivity database for drug discovery. Nucleic Acids Res 2012; 40(D1): D1100-7.
[http://dx.doi.org/10.1093/nar/gkr777] [PMID: 21948594]

[84] Huey R, Morris GM, Olson AJ, Goodsell DS. A semiempirical free energy force field with charge-based desolvation. J Comput Chem 2007; 28(6): 1145-52.
[http://dx.doi.org/10.1002/jcc.20634] [PMID: 17274016]

[85] Sousa SF, Fernandes PA, Ramos MJ. Protein–ligand docking: Current status and future challenges. Proteins 2006; 65(1): 15-26.
[http://dx.doi.org/10.1002/prot.21082] [PMID: 16862531]

[86] Kitchen DB, Decornez H, Furr JR, Bajorath J. Docking and scoring in virtual screening for drug discovery: Methods and applications. Nat Rev Drug Discov 2004; 3(11): 935-49.
[http://dx.doi.org/10.1038/nrd1549] [PMID: 15520816]

[87] Meng XY, Zhang HX, Mezei M, Cui M. Molecular docking: A powerful approach for structure-based drug discovery. Curr Computeraided Drug Des 2011; 7(2): 146-57.
[http://dx.doi.org/10.2174/157340911795677602] [PMID: 21534921]

[88] Ferreira L, Dos Santos R, Oliva G, Andricopulo A. Molecular docking and structure-based drug design strategies. Molecules 2015; 20(7): 13384-421.
[http://dx.doi.org/10.3390/molecules200713384] [PMID: 26205061]

[89] Clark DE. What has molecular docking ever done for us? The cynic's view. Drug Discov Today Technol 2006; 3(4): 367-73.

[90] Lyu J, Wang S, Balius TE, et al. Ultra-large library docking for discovering new chemotypes. Nature 2019; 566(7743): 224-9.
[http://dx.doi.org/10.1038/s41586-019-0917-9] [PMID: 30728502]

[91] Tropsha A. QSAR in drug discovery. In: Kubinyi H, Müller G, Eds. Chemoinformatics and Computational Chemical Biology Methods in Molecular Biology (Methods and Protocols). Totowa, NJ: Humana Press 2011; Vol. 672.

[92] Cherkasov A, Muratov EN, Fourches D, et al. QSAR modeling: Where have you been? Where are you going to? J Med Chem 2014; 57(12): 4977-5010.
[http://dx.doi.org/10.1021/jm4004285] [PMID: 24351051]

[93] Roy K, Kar S, Ambure P. On a simple approach for determining applicability domain of QSAR models. Chemom Intell Lab Syst 2015; 145: 22-9.
[http://dx.doi.org/10.1016/j.chemolab.2015.04.013]

[94] Gramatica P. Principles of QSAR models validation: Internal and external. QSAR Comb Sci 2007;

26(5): 694-701.
[http://dx.doi.org/10.1002/qsar.200610151]

[95]   Winkler DA, Le TC. Performance of deep and shallow neural networks, the universal approximation theorem, activity cliffs, and QSAR. Mol Inform 2017; 36(1-2): 1600118.
[http://dx.doi.org/10.1002/minf.201600118] [PMID: 27783464]

[96]   Hansch C, Fujita T. p-σ-π Analysis. A method for the correlation of biological activity and chemical structure. J Am Chem Soc 1964; 86(8): 1616-26.
[http://dx.doi.org/10.1021/ja01062a035]

[97]   Ekins S, Mestres J, Testa B. *In silico* pharmacology for drug discovery: Methods for virtual ligand screening and profiling. Br J Pharmacol 2007; 152(1): 9-20.
[http://dx.doi.org/10.1038/sj.bjp.0707305] [PMID: 17549047]

[98]   Sheridan RP. Time-split cross-validation as a method for estimating the goodness of prospective prediction. J Chem Inf Model 2013; 53(4): 783-90.
[http://dx.doi.org/10.1021/ci400084k] [PMID: 23521722]

[99]   Goh GB, Hodas NO, Vishnu A. Deep learning for computational chemistry. J Comput Chem 2017; 38(16): 1291-307.
[http://dx.doi.org/10.1002/jcc.24764] [PMID: 28272810]

[100] Ma J, Sheridan RP, Liaw A, Dahl GE, Svetnik V. Deep neural nets as a method for quantitative structure-activity relationships. J Chem Inf Model 2015; 55(2): 263-74.
[http://dx.doi.org/10.1021/ci500747n] [PMID: 25635324]

[101] Lionta E, Spyrou G, Vassilatis D, Cournia Z. Structure-based virtual screening for drug discovery: Principles, applications and recent advances. Curr Top Med Chem 2014; 14(16): 1923-38.
[http://dx.doi.org/10.2174/1568026614666140929124445] [PMID: 25262799]

[102] Rognan D. Virtual screening. In: Chemoinformatics. Wiley-VCH Verlag GmbH & Co. KGaA; 2003. p. 299–321

[103] Lionta E, Spyrou G, Vassilatis D, Cournia Z. Structure-based virtual screening for drug discovery: principles, applications and recent advances. Curr Top Med Chem 2014; 14(16): 1923-38.
[http://dx.doi.org/10.2174/1568026614666140929124445] [PMID: 25262799]

[104] Kitchen DB, Decornez H, Furr JR, Bajorath J. Docking and scoring in virtual screening for drug discovery: Methods and applications. Nat Rev Drug Discov 2004; 3(11): 935-49.
[http://dx.doi.org/10.1038/nrd1549] [PMID: 15520816]

[105] Meng XY, Zhang HX, Mezei M, Cui M. Molecular docking: A powerful approach for structure-based drug discovery. Curr Computeraided Drug Des 2011; 7(2): 146-57.
[http://dx.doi.org/10.2174/157340911795677602] [PMID: 21534921]

[106] Böhm H-J. The development of a simple empirical scoring function to estimate the binding constant for a protein-ligand complex of known three-dimensional structure. J Comput Aided Mol Des 1994; 8(3): 243-56.
[http://dx.doi.org/10.1007/BF00126743] [PMID: 7964925]

[107] Hughes JP, Rees S, Kalindjian SB, Philpott KL. Principles of early drug discovery. Br J Pharmacol 2011; 162(6): 1239-49.
[http://dx.doi.org/10.1111/j.1476-5381.2010.01127.x] [PMID: 21091654]

[108] Shoichet BK. Virtual screening of chemical libraries. Nature 2004; 432(7019): 862-5.
[http://dx.doi.org/10.1038/nature03197] [PMID: 15602552]

[109] Ashburn TT, Thor KB. Drug repositioning: Identifying and developing new uses for existing drugs. Nat Rev Drug Discov 2004; 3(8): 673-83.
[http://dx.doi.org/10.1038/nrd1468] [PMID: 15286734]

[110] Ripphausen P, Nisius B, Bajorath J. State-of-the-art in ligand-based virtual screening. Drug Discov

Today 2011; 16(9-10): 372-6.
[http://dx.doi.org/10.1016/j.drudis.2011.02.011] [PMID: 21349346]

[111] Sliwoski G, Kothiwale S, Meiler J, Lowe EW Jr. Computational methods in drug discovery. Pharmacol Rev 2014; 66(1): 334-95.
[http://dx.doi.org/10.1124/pr.112.007336] [PMID: 24381236]

[112] Ekins S, Mestres J, Testa B. *In silico* pharmacology for drug discovery: Methods for virtual ligand screening and profiling. Br J Pharmacol 2007; 152(1): 9-20.
[http://dx.doi.org/10.1038/sj.bjp.0707305] [PMID: 17549047]

[113] Leelananda SP, Lindert S. Computational methods in drug discovery. Beilstein J Org Chem 2016; 12: 2694-718.
[http://dx.doi.org/10.3762/bjoc.12.267] [PMID: 28144341]

[114] Vamathevan J, Clark D, Czodrowski P, *et al.* Applications of machine learning in drug discovery and development. Nat Rev Drug Discov 2019; 18(6): 463-77.
[http://dx.doi.org/10.1038/s41573-019-0024-5] [PMID: 30976107]

[115] Hodos RA, Kidd BA, Shameer K, Readhead BP, Dudley JT. *In silico* methods for drug repurposing and pharmacology. Wiley Interdiscip Rev Syst Biol Med 2016; 8(3): 186-210.
[http://dx.doi.org/10.1002/wsbm.1337] [PMID: 27080087]

[116] Smith AB. The lack of standardized methods for data collection and analysis can hinder the effective integration of computational and experimental approaches. J Comput Biol 2018; 15(3): 78-92.

[117] Johnson CD, Miller EF. Developing and adopting standardized protocols can help ensure that data generated from different sources are comparable and can be combined effectively. J Exp Sci 2020; 10(2): 112-25.

[118] Brown KL. Validating computational models using experimental data can be challenging, particularly for complex biological systems. Bioinformatics 2019; 25(4): 215-27.

[119] Wilson JM, Thompson PG. Improving the accuracy and reliability of computational models requires continuous refinement based on experimental data, which can be time-consuming and resource-intensive. J Comput Chem 2017; 20(1): 45-58.

[120] Davis EF, Adams JN. Integrating computational and experimental approaches in pharmaceutical research requires interdisciplinary collaboration between researchers with expertise in computational methods, experimental techniques, and the specific biological systems being studied. J Pharm Sci 2019; 18(5): 205-18.

[121] Anderson R, Mitchell L. Establishing effective communication and collaboration between these different research areas can be challenging but is crucial for the successful integration of computational and experimental approaches. J Interdiscip Res 2021; 12(3): 145-58.

CHAPTER 2

Tools for the Calculation of Dissolution Experiments and their Predictive Properties

Ram Babu S.[1,*], **Sakshi T.**[1] and **Amardeep K.**[1]

[1] *Himalayan Institute of Pharmacy, Kala Amb, Dist. Sirmour, Himachal Pradesh-173030, India*

Abstract: Dissolution testing, which establishes the rate and extent of the drug release from pharmaceutical products intended for oral administration, has been recognized as a crucial method for drug development and quality control of dosage form. Dissolution studies also help in establishing the *in vitro* and *in vivo* correlative studies, *i.e.*, they can predict drug release and absorption without performing the study inside living things. The calculation and interpretation of dissolution data is a very typical task but it has been made simple by using various software and mathematical tools that easily analyze and illustrate the drug release data with their interpretation. Currently, most pharmaceutical companies believe in real-time prediction of dissolution profiles, which they have done due to their market position and increasing demand. Because of their competitiveness and rising demand, the majority of pharmaceutical businesses now support real-time prediction of dissolution profiles. As a result, alternative methods have been added to acquire a rapid response, such as spectroscopic approaches, particularly near-infrared spectroscopy (NIRS), which gathers the data based on the physicochemical features of the dosage form. Advanced multivariate analytic approaches, such as principal component analysis (PCA), principal component regression, and classical least squares regression, are widely employed to extract such data for use in quantitative modelling. There is still a dearth of research into the combined impact of numerous critical factors and their interactions on dissolution, despite several studies showing that drug product dissolution profiles can potentially be predicted from material, formulation, and process information using advanced mathematical approaches.

Keywords: Dissolution studies, Mathematical model, NIRS, PCA, Quantitative modelling.

1. INTRODUCTION

A substantial solute penetrates a solution through a process known as dissolution. It can be characterized in the global market as the volume of a drug ingredient that

* **Corresponding author Ram Babu S.:** Himalayan Institute of Pharmacy, Kala-Amb, Dist. Sirmour, Himachal Pradesh-173030, India; E-mail: sharmaram77@ymail.com

Dilpreet Singh and Prashant Tiwari (Eds.)

dissolves in a time period under specified interfacial region, temperature, and fluid component conditions. The frequency, at which a capsule and tablet form distributes the medication, or active component, is essential for ensuring that the medication is supplied appropriately. The medication in solid dose forms is first dissolved in the biological medium and then absorbed into the body's circulation. The dissolution of a tablet can be schematically represented as shown in (Fig. **1A**). Drug dissolution evaluation is essential for characterizing product quality as part of standard quality assurance analysis and it is additionally crucial for the development of new drugs [1].

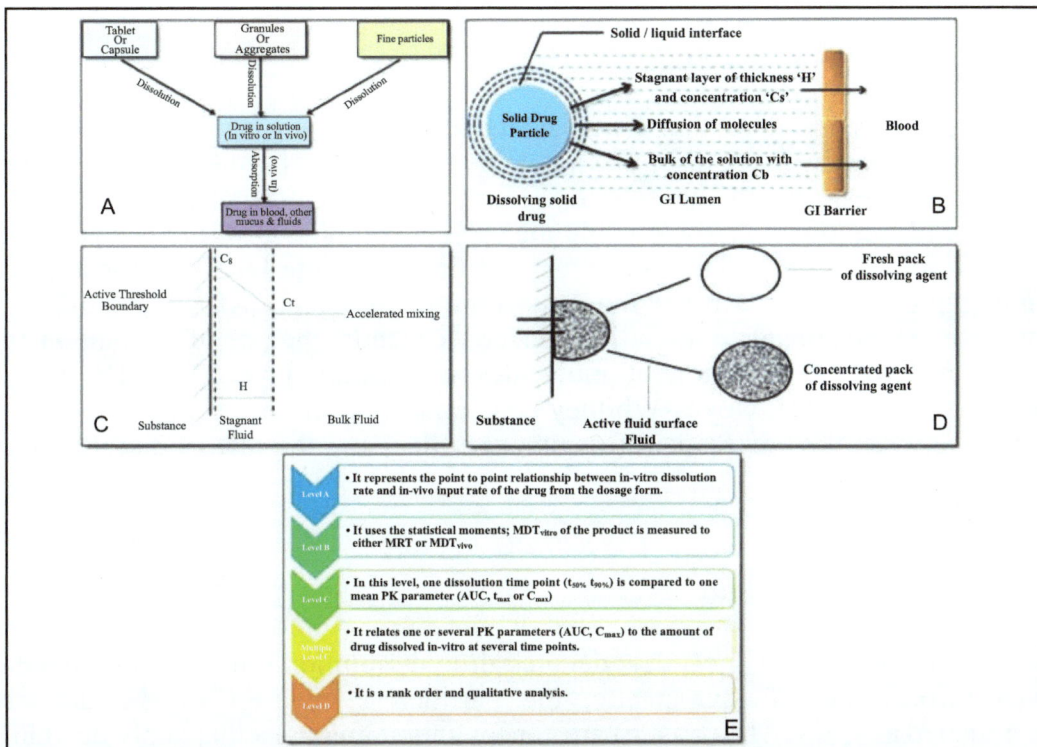

Fig. (1). A) Dissolution process; **B)** Diffusion layer model **C)** Interfacial barrier model **D)** The danckwert's model **E)** Levels of IVIVC studies.

Pharmacopeias utilize dissolution studies for the initial evaluation of how effectively drugs are released from solid and semi-solid dosage forms. Dissolution assays were originally developed to quantify the rate and amount of drug released from extended-release capsules, tablets, and other solid forms of oral administration [2]. Recently, dissolution has gained significance in evaluating drug release from dosage forms such as gums, soft gel capsules, suppositories, transdermal patches, aerosols, and semi-solids. The study of the dissolution process has been evolving since the final decade of the nineteenth century [3].

2. THEORIES OF DISSOLUTION

The stirring rate, temperature, rheology, pH, the content of the extraction solvent, and the inclusion or lack of wetting agents are some of the variables that can affect how quickly tablets dissolve. To explain the drug dissolution that has been reported, physiological simulations have been developed. The following three approaches can be employed to characterize dissolution processes, according to Higuchi.

2.1. The Diffusion Layer Model

This hypothesis postulates that a drug-saturated diffusion or stagnation layer at the liquid-solid interface, and also, that solute may mix in the liquid spread from this layer to the majority of the solution. Here, the solute is transported into the majority of the solution slowly while the solvent-solute association is controlled by diffusion. After the substances cross the fluid film interface, there is a quick mixing process that destroys the concentration gradient. At the solid-liquid interface, the rate of solute migration and, consequently, the rate of dissolution are determined by the diffusion of molecules in a liquid layer (Fig. **1B**) under Brownian motion [4]. Noyes-Whitney equation, depicted in Eq. (1), comes into existence when the process is diffusion controlled, and the rate of dissolution is presented as:

$$\frac{dC}{dt} = \frac{DWKo/w(Cs - Cb)}{Vh} \tag{1}$$

where, from equation [1] dC/dt = denoted drug dissolution rate, D represents the drug diffusion coefficient, while, the partition coefficient of the drug is represented as Ko/w. The (Cs-Cb) are concentration gradients that apply downhill driving force movement for solute without the expenditure of energy. V and h depict dissolution medium in the GIT and the thickness of the mucosal membrane which allows the solute the pass through it, respectively.

2.2. The Interfacial Barrier Model

The basis for the interface barrier model (Fig. **1C**), which depicts the gradual interaction between the solid and liquid interface, is a substantial active free energy threshold that must be overcome even before the substance can decompose. According to this model proposed by Higuchi, an intermediate concentration can be present at the solid/liquid interface due to the solvation mechanism, rather than diffusion, and this presence is attributed to solubility. When a crystal dissolves, the interfacial barrier will vary for each crystal face [5]. The interfacial barrier model places emphasis on solubility rather than dissolution, hence the rate-determining phase is the rate of solubility rather than the rate of dissolution. The equation of this model is as follows:

$$\frac{dm}{dt} \dots . Ki(Cs - Ct)/h \tag{2}$$

From Eq. (2), the rate of dissolution, or the change in solute content concerning time, is expressed as dm/dt. The dissolution rate constant, ki, is influenced by the solvent and solute characteristics. The concentration of solute in the bulk solution is given by Ct. The concentration at which the liquid is in balance (equilibrium) with the solid phase is known as the saturation concentration or Cs. h represents the thickness of the interfacial barrier that is determined by the solute and solvent characteristics.

2.3. Danckwert's Model

According to Danckwert's model, tiny, empty packets that are moved by eddies current fill up with medication solvent and migrate in the direction of the mucosal membrane (Fig. **1D**). Fresh new pockets were added to the vacant vacuoles in the full package to replace it. There won't be an equilibrium until the procedure is finished. The complete procedure is carried out in any order at the solid/liquid contact (Fig. **1D**) demonstrates the entire movement of new and saturated packets.

In accordance with the rules of diffusion, the packets might absorb the salt at the surface before being replenished by a new package of solution. The contaminant transport frequency and, consequently, the rate of dissolution are connected to this interface renewal process. Higuchi's model has explored rate theories that are predicted by the various techniques, which can be individual or in combination. However, Noyes and Whitney proposed an equation that quantitatively represents the rate of dissolution, which is as follows:

$$\frac{dc}{dt} = k(Cs - Ct) \tag{3}$$

Here dc/dt represents alteration in concentration rate in accordance with time and the reaction rate constant is defined by k. (Cs-Ct) is the concentration gradient which applies the driving force for moving the solute. On integration can write the same equation as:

$$\ln\left[\frac{Cs}{(Cs-Ct)}\right] = kt \tag{4}$$

Similar to the other kind of rate law equation, this one anticipates that it depends on the difference in concentration (Cs-Ct) by applying first-order kinetics, which was observed between the dynamic layer of liquid and bulk fluid [6].

3. IVIVC (*IN VITRO - IN VIVO* CORRELATION) STUDIES

The relationship between an *in vitro* parameter and the associated *in vivo* response for the same is described by a predictive mathematical model known as *in vitro in vivo* correlation (IVIVC). This model primarily serves as an alternative for bioequivalence (BE) evaluation in living beings. It has the potential to reduce the number of BE results needed for initial approval, as well as for specific scale-up and post-approval adjustments [7]. The usage of IVIVC could additionally be supported and/or validated by the establishment of dissolution requirements and by:

• Procedure for dissolution of a drug.

• Methods to control quality.

• Disintegration of solid dosage forms.

• Use of modern analytical tools for analysis.

• Dissolution speed evaluation.

• Rate at which absorption of drug occurs.

• Dissolution outline variables.

• Performance during *in vivo* studies.

• Proper rate at which a drug gets dissolved in a medium when observed *in vitro*.

• Bioavailability correlation with data.

The United States Food and Drug Administration (USFDA) issued regulatory guidance in September 1997, which covers the development, evaluation, and application of *in vitro-in vivo* correlation (IVIVC) for oral dosage forms with extended-release (ER). However, despite these guidelines being issued, the Division of Bioequivalence continues to identify flaws related to IVIVC during the examination of generic drug approval applications, specifically Abbreviated

New Drug Applications (ANDA). Data from animal studies conducted *in vivo* and *in vitro* can be easily connected, simplifying result comprehension. IVIVC is often used to obtain ANDA clearance for generic medication applications, but this does not imply a causal connection [8]. There are five main levels, each characterized by the type of data used to establish the relationship (Fig. **1E**).

IVIVC is extensively studied and validated in terms of general principles, theory, modeling methodology, evaluation, and implementation. However, gaps still exist in the scientific community's understanding of specific model design and assessment details. The current knowledge level lacks a standardized *in vitro* assay that can replicate the complex and dynamic conditions of the gastrointestinal (GI) environment or predict the *in vivo* behavior of solid oral dose forms. Since various factors like product properties, API features, and interactions with *in vitro* assessment techniques need consideration when constructing an IVIVC, it must be approached on a case-by-case basis.

As we understand, the absorption of drugs in immediate-release (IR) dosage forms is significantly faster compared to extended-release (ER) dosage forms. In this context, IVIVC is more applicable to the ER type of dosage form. Therefore, when evaluating the IVIVC viability of *in vivo* and *in vitro* evaluation results, it is crucial to consider the development of dosage forms, their interactions with the GI environment, and *in vitro* test settings. This comprehensive approach enhances the likelihood of success. Additionally, incorporating the IVIVC approach as a critical step in the dosage formulation process is essential. An organized and validated *in vitro* test can replace *in vivo* research, act as a template for convincing product specifications, and be a trustworthy instrument for quality assurance. Every time it is feasible in the QC laboratories, an IVIVC-based *in vitro* testing technique must be used to ensure the quality and efficacy of each commercial sample. This will make it possible to identify any potential or unforeseen changes in a product's *in vivo* behavior during production [9].

4. EXPERIMENTAL TECHNIQUES FOR DISSOLUTION

The United States Pharmacopeia (USP) currently recognizes seven distinct kinds of dissolution apparatus: Basket, reciprocating cylinder, paddle, paddle over the disc, flow-through cell, reciprocating disc and rotating cylinder. Mostly, for oral dosage form dissolution testing, USP-I (basket type) and USP-II (paddle type) are being-used. However, the apparatus specified in the USP may be utilized to evaluate a variety of formulation types. They are adaptable and enable the creation of an extensive variety of dissolution techniques, ranging from all of those used mostly for method development to those utilized for QC assessment of industrial batches. As medicinal products become a little more complicated as well as the pursuit for an improved bio-predictive methodology continues, there are additionally a number of increasingly specialized dissolution equipment being developed and deployed [10].

Despite the intricate physicochemical properties of the active pharmaceutical ingredient (API) and the stringent regulations imposed by regulatory authorities for drug approval, the development of a novel, state-of-the-art *in vitro* technique that meets the criteria for predicting *in vivo* outcomes remains essential. A rapidly expanding toolkit of certain other unique ways can be provided by academic results, like physisorption or even other specified industries because of certain particular problems, furthermore Apparatus III and IV, seem to be choices for the estimation of the comprehensive gastrointestinal tract with various test mediums or bioequivalent amounts. When selecting the best dissolving test to utilize, volumes may present a difficulty. Because the most frequently used equipment is capable of being utilized with media quantities around 500 mL and 1000 mL, the therapeutic significance may be limited. Yet, a statistical exaggeration *in vivo* may result from employing large amounts for dissolving testing. The use of mini-paddles along with smaller containers can sometimes prove beneficial in better simulating conditions within the human gastrointestinal tract. However, this approach has not yet been taken into consideration by pharmacopeias. Reports suggest that the use of modified and non-compendial equipment in the field of dissolution testing has seen an increase over the past several years.

In the interest of rendering dissolution outcomes increasingly bio-relevant, these instruments are being used to provide fresh insights into various dosage forms, delivery systems, and formulations. The use of formerly "fringe" approaches including inherent dissolution, comparatively tiny dissolution, or dissolutions utilizing enhancers gets submerged, or extraction cells are expanded as a consequence. Moreover, improvements in the detection methods are making it possible to test increasingly complicated, multi-component compositions digitally or non-digitally. Manufacturers can obtain an extensive data set that offers a better knowledge of the combinations of APIs and inactive ingredients in pharmaceutical preparations thanks to the enhanced possibilities provided by such innovative detection methods [11].

Globally, various pharmacopeias provide detailed guidelines for tools, procedures, and evaluations that aid developers in fulfilling regulatory requirements for dissolution testing. Since 2014, Europe has started publishing distinct formulation monographs that encompass dissolution procedures and criteria for acceptability, following the approach outlined in the United States Pharmacopeia (USP). Additionally, the FDA's database contains more openly accessible data on dissolution processes in the United States. Further details can be obtained on the physical operating circumstances of the dissolution analyzers, including instructions for dissolution testing of dosage forms with imminent release, staggered release, and sustained release. Nevertheless, it is occasionally challenging to only rely upon that dissolving study as a means of forecasting how well a pharmaceutical dosage form might operate *in vivo* considering the intricate nature of the human anatomy, metabolism, and the biological reactions that occur. bio-relevant media can assist in such evaluations. However, clinical trials are essential to comprehensively assess how accurately the dissolution test can predict *in vivo* effectiveness [12].

On how to investigate bioequivalence, the European Medicines Agency (EMA) also offers instructions. These recommendations cover the comparison of optimized formulations as well as utilizing dissolution evaluations to avoid bioequivalence research when appropriate. The EMA's recommendation on the pharmacological and therapeutic assessment of dose modification forms also explains the way that utilizing dissolution data in IVIVC techniques works. Dissolution analysis is essential for evaluating the effectiveness of oral dosage forms and is utilized extensively worldwide, therefore significant effort was put into developing a globally standardized technique. The USP I and II Apparatus in particular have been standardized throughout numerous locations thanks to the collaboration between the International Council for Harmonization (ICH) and numerous pharmacopeias. Dissolution is a standardized procedure utilized in many pharmacopeias where the equipment's specifications and operating conditions are specified and documented. This harmonization makes it reasonably simple to successfully transfer validated dissolve procedures from one laboratory to another [13].

5. FUNDAMENTALS OF DISSOLUTION TESTING

One of the most flexible and efficient treatment alternatives for patients remains oral dose forms. The quality of solid oral medicinal formulations has traditionally been assessed largely using the dissolving test. In actuality, this is the only QC test to quantify the rate at which a pharmaceutical product releases a medication.

An essential element in the product release and stability testing process during the product development cycle is the dissolution of solid dosage forms. It is a crucial analytical procedure used to spot physical alterations in a pharmaceutically active component and the finished product after formulation. Is it feasible to carry out *in vitro* trials during the early phases of drug development? In order to maximize medication release from a certain formulation, dissolution testing is employed [14].

5.1. Signifance of Dissolution Testing

Normally, this testing purpose is used to ensure consistency between batches of commercial products as well as to support scaling up and changes to the manufacturing process after product approval. Dissolution testing can be used to achieve the following goals:

• A pivotal tool for the development of generic drug dosage forms.

• Aids in formulating prototypes and determining optimal ingredient concentrations to achieve the desired cumulative drug release from dosage forms, particularly in the context of extended-release (ER) formulations.

• Ensures the conduct of bioequivalence studies for generic drug products, facilitating their approval for market distribution. Additionally, it lays the groundwork for supporting *in vivo* bioavailability trials of branded (innovative) products [15, 16].

• Facilitates the maintenance of uniformity and consistency across all batches. It also serves as a means to assess the batch-to-batch performance of a single batch.

5.2. Product Stability

In vitro dissolution is also used to determine the stability and shelf-life of a pharmaceutical product. As a drug product gets older, physiological changes may occur that cause it to lose some of its safety, efficacy, and quality, and such parameters of the product when they are evaluated for dissolution studies. Changes in the safety, efficacy, and quality of the product with time are due to changes in the physiochemical properties of the drug and its excipients. It has also been observed that some of the APIs, which exist in more than one form (polymorphs), change to another form due to chemical or physical changes, resulting in a change in the dissolution parameters.

5.3. Comparability Assessment

In vitro dissolution studies are crucial for understanding the impact of changes made during pre or post-approval in manufacturing, formulation, and processes on the product's quality. Additionally, in order to guarantee product similarity and equivalence in performance, *in vitro* comparability testing plays a vital role.

It is also helpful for determining whether changes made to the formulation and production of medicinal products before or after clearance will have an impact on the dissolution tests. Therefore, it is essential to conduct *in vitro* comparability testing in order to maintain product uniformity and performance comparable [17, 18].

5.4. Noyes-Whitney Rule

The equation was created by Noyes and Whitney in 1897 and serves as the cornerstone for analyzing the kinetics of drug release.

$$KS=dM/dt \; (Cs\text{-}Ct) \tag{5}$$

As per Eq. (5), dm/dt stands for a change in mass with respect to time which is in contact with the dissolution media; it is also known as the dissolution rate. Cs stands for solubility of solute at the equilibrium stage while Ct stands for the concentration of solute with respect to time [19].

5.5. Nernst and Brunner Film Theory

According to Nernst and Brunner's theory, surface processes happen much more quickly than transport processes and linear concentration gradients only appear in the layer of solution that clings to a solid surface. The theory states that the dissolving process of the diffusion layer follows the first-order rate kinetics. Using the Fick first law, Nernst and Brunner created an equation that established the relation between K (the Noyes-Whitney equation) and the solute diffusion constant as given below.

$$K = \frac{DS}{h\gamma} \tag{6}$$

In Eq. (6), S depicts the surface area of the diffusion layer, D is the diffusion coefficient, h denotes the thickness, and γ indicates the volume of the solvent being contacted by the diffusion layer [20, 21].

6. MATHEMATICAL AND STATISTICAL TOOLS FOR IN VITRO DISSOLUTION METHODS

6.1. Kinetics in Dosage Form

The dissolution patterns are illustrated by the model-dependent methods based on different mathematical roles [22]. First order, zero order, Hixon, Higuchi, Baker Lonsdale, Hopfenberg, Weibull, Korsmeyer-Peppas *etc.* are some of the model-dependent variables [23].

6.2. Empirical and Semi-empirical Mathematical Modeling

6.2.1. Zero order Kinetics

Zero-order kinetics describes a type of chemical reaction that is independent of the concentration of reactants. In simpler terms, the reaction progresses at a constant rate irrespective of the amount of reactants present (Fig. **2A**). In zero-order kinetics, it is the rate constant (k) that determines the rate of a reaction, not the concentration of the reactants.

$$Rate = k$$

In the reaction, k represents the rate constant. Mol/L/s or M/s are common units of k for this reaction. The decomposition process takes place in the liver by booze and the oxidation process of some substances initiated in the body are two examples of zero-order reactions. During this reaction, a constant rate of reaction was observed despite the difference in concentrations of the reactant (alcohol or drug) [24].

6.2.2. First-order Kinetics

The first-order rate of a chemical reaction is a concentration-dependent chemical reaction, which depends upon the concentration of the reactants. Most of the drug released from the dosage form in the body ideally follows the first-order rate of reaction. The first-order rate kinetics are expressed by the following equation:

$$\log Qt = log Q^0 + \frac{Kt}{2.303} \tag{7}$$

Where Q^0 represents the drug's starting concentration; Qt represents the quantity of drug released at time t. K is the first order rate constant. A straight line will appear on a graph showing the cumulative logarithm percentage release over time (Fig. **2B**). The cumulative percentage release of the logarithm was displayed against time in graphical form. For example, water-soluble medications are used as an example to show how drugs dissolve in pharmaceutical dosage forms using first-order kinetics [25, 26].

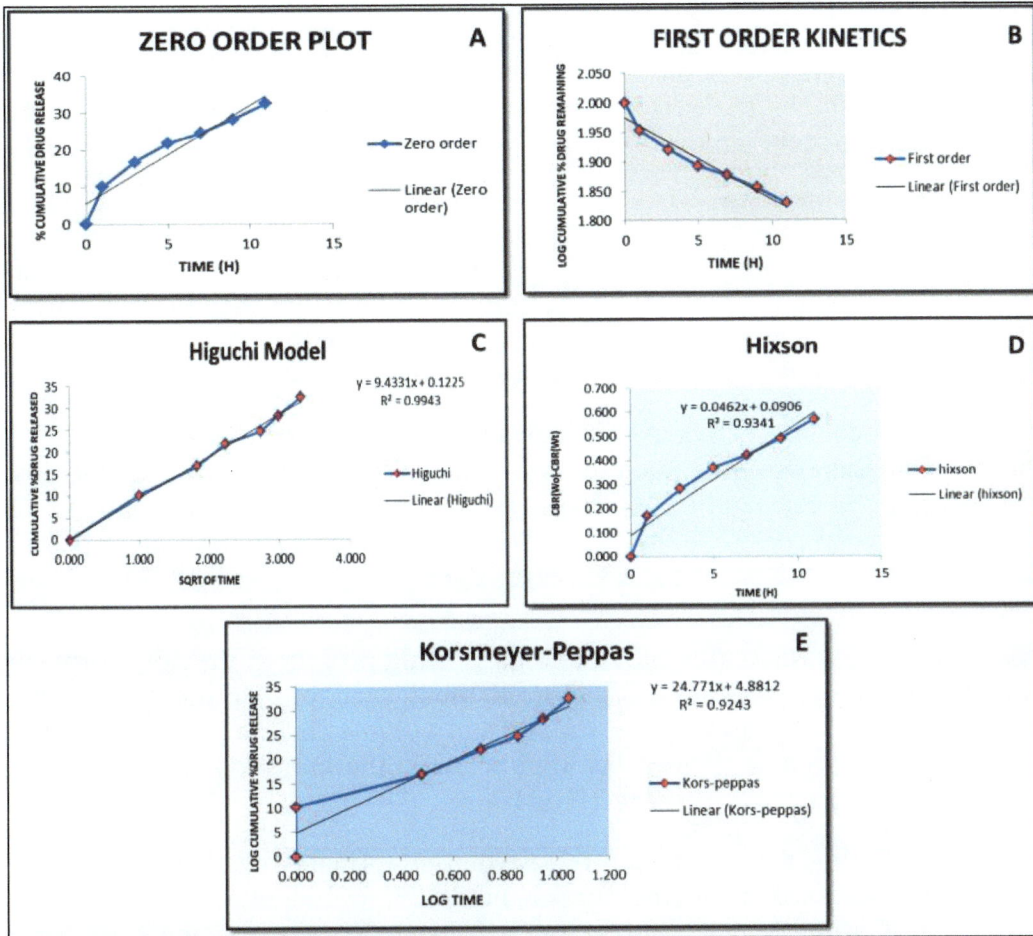

Fig. (2). Graphical representation of: **A**) Zero order kinetics; **B**) First order kinetics; **C**) Higuchi model; **D**) Hixson model; **E**) Korsmeyer peppas model.

6.2.3. Higuchi Model

The Higuchi equation explains how the drug is released in dose forms from a matrix system. This model can be represented by the following equation:

$$Q = [Dt/T(2A - TCS)CSt)]^{1/2} \qquad (8)$$

where Q is the amount of medicine that is discharged per unit area over time t.

In Fig. (**2C**), A represents the initial concentration of the drug, Cs, is the solubility in the matrix media and D shows the diffusion coefficient in the matrix substance [23]. Some regulated and modified-release dosage forms, such as sustained-release matrix tablets, transdermal drug delivery systems, nanotechnology, *etc.* typically show Higuchi's model [4].

6.2.4. Hixson- Crowell Model

The model states that the rate of dissolution of a solid drug particle is proportional to the surface area of the particle that remains. The rate of change in the diameter of the particle is proportional to the square root of the initial diameter minus the cube root of the diameter at any given time.

The Hixson model shows that as the diameter and surface area of the solute in the solution decreases over time, the rate at which the drug is released from the dose form also changes constantly. The Hixson paradigm is represented by Eq. (9).

$$W_0{}^{1/3} - Wt^{\frac{1}{3}} = \kappa t \tag{9}$$

In Eq. (9), W_0 represents the initial amount of drug present in the dose form, W_t represents the amount of drug still present in the dosage form at time t, and kappa (κ) is a constant that preserves the surface and volume relations [27]. A straight line is obtained when a graph is plotted between the cube root of the % drug remaining in the matrix versus time (Fig. **2D**).

The release of drugs from sustained-release dosage forms, like tablets or capsules, is frequently assessed using the Hixson model. It is believed that the rate of dissolution of drugs, rather than its rate of dispersion through the dose form, determines how quickly it is released into the body [4].

6.2.5. Korsmeyer-Peppas Model

The model is founded on the observation that a polymeric system, which is described by the following equation, is used to deliver the drug in a dosage form:

$$Mt/M\infty = Kt^n \tag{10}$$

Where $Mt/M\infty$ stands for the percentage of the substance released at time t, k is the Korsmeyer release rate constant, and n denotes the release exponent (Fig. **2E**). When describing various releases for matrices with a cylindrical form, the n number is used [27]. The graphical representation of the cumulative proportion of drug release versus the logarithm of time was used to illustrate the model using data from *in vitro* drug release experiments. A Korsmeyer-Peppas model was employed to linearize the release data for different compositions of microspheres and microcapsules [28].

6.2.6. Baker Lonsdale Model

This model is expressed as the following equation to show controlled release of drugs from a spherical matrix.

$$f_1 = 3/2[1 - (1 - \frac{M_t}{M_\infty})^{\frac{2}{3}}]\frac{M_t}{M_\infty} = k_t \qquad (11)$$

Here, drug release data designated as [d (Mt / M)] / dt for inverse of the root of time *in vitro*. For generating linearity reports of release data from microspheres and microcapsules, Baker Lonsdale models are helpful [25].

6.2.7. Weibull Model

Drug release curves can be explained using the Weibull equation. Most dissolution curve types can be fitted by this equation. The Weibull equation increases the drug fraction when it is used to solve drug release (m).

$$M = M_0[1 - e^{-(t-T)^{b/a}}] \qquad (12)$$

Here, M shows the drug quantity dissolved at time t. The net amount of medication released throughout the breakdown process is M_0. Parameter b designates the dissolution curve's shape. Alternatively, a parameter designates a time-dependent scale parameter [29].

6.2.8. Hopfenberg Model

If there is no change in the surface area, Hopfenberg described a mathematical model for forecasting the drug release data from the polymers that are breaking down at their surfaces. Based on the equation below, we can calculate the cumulative fraction of medication released as a function of time t:

$$\frac{M_t}{M_\infty} = 1 - [1 - k_0 t/CL\ a]^n \qquad (13)$$

In this context, where k0 represents the zero-order rate constant associated with polymer degradation, "a" signifies half the thickness of the system, and the geometric exponent "n" assumes values of 1, 2, and 3, corresponding to slab (flat), cylindrical, and spherical shapes, respectively [30]. In this model, the release mechanism from the optimized oil sphere is conceptualized by utilizing a composite profile with site-explicit biphasic release kinetics.

7. PREDICTION OF *IN VITRO* DISSOLUTION STUDIES USING ADVANCED MEASUREMENTS TECHNIQUES

More and more data generated during pharmaceutical research is handled and processed by computers. Computer programmes with a wide range of therapeutic uses have been developed by many people. Among these programmes, only a few are designed to evaluate drug release from various oral dosage forms [31, 32]. Due to the ever-increasing complexity of their job processes, pharmaceutical professionals still need innovative user-interactive software.

In 1998, Singh *et al.* created ZOREL, a software tool for studying the drug release kinetics from a variety of drug delivery systems.

The FORTRAN-based application accepts raw dissolution data as input. In the first step, this application modifies the dissolution data for the drug using WR (Weighting with Replacement) and/or WOR (Weighting without Replacement) sampling algorithms based on the dosage form's weight and drug content at sample times [33]. Subsequently, it calculates the estimated values for the quantity and percentage of drug released at different time intervals for each formulation unit. By applying log mean fractions released and log time regression, the application computes these values. Then, using the Korsmeyer-Peppas relationship proposed for swelling matrices, it calculates the mean dissolution time (MDT). This calculation provides the release exponent (n) and kinetic constant (k) according to the Peppas and Higuchi models.

The constants kl and k2, as well as the corresponding contributions of diffusion and polymer relaxation, are too estimated by analyzing phenomenological data, the software identifies Fickian, non-Fickian (anomalous), and zero-order releases. This programme is employed for determination of the drug release rates at 50%, 60%, and 90%.

Many alternative mathematical models have been developed to fit the drug dissolution data, but the majority of these are nonlinear equations. There have only been five release models implemented, and MSFIT is the unique programme that has been described mainly for fitting dissolution profile data. Just a small subset of data could be used with these models [33]. Sigma Plot [34, 35], GraphPad Prism [36], and SYSTAT [37] are a few more expert statistical software programmes that can be utilized in addition to Micro Math Scientist [35] to fit dissolution data nonlinearly. On other systems, the equation must be manually entered and each parameter must have a primary value. However, DD Solver adds a menu-driven interface made in Visual Basic for Applications that work with the Microsoft Excel spreadsheet software. This program can be used to

• Examine dissolution profiles using a variety of popular methods for assessing similarity.

• Simplify dissolution data modeling by utilizing nonlinear optimization methodologies by using 40 dissolution models built into the model library.

• Reduce user errors, speed up the calculation, and offer an appropriate mode to report dissolution data easily and quickly [38, 39].

An evaluation of the geometrical differences between swellable polymeric matrix devices, MATLAB software is a programming language employed for characterization and stimulation of drug release rate and kinetics by illustrating, studying and distinguishing the effects of several parameters of mathematical equations on drug release profiles [40].

KinetDS is a curve-fitting programme that was specifically created to use a single or group of equations to represent the cumulative dissolution curve.

Michaelis-Menten and Hill equations, Weibull (3- and 2-parameter), Hixson-Crowell, Higuchi, and Korsmeyer-Peppas equations were among the most liked empirical and mechanistic models used to describe the drug dissolution curves [41].

CONCLUSION

The dissolution study is one of the most important aspects of any dosage form in terms of its efficacy, efficiency and therapeutic value. It has been suggested that the release of drugs may reflect the release of drugs *in vivo*. There are several techniques available for calculating and predicting dissolution properties, including mathematical models, computational methods, and experimental techniques which can further assist in the prediction of *in vitro* and *in vivo* values which further avoids animal curability for experimentation. As a result, these tools can also assist in understanding how drugs dissolve in various media, which can facilitate formulation design and optimization for optimal therapeutic effect. Mathematical models, such as the Higuchi equation and the Noyes-Whitney equation, are widely used for predicting dissolution rates based on experimental data. A computer simulation, such as a molecular dynamics simulation, or a quantum chemical calculation, can offer valuable insight into the molecular-level interactions that govern drug dissolution. A combination of empirical, computational, and experimental approaches can help researchers gain a more comprehensive understanding of drug dissolution behavior and design formulations that maximize the therapeutic efficacy. Prediction values and modern technology come into existence making dissolution studies more reliable, and more accurate in terms of their scientific significance.

CONSENT FOR PUBLICATON

I certify that I am the rightful owner of the intellectual property of the above-mentioned content and have the necessary rights to grant permission for its publication.

CONFLICT OF INTEREST

I am writing this letter to confirm that I have no conflict of interest regarding my contribution to the book chapter titled, 'Tools for calculations of dissolution experiments and its prediction properties' for the upcoming book, titled, **Softwares and Programming Tools in Pharmaceutical Research**.

ACKNOWLEDGEMENT

I would like to show my sincere gratitude to Dr. Sakshi and Ms. Amardeep Kaur for their invaluable contribution and insights to this book chapter. Their expertise and support have been instrumental in shaping the ideas presented this work.

REFERENCES

[1]   Diaz DA, Colgan ST, Langer CS, Bandi NT, Likar MD, Van Alstine L. Dissolution similarity requirements: How similar or dissimilar are the global regulatory expectations? AAPS J 2016; 18(1): 15-22.
[http://dx.doi.org/10.1208/s12248-015-9830-9] [PMID: 26428517]

[2]   Chen Y, Gao Z, Duan JZ. Dissolution Testing of Solid Products. In: Yu LX, Li H, Eds. Developing Solid Oral Dosage Forms: Pharmaceutical Theory and Practice. 2nd ed. Elsevier 2017; pp. 203-35.
[http://dx.doi.org/10.1016/B978-0-12-802447-8.00013-3]

[3]   Khan KA, Khan GM, Rehman AUR, Shah KU. Studies on drug release kinetics of controlled release matrices of flurbiprofen and comparison with market product. Lat Am J Pharm 2013; 32(4): 526-31.

[4]   Baishya H. Application of mathematical models in drug release kinetics of carbidopa and levodopa ER tablets. J Dev Drugs 2017; 6(2): 1-7.
[http://dx.doi.org/10.4172/2329-6631.1000171]

[5]   Ma T, Jivkov AP, Li W, *et al.* A mechanistic model for long-term nuclear waste glass dissolution integrating chemical affinity and interfacial diffusion barrier. J Nucl Mater 2017; 486: 70-85.
[http://dx.doi.org/10.1016/j.jnucmat.2017.01.001]

[6]   Anal AK, Bhowmik D, Gopinath H, Kumar BP, Duraivel S, Kumar KPS, *et al.* Overview on controlled release dosage forms. Int J Pharma Sci 2013; 3(2): 14-9.

[7]   McDonald TO, Giardiello M, Martin P, *et al.* Antiretroviral solid drug nanoparticles with enhanced oral bioavailability: Production, characterization, and *in vitro-in vivo* correlation. Adv Healthc Mater 2014; 3(3): 400-11.

[http://dx.doi.org/10.1002/adhm.201300280] [PMID: 23997027]

[8]   Kaur P, Jiang X, Duan J, Stier E. Applications of *in vitro–in vivo* correlations in generic drug development: Case studies. AAPS J 2015; 17(4): 1035-9.
[http://dx.doi.org/10.1208/s12248-015-9765-1] [PMID: 25896303]

[9]   Qiu Y, Duan JZ. *In vitro/in vivo* Correlations: Fundamentals, Development Considerations, and Applications In: Yu LX, Li H. Developing Solid Oral Dosage Forms: Pharmaceutical Theory and Practice: Second Edition. Elsevier 2017; pp. 237-59.

[10]   Klein S. Multiparticulate Drug Delivery: Formulation, Processing and Manufacturing 2017. Woodhead Publishing 2017.

[11]   Deng J, Staufenbiel S, Hao S, Wang B, Dashevskiy A, Bodmeier R. Development of a discriminative biphasic *in vitro* dissolution test and correlation with *in vivo* pharmacokinetic studies for differently formulated racecadotril granules. J Control Release 2017; 255: 202-9.
[http://dx.doi.org/10.1016/j.jconrel.2017.04.034] [PMID: 28450206]

[12]   Schneider F, Koziolek M, Weitschies W. *in vitro* and *in vivo* test methods for the evaluation of gastroretentive dosage forms. Pharmaceutics 2019; 11(8): 416.
[http://dx.doi.org/10.3390/pharmaceutics11080416] [PMID: 31426417]

[13]   Chaturvedi K, Shah HS, Sardhara R, Nahar K, Dave RH, Morris KR. Protocol development, validation, and troubleshooting of *in-situ* fiber optic bathless dissolution system (FODS) for a pharmaceutical drug testing. J Pharm Biomed Anal 2021; 195: 113833.
[http://dx.doi.org/10.1016/j.jpba.2020.113833] [PMID: 33358085]

[14]   Kapoor D, Maheshwari R, Verma K, Sharma S, Pethe A, Tekade RK. Fundamentals of diffusion and dissolution: Dissolution testing of pharmaceuticals.Elsevier. Grumezescu, AM 2019; pp. 131-63.

[15]   Adeline Siew. Dissolution testing | pharmaceutical technology. PharmTech.com. 2016. Available from: https://www.pharmtech.com/view/dissolution-testing

[16]   Grady H, Elder D, Webster GK, *et al.* Industry's view on using quality control, biorelevant, and clinically relevant dissolution tests for pharmaceutical development, registration, and commercialization. J Pharm Sci 2018; 107(1): 34-41.
[http://dx.doi.org/10.1016/j.xphs.2017.10.019] [PMID: 29074376]

[17]   Sun F, Hou G, Cong W, Li HH, Zhang X, Ren Y, *et al.* DMS-Logo.pdf. J Pharm Sci 2018; 107(11): 2862.

[18]   Muthappa R, Purushothaman BK, Meera Sheriffa Begum KM, Maheswari PU. Kinetic modeling and optimization of the release mechanism of curcumin from folate conjugated hybrid BSA nanocarrier. Chemical Product and Process Modeling 2020; 15(1): 20190026.
[http://dx.doi.org/10.1515/cppm-2019-0026]

[19]   Siepmann J, Siepmann F. Mathematical modeling of drug dissolution. Int J Pharm 2013; 453(1): 12-24.
[http://dx.doi.org/10.1016/j.ijpharm.2013.04.044] [PMID: 23618956]

[20]   Haidar ZS. Mathematical modeling for pharmacokinetic predictions from controlled drug release nano systems: A comparative parametric study. Biomed Pharmacol J 2018; 11(4): 1801-6.
[http://dx.doi.org/10.13005/bpj/1552]

[21]   Li X, Liu Y, Yu Y, Chen W, Liu Y, Yu H. Nanoformulations of quercetin and cellulose nanofibers as healthcare supplements with sustained antioxidant activity. Carbohydr Polym 2019; 207: 160-8.
[http://dx.doi.org/10.1016/j.carbpol.2018.11.084] [PMID: 30599995]

[22]   Sopyan I, Alvin B, Insan Sunan KS, Cikra Ikhda NHS. Systematic review: Cocrystal as efforts to improve physicochemical and bioavailability properties of oral solid dosage form. Int J Appl Pharm 2021; 13(3): 17-26.
[http://dx.doi.org/10.22159/ijap.2021v13i1.39594]

[23]   Release Kinetics – Concepts and Applications. Int J Pharm Res Technol 2019; 3(1): 58-61.

[24]   Sahoo CK, Rao SRM, Sudhakar M, Satyanarayana K. The kinetic modeling of drug dissolution for drug delivery systems: An overview. Pharm Lett 2015; 7(11): 26-38.

[25]   R KA. Mathematical models of drug dissolution: A review. (sch acad j pharmacyonline) sch acad. J Pharm 2014; 3(9): 706-10.

[26]   Mohamed Rizwan I, Damodharan N. Mathematical modelling of dissolution kinetics in dosage forms. Res J Pharm Technol 2020; 13(11): 5291.

[27]   Permanadewi I, Kumoro AC, Wardhani DH, Aryanti N. Modelling of drug dissolution: A comparison of dissolution profile methods. Int J Pharma Bio Sci 2013; 4(1): 174-83.

[28]   Senthila S, Manojkumar P, Venkatesan P. Characterisation of preformulation parameters to design, develop and formulate silymarin loaded PLGA nanoparticles for liver targeted drug delivery. Res J Pharm Technol 2021; 14(8): 3927-37.

[29]   Muselík J, Komersová A, Kubová K, Matzick K, Skalická B. A critical overview of FDA and EMA statistical methods to compare *in vitro* drug dissolution profiles of pharmaceutical products. Pharmaceutics 2021; 13(10): 1703.
[http://dx.doi.org/10.3390/pharmaceutics13101703] [PMID: 34683995]

[30]   Vigoreaux V, Ghaly ES. Fickian and relaxational contribution quantification of drug release in a swellable hydrophillic polymer matrix. Drug Dev Ind Pharm 1994; 20(16): 2519-26.
[http://dx.doi.org/10.3109/03639049409042655]

[31]   Gao P, Skoug JW, Nixon PR, Robert Ju T, Stemm NL, Sung KC. Swelling of hydroxypropyl methylcellulose matrix tablets. 2. Mechanistic study of the influence of formulation variables on matrix performance and drug release. J Pharm Sci 1996; 85(7): 732-40.
[http://dx.doi.org/10.1021/js9504595] [PMID: 8818998]

[32]   Lu DR, Abu-Izza K, Mao F. Nonlinear data fitting for controlled release devices: An integrated computer program. Int J Pharm 1996; 129(1-2): 243-51.
[http://dx.doi.org/10.1016/0378-5173(95)04356-X]

[33]   El-Nabarawi MA, Elshafeey AH, Mahmoud DM, El Sisi AM. Fabrication, optimization, and *in vitro/in vivo* evaluation of diclofenac epolamine flash tablet. Drug Deliv Transl Res 2020; 10(5): 1314-26.
[http://dx.doi.org/10.1007/s13346-020-00709-4] [PMID: 32072473]

[34]   Papadopoulou V, Kosmidis K, Vlachou M, Macheras P. On the use of the Weibull function for the discernment of drug release mechanisms. Int J Pharm 2006; 309(1-2): 44-50.
[http://dx.doi.org/10.1016/j.ijpharm.2005.10.044] [PMID: 16376033]

[35]   Desai P, Thakkar A, Ann D, Wang J, Prabhu S. Loratadine self-microemulsifying drug delivery systems (SMEDDS) in combination with sulforaphane for the synergistic chemoprevention of pancreatic cancer. Drug Deliv Transl Res 2019; 9(3): 641-51.
[http://dx.doi.org/10.1007/s13346-019-00619-0] [PMID: 30706304]

[36]   Mobarak D, Salah S, Ghorab M. Improvement of dissolution of a class II poorly water-soluble drug, by developing a five-component self-nanoemulsifying drug delivery system. J Drug Deliv Sci Technol 2019; 50: 99-106.
[http://dx.doi.org/10.1016/j.jddst.2018.12.018]

[37]   Zuo J, Gao Y, Bou-Chacra N, Löbenberg R. Evaluation of the DDSolver software applications. BioMed Res Int 2014; 1-9.
[http://dx.doi.org/10.1155/2014/204925] [PMID: 24877067]

[38]   Knöös P, Svensson AV, Ulvenlund S, Wahlgren M. Release of a poorly soluble drug from hydrophobically modified poly (acrylic acid) in simulated intestinal fluids. PLoS One 2015; 10(10): e0140709.

[http://dx.doi.org/10.1371/journal.pone.0140709] [PMID: 26473964]

[39]   Sou T, Kukavica-Ibrulj I, Levesque RC, Friberg LE, Bergström CAS. Model-informed drug development in pulmonary delivery: Semimechanistic pharmacokinetic–pharmacodynamic modeling for evaluation of treatments against chronic *Pseudomonas aeruginosa* lung infections. Mol Pharm 2020; 17(5): 1458-69.
[http://dx.doi.org/10.1021/acs.molpharmaceut.9b00968] [PMID: 31951139]

[40]   Mendyk A, Jachowicz R, Fijorek K, *et al.* An open source software for dissolution test data analysis. Dissolut Technol 2012; 19(1): 6-11.
[http://dx.doi.org/10.14227/DT190112P6]

[41]   Smith J, Brown A, Johnson C, *et al.* Enhanced oral bioavailability of atorvastatin using self-nanoemulsifying drug delivery systems. Eur J Pharm Sci 2022; 168: 106036.

CHAPTER 3

The Role of Principal Component Analysis in Pharmaceutical Research: Current Advances

Diksha Sharma[1], Anjali Sharma[1], Punam Gaba[1], Neelam Sharma[1,*], Rahul Kumar Sharma[1] and Shailesh Sharma[1]

[1] Department of Pharmaceutics, Amar Shaheed Baba Ajit Singh Jujhar Singh Memorial College of Pharmacy, Bela, (An Autonomous College) Ropar, India

Abstract: Karl Pearson developed Principal Component Analysis (PCA) in 1901 as a mathematical equivalent of the principal axis theorem. Later on, it was given different names according to its application in various fields. Principal Component Analysis provides a foundation for comprehending the fundamental workings of the system under examination. It has various applications in different fields such as signal processing, multivariate quality control, psychology, biology, meteorological science, noise and vibration analysis (spectral decomposition), and structural dynamics. In this chapter, we will discuss its application in pharmaceutical research and drug discovery. This technique allows for the representation of multidimensional data and the evaluation of large datasets to improve data interpretation while retaining the maximum amount of information possible. PCA is a technique that does not require extensive computations and offers reduced memory and storage requirements. PCA can be conceptualized as an n-dimensional ellipsoid fitted to the data, with each axis representing a principal component. The ellipse's axes are determined by subtracting the mean of each variable from the datasheet. In the pharmaceutical research field, original variables are often expressed in various measurement units. Therefore, the original variables are divided by their standard deviation once the mean has been subtracted. This step is taken to work with z-scores, which are further used for extracting the eigenvalues and eigenvectors of the original data.

Keywords: PCA, Principal component, Research, Statistical method.

1. INTRODUCTION

Principal Component Analysis (PCA) is a widely used method that reduces a large number of potentially correlated variables to a smaller set of variables called principal components, achieved through the application of complex mathematical principles [1]. The primary goal of PCA is to minimize a dataset comprising num-

---

* Corresponding author Neelam Sharma: Department of Pharmaceutics, Amar Shaheed Baba Ajit Singh Jujhar Singh Memorial College of Pharmacy, Bela, (An Autonomous College) Ropar, India;
E-mail: pharmneelam@gmail.com

Dilpreet Singh and Prashant Tiwari (Eds.)

erous interrelated variables while retaining as much of the dataset's variance as possible [2]. Karhunen-Loeve expansion is another term for the representation method known as principal component analysis (PCA). This method finds applications in pattern recognition and computer vision, such as face identification [3]. This reduction is achieved by transforming the original variables into a new set of uncorrelated variables called principal components. These components are arranged in a way that the first few components capture the majority of the variance present in the original variables. While the computation of principal components may seem straightforward, this method's apparent simplicity masks its wide range of potential applications and diverse derivations [4]. The principal components optimally account for the total variance of the original variables, serving as a measure of information. The geometric properties of these components facilitate a structured and intuitive interpretation of key features within complex multivariate datasets [5]. A range of diagnostic techniques, including mass spectrometry, Raman spectroscopy, near-infrared spectroscopy, infrared spectroscopy, nuclear magnetic resonance (NMR), laser-induced breakdown spectroscopy (LIBS), ultraviolet-visible spectroscopy, and X-ray absorption spectroscopy, have found utility in conjunction with PCA [6 - 8].

1.1. Definition of PCA

1.1.1. Definition

Principal Component Analysis (PCA) is a statistical technique used for dimensionality reduction and data transformation. It involves transforming a high-dimensional dataset into a lower-dimensional one while retaining as much of the original data's variability as possible. PCA achieves this by identifying new orthogonal axes, called principal components, along which the data varies the most. These components are linear combinations of the original variables, and they are sorted in order of decreasing variance.

1.1.2. Goals

The primary goals of PCA include:

1.1.2.1. Dimensionality Reduction

One of the main objectives of PCA is to reduce the number of dimensions (features or variables) in a dataset while maintaining the most important information. This is particularly useful when dealing with datasets that have many variables, as it simplifies analysis and visualization.

1.1.2.2. Variance Maximization

PCA aims to capture as much of the variance present in the original data as possible in the lower-dimensional representation. The first principal component accounts for the most variance, followed by the second component, and so on. By retaining the top few principal components, a significant portion of the dataset's variability can often be preserved.

1.1.2.3. Feature Interpretation

PCA allows for the interpretation of patterns and relationships within the data. Principal components are orthogonal to each other, and each component represents a linear combination of the original variables. These components can correspond to meaningful aspects of the data, making it easier to understand the underlying structure.

1.1.2.4. Data Visualization

By reducing data dimensions, PCA can facilitate the visualization of complex datasets in two or three dimensions. This aids in gaining insights and identifying patterns that might not be apparent in the original high-dimensional space.

1.1.2.5. Data Compression

PCA can be used for data compression, as the lower-dimensional representation requires less storage space. This is particularly useful in cases where storage or processing resources are limited.

1.1.2.6. Data Preprocessing

PCA can also be used as a preprocessing step to decorrelate variables, making subsequent analyses, such as regression or clustering, more effective.

1.2. History of PCA

Tracing the origins of statistical methods can sometimes be challenging. In a two-way study, Fisher and Mackenzie utilized Singular Value Decomposition (SVD). However, it is widely acknowledged that the first descriptions of the approach now known as Principal Component Analysis (PCA) were provided by Pearson and Hotelling [5]. Hotelling's approach to PCA distinguished it as having a unique character separate from factor analysis. Hotelling's theory is based on the idea that the initial set of p variables might be represented by a "fundamental set of independent variables." The components that Hotelling selects are termed "primary components" because they are derived to maximize their subsequent variance. Another work by Hotelling introduced an expedited method for determining principal components. In the same year, Girshick presented several alternative derivations of principal components and introduced the concept that sample principal components are maximal. PCA's development is in line with the growth of statistical literature as a whole. However, PCA gained popularity in tandem with the widespread adoption of electronic computers since it demands substantial processing power [7].

2. TERMINOLOGY IN THE PCA ALGORITHM

2.1. Dimensionality

Dimensionality refers to the number of factors or variables within the dataset. This can be readily determined by counting the columns in the dataset [8].

2.2. Correlation

Correlation signifies the degree of association between two factors or variables. For instance, when a change occurs in one variable, corresponding changes also take place in another variable. The correlation score ranges between -1 and +1. A value of -1 indicates an inverse relationship between the variables, while a value of +1 indicates a direct or straight relationship between the variables [9].

2.3. Orthogonal

This matrix summarizes the relationships between different features in the dataset. Off-diagonal elements represent the covariances between pairs of features, while diagonal elements represent the variances of individual features [10, 11].

2.4. Eigenvectors

Principal components are derived from the eigenvalues and eigenvectors of the covariance matrix of the original data. Eigenvalues represent the amount of variance explained by each principal component, while eigenvectors are the directions of those components [12].

2.5. Covariance Matrix

This matrix summarizes the relationships between different features in the dataset. Off-diagonal elements represent the covariances between pairs of features, while diagonal elements represent the variances of individual features [13, 14]. This is the ratio of the variance explained by a principal component to the total variance in the data. It helps you understand how much information each principal component retains [15, 16].

2.6. The PCA Algorithm's Steps

2.6.1. Getting the Dataset

First, we must split the input information into 2 sections, x, and y.

x: training set

y: validation set

2.6.2. Structure of Representing Data

A structure is created to describe our dataset. We will use the 2-dimensional matrix of independent variable X as an example. Each row represents a data object, and each column represents a feature. The number of sections in the dataset can also be determined. [17].

2.6.3. Standardizing the data

This stage involves standardizing our dataset. The features with higher variance, for example, are more significant than the features with smaller variance in a given column. Each piece of information in a column will be divided by the SD of the column if the significance of features is independent of the variation of the feature. We will call the matrix Z in this case [18].

2.6.4. Covariance of Z

Z matrix was transposed in order to determine Z's correlation. We will transpose it and then increase the result by Z. This matrix for Z will be the output matrix [19].

2.6.5. Eigenvalues and Eigenvectors

For the resulting correlation matrix Z, we must now compute the eigenvalues and eigenvectors. The orientations of axes having the most data are represented by an eigenvector. Additionally, the eigenvalues of these eigenvectors are specified as their coefficients [20].

2.6.6. Sorting the Eigenve+-ctors

We will take all of the eigenvalues and sort them in this phase in the descending order. In addition, we concurrently sort the eigenvectors in the matrix P of eigenvalues in accordance. P* is the designation of the resulting matrix [21].

2.6.7. Calculating the New Features

Here, the novel features have been measured to the final P* matrix by the Z. Each observation is the linear combination of the initial features in the resulting matrix Z*. The Z* matrix's individual columns are unrelated to one another [22 - 24].

2.6.8. Unimportant Features from the New Data

Since the new feature set has been implemented, we will decide here what to retain and what to remove from the data. It indicates that we will only retain pertinent or significant features in the new dataset and will eliminate irrelevant features [25 - 27].

2.7. PCA for Feature Engineering

There are two ways you could use PCA for feature engineering. The first method is to employ it as an expository strategy. You can compute the MI scores for the components and determine which type of variation is the most predictive of your goal since the components inform you about the variation [28]. This can provide you with ideas for designing features, such as a ratio of height to diameter if shape is essential or the product of height and diameter if size matters. You might even consider grouping based on one or several of the high-scoring elements [29]. The second approach is to utilize the components as features directly. The components often provide more information than the original features because they immediately reveal the variation structure of the data

2.7.1. Dimensionality Reduction

When your features are extremely redundant (more specifically, multicollinear), PCA will divide the redundancy into one or more near-zero variance components, which you can then discard because they will be information-poor or lacking altogether [30].

2.7.2. Anomaly Detection

The low-variance components frequently reveal unusual variations that are not visible from the initial features. These elements might be very helpful when performing an anomaly or outlier detection job [31].

2.7.3. Noise Reduction

Background noise is often present in a collection of sensor data. PCA can improve the signal-to-noise ratio when informative information can be condensed into fewer features while leaving the noise unaffected [32].

2.7.4. Decorrelation

Highly correlated features can be challenging for some machine learning methods. Your algorithm might find it simpler to deal with uncorrelated components created by PCA, which transforms correlated features [33, 34].

2.8. Role of Principal Component Analysis in Pharmaceutical Research

Biology depends heavily on statistics since experiments rarely yield precise results and are prone to measurement errors. Because, nothing can be measured with absolute precision and because measurement errors might be random or biased, measure errors in experiments [35]. Three variable types in principal component research are depicted in Fig. (**1**).

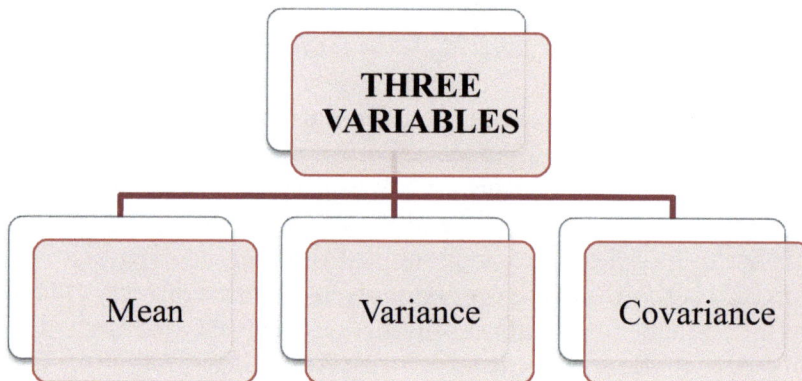

Fig. (1). Types of variable.

The mean is the average of the given numbers and is calculated by dividing the sum of the given numbers by the total number of numbers [36]. The data is provided after conducting five different experiments to measure the relative abundance of different proteins following treatment with a specific drug. We can compute the mean if we want to get an idea of how much protein is typically present. It can be explained by the following example:

1. The mean of any set is given by, $S = \frac{1}{n} \sum_{i=0}^{n} S_i$

$S_i = (1.4, 0.6, 0.4, 0.8, 1.5)$

Where,

S = set of numbers

$\underline{S}$= mean of the set of numbers

S_i= individual numbers in a set of numbers

$\underline{S}=\frac{1}{5}(1.4+0.6+0.4+0.8+1.5)$

$=1$

value 1.48: the average abundance of the protein.

Variance is the expected value of the squared variation of a random variable from its mean value, in probability and statistics. It quantifies the spread of data.

2. Variance (S) $=\frac{1}{n-1}\sum_{i=0}^{n}$ $(Si\text{-}\underline{S})^2$

$\qquad$ S_i= (1.4, 0.6, 0.4, 0.8, 1.5)

Where,

n-1= gives closer approximation to the true value

Variance (S)$=\frac{1}{5-1}[$ $(1.4-1)^2+(0.6-1)^2+(0.4-1)^2+(0.8-1)^2+(1.5-1)^2$

$\qquad =\frac{1}{4}[$ $(0.4)^2+(-0.4)^2+(-0.6)^2+(-0.2)^2+(1)^2)$]

$\qquad =\frac{1}{4}[$ $(0.16+0.16+0.36+0.4+1)$]

$\qquad =\frac{2.08}{4}$

$\qquad = 0.52$

2.8.1. Covariance

Covariance measures the degree of interdependence between two variables (Fig 2). In this case, two scenarios arise where the measure of central tendency is not dependent on the covariance and can be used as a substitute [37].

3. Two variables (x,y) can be,

$\qquad$ **Cov(x,y)**$=\frac{1}{n-1}\sum_{i=1}^{n}$ $(x_i\text{-}\underline{x})(y_i\text{-}\underline{y})$

3 dimensional data sets with three variables x,y,z can be given as:

$$[Cov(x,x)Cov(x,y)Cov(x,z)Cov(y,x)Cov(y,y)Cov(y,z)Cov(z,x)Cov(z,y)Cov(z,z)]$$

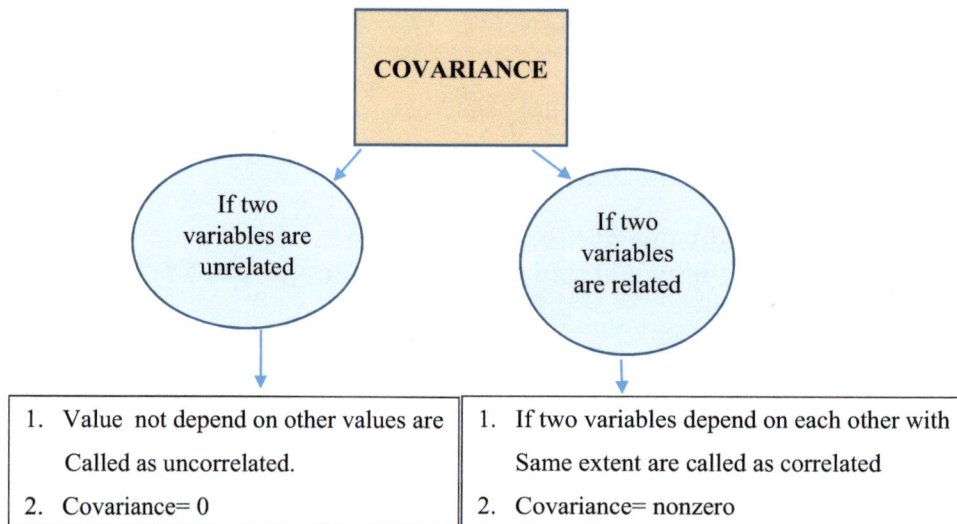

Fig. (2). Depiction of measurement of covariance.

2.8.2. Eigen vectors and eigen values

When linear transformation T is applied to the vectors u1, u2, u3,......u_n, new vectors b1, b2, b3,...bn will be obtained.

$$Tu1=b1$$

$$Tu2=b2$$

$$............$$

$$Tu_n=bn$$

The vectors that change the length but not the direction on applying transformation (T) are called eigenvectors and the scalar which represents the multiple of eigenvectors are called eigenvalues.

$$Tui=\lambda ui$$

Eigenvectors and eigenvalues together provide a matrix's eigen-decomposition, which allows us to examine the matrix's structure. While the eigendecomposition is not applicable to square matrices, it offers a particularly straightforward expression for matrices such as correlation and covariance when seeking maximum functions.

Example: A=2321

Matrix A is the vector u if the length of the vector is changed when it is multiplied by A.

Eigenvector   u1 = [32]   with eigenvalues $\lambda 1 = 4$

And   u2 = [-11]   with eigenvalues $\lambda 2 = 1$

To normalize the eigenvectors linear transformation is applied for most applications: length is equal to one

$u^T u = 1$

$AU = U\Lambda U^{-1}$

Where,

U = set of A in a matrix

Λ = diagonal matrix [10].

2.9. PCA in Drug Excipients Interaction Studies

The dosage form of the drug contains various excipients that enhance the drug's release and administration effectiveness. Although all the excipients in the formulation are inherently inert, they can interact with the drug in the dosage form. Therefore, studies on drug-excipient interactions play a crucial role in formulating new dosage forms and saving valuable time and expensive raw materials. There are various methods to study drug-excipients interactions, like isothermal microcalorimetry and high-performance liquid chromatography [38]. Overlapping of peaks of drugs and excipients makes the interpretation of data difficult. PCA is one of the factor analysis methods used to find relationships among variables with high variance. PCA obeys the following:

$X = T_A - P_A + E_A$

Where,

X = M ^{x} N matrix of data

T = N ^{x} A scores matrix

E = M ^{x} A loading matrix

2.10. Role of PCA in Various Pharmaceutical Fields

In order to identify trends, jumps, clusters, and outliers, it is crucial to describe a multivariate data table as a smaller number of variables that are summary indices, and this is done with the help of PCA [39, 40]. Table **1** provides critical analysis of various roles of PCA in pharmaceutical research.

Table 1. Role of PCA in various pharmaceutical research fields.

Pharmaceutical Research Field	Role of PCA
Clinical research	Independent variables in multivariate analysis are frequently associated with one another, which can cause multicollinearity in the regression models. Using principal component analysis (PCA) on these variables is one way to fix this issue.
Drug discovery and biomedical data	PCA is a 'hypothesis-generating' method that constructs a statistical mechanics framework for modeling biological systems without the need for significant prior theoretical assumptions. This feature makes PCA essential for overcoming overly constrained reductionist techniques and adopting a systematic approach to drug discovery.
QSAR Studies	With the help of PCA, interrelationships between descriptors and molecules are being highlighted. Biological activity can also be connected to a handful of the first main components of a pool of descriptors using PCA.
Accelerating autism drug research	The computer analysis employs a model known as principal component analysis (PCA), distinct from clustering, to group data based on their properties and relationships. This approach can be used to amalgamate studies on autism, potentially expediting learning, or for practical applications such as utilizing currently available medications for autism treatment.

2.10.1. Adaptations

PCA is employed as an exploratory tool for data analysis in various fields, including medicine and finance. Consequently, several scientists have modified PCA to create simple two-dimensional representations of large datasets. These modified forms of PCA are commonly called adaptations of PCA. These updates make the data analysis simple. There are many such adaptations available in literature. Four such adaptations are functional adaptation, simplified adaptation, robust adaptation, and symbolic data adaptation [41].

2.10.2. Functional PCA

Recent advancements in computation and the capacity to collect and store high-dimensional data have enabled statistics to analyze high-dimensional data models. Functional Data Analysis (FDA) is a method that allows for the incorporation of independent functional features obtained from stochastic processes into the analysis.

There are three methods for computing functional principal component:

2.10.2.1. Discretization

Similar to PCA, FPCA is performed with the exception that the eigenvectors must be renormalized and interpolated using an appropriate smoother. The first way to compute functional principal components used a discretization methodology.

2.10.2.2. Discretization

Basis function expansion: In the second method, a stochastic process is expressed as a linear combination of basis functions, that is $X^c(t)=\sum_{k=1}^{\infty} \beta_k \phi_k(t)$

The basis function expansion offers an advantage over previous discretization strategies in that it allows for the application of roughness penalty approaches to enforce the smoothness of $\phi k(t)$. Among the numerous possible basis functions, the most popular ones include polynomial basis functions, Bernstein polynomial basis functions, Fourier basis functions, radial basis functions, wavelet basis functions, and orthogonal basis functions.

2.10.2.3. Numerical Approximation

This method uses quadrature methods to approximate functional [42].

2.10.3. Simplified PCA

In the quest to maximize the variance in q dimensions, we aim to achieve the best possible representation of p-dimensional datasets in q dimensions. Several adaptations have been proposed to simplify the interpretation of these q dimensions. During this step, the transformed data matrix undergoes Principal Component Analysis (PCA). PCA calculates the principal components, which are orthogonal directions in the data space that capture the most variance in the data. The PCA algorithm generates principal components, with each one representing a linear combination of the original features [43]. The first principal component captures the most variance, the second captures the second most, and so on. The flow chart of disintegration of data matrix of PCA is given in Fig (**3**).

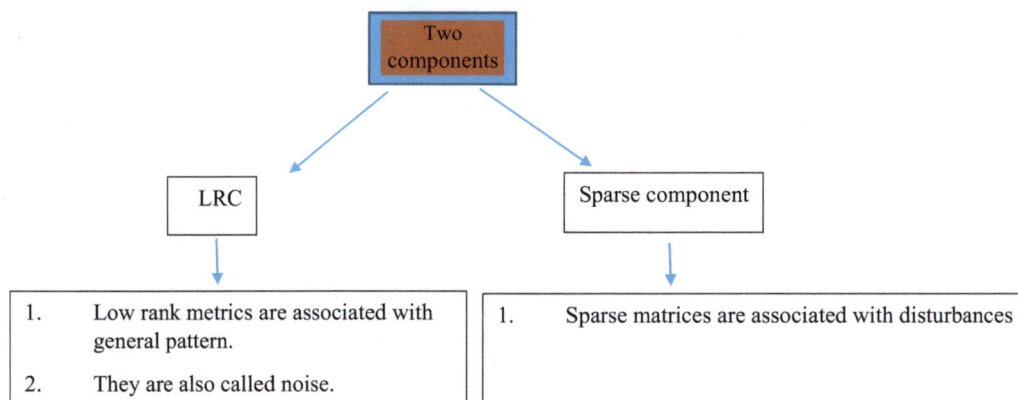

Fig. (3). Flowchart of the disintegration of the data matrix Y into two components.

2.11. Symbolic Data Principal Component Analysis

General designation for more complex data structures. In order to retain the variability in observations, interval data is created [44].

2.11.1. Advantages of PCA

2.11.1.1. Dimensionality Reduction

One of PCA's main advantages is that it decreases the dimensionality by identifying the important components. When the initial data has a lot of variables and is difficult to visualize or understand this can be useful.

2.11.1.2. Data Visualization

PCA can be used to visualize data in 2-3 dimensions. Finding data clusters or patterns that may not have been obvious in the initial high-dimensional space can be made easier using this.

2.11.1.3. Noise Reduction

PCA can be used to reduce the effects of noise or measurement errors in the data by identifying the underlying signal or pattern in the data.

2.11.1.4. Multicolinearity

Multicollinearity in the data occurs when two or more variables have a strong correlation, PCA can handle this. By highlighting the most important characteristics or elements, PCA can reduce the effects of multicollinearity on the analysis [45].

2.11.2. Disadvantages of PCA

2.11.2.1. Information Loss

PCA reduces data dimensionality by selecting a subset of the most crucial features or components. While this simplifies the data and reduces noise, it can lead to information loss if vital aspects are excluded from the chosen components.

2.11.2.2. Scaling

PCA assumes that the data is scaled and centralized, which may not always be suitable. If the data is not properly scaled, the resulting principal components might fail to accurately represent underlying data patterns.

2.11.2.3. Outliers

Abnormalities in the data can significantly impact the principal components derived from PCA. Outliers can distort the covariance matrix, making it more challenging to identify the most significant characteristics [46].

2.12. Software's used to Perform Principal Component Analysis

There are several software packages available for principal component analysis. These include XLSTAT, BioVinci, R software, Unscrambler X, SPSS 23.0, Minitab, Rapid Miner Studio, Solo Simca-P, Origin Pro, Statgraphics, SimcaCemetrics, Statistca, SAS, Canoco5, PAST (Palentologica Statistical Software), C2, OriginLab, Python 3, and GRETL [47 - 50]. Table. (2) provides a list of softwares available in PCA for pharmaceutical research.

Table 2. List of commonly used software's for principal component analysis.

S. No.	Software	Developed	Key Features	Advantages
1.	XLSTAT	Developed in 1993 by Addinsoft.	Modeling data, machine learning sensory data analysis, path modeling, design of experiments, method validation, dose effect analysis.	• Allows parallel excel analysis alongside statistical analysis. • Simplifies the computation of statistical tests and the creation of figures. • User-friendly and robust for various statistical analyses, including sensory data analysis.

S. No.	Software	Developed	Key Features	Advantages
				• Provides ease of sample size and power estimates in XLSTAT.
2.	BioVinci (version 3.0.0)	Developed in December 2020	Dimensionality reduction analysis including PCA in 2D/3D, Isometric feature mapping	• It quickly creates heat maps for datasets as large as 10^5 rows $\times$ 10^5 columns. • It is a powerful web application that produces high quality scientific figures in seconds. • Any statistical analysis from basic to sophisticated can run.
3.	Matlab (version1.0.0.0)	Developed in June 2009	Feature selection, feature extraction t-SNE multidimensional visualization, PCA and cannonocal correlation, factor analysis, non negative matrix factorization.	• Develop the computational codes easily. • Debug easily. • Symbolic computation can be easily done.
4.	SPSS 24.0	Initially launched in 1968 by SPSS Inc. Later on by IBM in 2009	Text analysis, visualization design, handy data management system and editing tools.	• It offers reliable and fast answers. • The chances of errors are little with SPSS. • Effective data management.
5.	Canoco5 (Old versions- 4.0, 4.5)	Developed by CajoterBraak and PetrSmilauer and its latest version was released in 2012	Straightforward testing of all constrained axes, graphing of Principal Response Curves, enhanced visualization capabilities.	• Data can be easily imported from excel. • Multiple methods can be combined in a single analysis

3. APPLICATIONS OF PCA IN PHARMACEUTCAL RESEARCH

3.1. Neuroscience

Spike-triggered covariance analysis is a method that identifies the stimulus characteristics leading to the generation of an action potential by adapting PCA in neuroscience. Principal Component Analysis can also be employed to distinguish neurons based on the appearance of their action potentials. Large neural ensembles utilize dimension reduction techniques like PCA to detect coordinated activities. Additionally, PCA has been employed to establish order parameters or collective variables during phase transitions in the brain [51].

3.2. Role of PCA in Drug Discovery

Principal Component Analysis (PCA) finds applications in various fields, including quantitative Structure-Activity Relationship (SAR), data mining, and 'omics' techniques within pharmacological and biomedical sciences. PCA's relevance in pharmaceutical and biological research is growing, thanks to the data explosion driven by high-tech methods [52]. In the context of drug discovery, the emerging field of 'network pharmacology' is reshaping our understanding of therapeutic targets. The continuous challenges in developing new drugs are prompting researchers to reevaluate the concept of an 'ideal biological profile.' For years, a reductionist approach focused on identifying the fundamental molecular causes of diseases, often involving protein molecules as drug targets. PCA primarily considers uncertainty related to 'y' in its analysis (left panel), where the distances to minimize in the classical regression model (right panel) are perpendicular to the 'x' axis. This technique combines 'n' variables within a studied system linearly to create principal components, as outlined in the formula.

$$PC = ax1 + bx2 + cx3........+kxn.$$

The experimental/observational variables X1–Xn define the statistical units, and least squares optimization is used to determine the coefficients a–k. In a least squares sense, principal components are both the "best summary" of the data [40].

3.3. Image Recognition

Following normalisation, it would take a lot of processing time to execute facial recognition directly. This is because every pixel in the original image contains every bit of information, necessitating a reduction in the image's size. The right features are then recorded in order to express a lot of information. With the use of PCA and a few mutually independent linear combinations in place of the original data, it was possible to significantly reduce the number of variants for the data. The difference in the variation had a significant impact on the data through linear combination computing. The data were analysed to show the most significant individual differences.

3.3.1. Advantages

• The outcome may be calculated quickly and simply.

• In the linear projection, PCA could maintain the most projection data's information.

• PCA used the entire face to extract features that may compensate for the wearer's glasses and shifting facial expressions. The PCA's operational process is listed below [53].

3.4. QSAR Studies

PCA is an effective technique for addressing challenges related to the unfavorable greater descriptor/molecule ratio and collinearity. Figures can be employed to illustrate the overall concept of PCA. It is particularly useful when dealing with large datasets that are challenging to evaluate directly and when intricate correlations between independent variables are difficult to comprehend and visualize. Such situations are common in typical QSAR investigations, where a large number of theoretical descriptors are calculated. These newly created variables, referred to as principal components (PC), are linear combinations of the basic descriptors, designed to capture the majority of the variance in the original data [54].

4. MEDICAL DATA IN PCA REPOSITORIES

PCA is used in a stepwise format to find the correlation between variables. The effects of these chemical tests, including cholesterol, lipoprotein, triglycerides, Apo protein A-1, Apoprotein B, LDLR, phospholipids, total cholesterol, glucose, and uric acid, in patients with ischemic heart disease were investigated using step-by-step principal component analysis [55].

4.1. As Sensory Assessment Tool for Fermented Food Products

Fermented food products, such as yogurt, soymilk, and ice cream, are produced when microorganisms (probiotics) invade the food and hydrolyze lipids and proteins into non-toxic substances. This process is known as fermentation and leads to changes in the physical properties of the food. The acceptability of fermented food products depends on various attributes and is often analyzed using principal component analysis [56].

4.2. In Nanomaterials

PCA proved to be an important tool for constructing friction-free intertube rotation and translation devices. PCA provides mechanical dynamics-based approaches for investigating the characteristics of nanomachines like carbon nanotubes, mechanical oscillators, resonators, rotational bearings, and actuators. In the field of nanotoxicology, which explores the effects of nanomaterials on living systems, PCA plays a crucial role in constructing QSAR models by identifying the relevant features contributing to toxicity [57]. PCA serves as a versatile tool that extracts valuable insights from complex and multidimensional nanomaterials data. Its capacity to streamline data while preserving critical information makes it an indispensable technique for researchers in nanotechnology and materials science

4.3. Bimolecular Molecule Dynamics

For molecular dynamics simulations of large macromolecules, PCA is essential. Both PCA and the mathematically equivalent essential dynamics have proven successful in identifying significant motions in various biomolecules, including proteins and nucleic acids. Computational chemists can find PCA invaluable for managing extensive and complex datasets, as long as they take proper precautions when using these approaches to identify biochemically significant molecular movements [58].

4.4. ECG Signal Determination

Principal Component Analysis (PCA) emerges as a valuable technique in the realm of cardiovascular health, contributing significantly to the determination and diagnosis of heart diseases. Its versatility extends across various facets of cardiac care, including remote patient monitoring, electronic cardiac pacemakers, and intensive care units (ICUs) [58]. In the context of remote patient monitoring, PCA serves as a robust analytical tool for processing vast amounts of patient data, offering insights into patterns and anomalies that might signal cardiac issues. This technology enables healthcare providers to monitor patients' heart health remotely, enhancing early detection and intervention. Furthermore, PCA plays a pivotal role in optimizing the performance of electronic cardiac pacemakers. By analyzing patient-specific data, it allows for the customization and fine-tuning of pacemaker settings to ensure precise and adaptive control over cardiac rhythms, thus improving patient outcomes and quality of life. In the high-stress environment of intensive care units, PCA aids clinicians in processing intricate cardiac data swiftly and effectively. By identifying key variables and patterns, it assists in the early detection of critical events and guides timely interventions, ultimately saving lives.

CONCLUSION

There is a pressing need to define and explore the diverse fields within pharmaceutical research where statistical challenges can be effectively addressed by leveraging powerful tools like Principal Component Analysis (PCA). PCA offers a versatile approach to simplifying complex data structures and revealing valuable insights across various pharmaceutical domains. Its applications span from drug discovery to formulation development, clinical trials, and pharmacovigilance. PCA can be readily executed using a variety of software packages, such as XLSTAT, BioVinci, Canoco5, and many others, making it accessible and adaptable for researchers across the pharmaceutical landscape. These tools empower scientists to efficiently analyze large datasets, extract key information, and identify underlying patterns. This chapter aims to provide a comprehensive overview of principal components and the methods for reducing dimensionality in pharmaceutical research. By delving into the fundamentals of PCA, researchers can enhance their analytical capabilities, leading to more robust and informed decision-making processes. As the pharmaceutical industry continues to evolve and generate vast amounts of data, PCA remains a valuable ally in uncovering actionable insights and advancing drug development and safety.

REFERENCES

[1]   Bezdek JC, Ehrlich R, Full W. FCM: The fuzzy c-means clustering algorithm. Comput Geosci 1984; 10(2-3): 191-203.
        [http://dx.doi.org/10.1016/0098-3004(84)90020-7]

[2]   Borbély-Kiss I, Koltay E, Szabó GY. Apportionment of atmospheric aerosols collected over Hungary to sources by target transformation factor analysis. Nucl Instrum Methods Phys Res B 1993; 75(1-4): 287-91.
        [http://dx.doi.org/10.1016/0168-583X(93)95660-W]

[3]   Karamizadeh S, Abdullah SM, Manaf AA, Zamani M, Hooman A. An overview of principal component analysis. J Signal Inf Process 2013; 4(3B): 173.

[4]   Deane JM. Data reduction using principal components analysis. In: Brereton RG, Ed. Multivariate Pattern Recognition in Chemometrics. Elsevier 1992; pp. 125-77.

[5]   Wold S, Esbensen K, Geladi P. Principal component analysis. Chemom Intell Lab Syst 1987; 2(1-3): 37-52.
        [http://dx.doi.org/10.1016/0169-7439(87)80084-9]

[6]   Beattie JR, Feskanich D, Caraher MC, Towler MR. A preliminary evaluation of the ability of keratotic tissue to act as a prognostic indicator of hip fracture risk. Clin Med Insights Arthritis Musculoskelet Disord 2018; 11.
        [http://dx.doi.org/10.1177/1179544117754050] [PMID: 29371785]

[7]   Yang J, Zheng N, Soyeurt H, Yang Y, Wang J. Detection of plant protein in adulterated milk using nontargeted nano-high-performance liquid chromatography–tandem mass spectroscopy combined with principal component analysis. Food Sci Nutr 2019; 7(1): 56-64.
        [http://dx.doi.org/10.1002/fsn3.791] [PMID: 30680159]

[8]   Moreira LP, Silveira L Jr, Pacheco MTT, da Silva AG, Rocco DDFM. Detecting urine metabolites related to training performance in swimming athletes by means of Raman spectroscopy and principal component analysis. J Photochem Photobiol B 2018; 185: 223-34.
        [http://dx.doi.org/10.1016/j.jphotobiol.2018.06.013] [PMID: 29966989]

[9]   Draper NR, Smith H. Applied Regression Analysis. John Wiley & Sons 1981; p. 709.

[10]   Abdi H, Williams LJ. Principal component analysis. Wiley Interdiscip Rev Comput Stat 2010; 2(4): 433-59.
        [http://dx.doi.org/10.1002/wics.101]

[11]   Eastment HT, Krzanowski WJ. Cross-validatory choice of the number of components from a principal components analysis. Technometrics 1982; 24(1): 73-7.

[http://dx.doi.org/10.1080/00401706.1982.10487712]

[12]   Tauler R, Barcelo D, Thurman EM. Multivariate correlation between concentrations of selected herbicides and derivatives in outflows from selected U.S. midwestern reservoirs. Environ Sci Technol 2000; 34(16): 3307-14.
[http://dx.doi.org/10.1021/es000884m]

[13]   Rojas-Valverde D, Gómez-Carmona CD, Gutiérrez-Vargas R, Pino-Ortega J. From big data mining to technical sport reports: The case of inertial measurement units. BMJ Open Sport Exerc Med 2019; 5(1): e000565.
[http://dx.doi.org/10.1136/bmjsem-2019-000565] [PMID: 31673403]

[14]   Parmar N, James N, Hearne G, Jones B. Using principal component analysis to develop performance indicators in professional rugby league. Int J Perform Anal Sport 2018; 18(6): 938-49.
[http://dx.doi.org/10.1080/24748668.2018.1528525]

[15]   Oliva-Lozano JM, Rojas-Valverde D, Gómez-Carmona CD, Fortes V, Pino-Ortega J. Impact of contextual variables on the representative external load profile of Spanish professional soccer match-play: A full season study. Eur J Sport Sci 2020.
[PMID: 32233969]

[16]   Rojas-Valverde D, Sánchez-Ureña B, Pino-Ortega J, *et al.* External workload indicators of muscle and kidney mechanical injury in endurance trail running. Int J Environ Res Public Health 2019; 16(20): 3909.
[http://dx.doi.org/10.3390/ijerph16203909] [PMID: 31618865]

[17]   Casamichana D, Castellano J, Gómez Díaz A, Martín-García A. Looking for complementary intensity variables in different training games in football. J Strength Cond Res 2019; Publish Ahead of Print.
[http://dx.doi.org/10.1519/JSC.0000000000003025] [PMID: 30844980]

[18]   Svilar L, Castellano J, Jukic I, Casamichana D. Positional differences in elite basketball: Selecting appropriate training-load measures. Int J Sports Physiol Perform 2018; 13(7): 947-52.
[http://dx.doi.org/10.1123/ijspp.2017-0534] [PMID: 29345556]

[19]   Maskey R, Fei J, Nguyen HO. Use of exploratory factor analysis in maritime research. Asian Journal of Shipping and Logistics 2018; 34(2): 91-111.
[http://dx.doi.org/10.1016/j.ajsl.2018.06.006]

[20]   Zago M, Codari M, Grilli M, Bellistri G, Lovecchio N, Sforza C. Determinants of the half-turn with the ball in sub-elite youth soccer players. Sports Biomech 2016; 15(2): 234-44.
[http://dx.doi.org/10.1080/14763141.2016.1162841] [PMID: 27111261]

[21]   Leiva Deantonio JH, Amú-Ruiz FA. Características morfofuncionales y motoras de los seleccionados deportivos de la Universidad del Valle. Revista Científica General José María Córdova 2016; 14(18): 169-93.
[http://dx.doi.org/10.21830/19006586.48]

[22]   Torrents C, Ric A, Hristovski R, Torres-Ronda L, Vicente E, Sampaio J. Emergence of exploratory, technical and tactical behavior in small-sided soccer games when manipulating the number of teammates and opponents. PLoS One 2016; 11(12): e0168866.
[http://dx.doi.org/10.1371/journal.pone.0168866] [PMID: 28005978]

[23]   Ric A, Torrents C, Gonçalves B, Sampaio J, Hristovski R. Soft-assembled multilevel dynamics of tactical behaviors in soccer. Front Psychol 2016; 7: 1513.
[http://dx.doi.org/10.3389/fpsyg.2016.01513] [PMID: 27761120]

[24]   Abdullah MR, Maliki ABHM, Musa RM, Kosni NA, Juahir H. Intelligent prediction of soccer technical skill on youth soccer player's relative performance using multivariate analysis and artificial neural network techniques. Int J Adv Sci Eng Inf Technol 2016; 6(5): 668-74.
[http://dx.doi.org/10.18517/ijaseit.6.5.975]

[25]   Abdullah MR, Musa RM, Azura N. Similarities and distinction pattern recognition of physical fitness

related performance between amateur soccer and field hockey players. Int J Life Sci Pharma Res 2016; 6: 12.

[26]   Negra Y, Chaabene H, Hammami M, *et al.* Agility in young athletes: Is it a different ability from speed and power? J Strength Cond Res 2017; 31(3): 727-35.
[http://dx.doi.org/10.1519/JSC.0000000000001543] [PMID: 28186497]

[27]   Abdullah MR, Maliki ABHM, Musa RM, Kosni NA, Juahir H, Mohamed SB. Identification and comparative analysis of essential performance indicators in two levels of soccer expertise. Int J Adv Sci Eng Inf Technol 2017; 7(1): 305-14.
[http://dx.doi.org/10.18517/ijaseit.7.1.1150]

[28]   Williams S, Trewartha G, Cross MJ, Kemp SPT, Stokes KA. Monitoring what matters: A systematic process for selecting training-load measures. Int J Sports Physiol Perform 2017; 12(2-101): 2-106.
[http://dx.doi.org/10.1123/ijspp.2016-0337]

[29]   Svilar L, Jukić I, Jukic I. Load monitoring system in top-level basketball team. Kinesiology 2018; 50(1): 25-33.
[http://dx.doi.org/10.26582/k.50.1.4]

[30]   Teramoto M, Cross CL, Rieger RH, Maak TG, Willick SE. Predictive validity of national basketball association draft combine on future performance. J Strength Cond Res 2018; 32(2): 396-408.
[http://dx.doi.org/10.1519/JSC.0000000000001798] [PMID: 28135222]

[31]   Floría P, Sánchez-Sixto A, Harrison AJ. Application of the principal component waveform analysis to identify improvements in vertical jump performance. J Sports Sci 2019; 37(4): 370-7.
[http://dx.doi.org/10.1080/02640414.2018.1504602] [PMID: 30058950]

[32]   Weaving D, Dalton NE, Black C, *et al.* The same story or a unique novel? Within-participant principal-component analysis of measures of training load in professional rugby union skills training. Int J Sports Physiol Perform 2018; 13(9): 1175-81.
[http://dx.doi.org/10.1123/ijspp.2017-0565] [PMID: 29584514]

[33]   Figueiredo DH, Gonçalves HR, Stanganelli LCR, Dourado AC. Análise de componentes principais na identificação de características primordiais em esportes coletivos. Revista Brasileira de Ciência e Movimento 2019; 27(3): 41-51.
[http://dx.doi.org/10.31501/rbcm.v27i3.9881]

[34]   Welch N, Richter C, Moran K, Franklyn-Miller A. Principal component analysis of the associations between kinetic variables in cutting and jumping, and cutting performance outcome. J Strength Cond Res 2019.
[PMID: 30741857]

[35]   Welch N, Richter C, Franklyn-Miller A, Moran K. Principal component analysis of the biomechanical factors associated with performance during cutting. J Strength Cond Res 2019.
[PMID: 30664108]

[36]   Verheul J, Warmenhoven J, Lisboa P, Gregson W, Vanrenterghem J, Robinson MA. Identifying generalised segmental acceleration patterns that contribute to ground reaction force features across different running tasks. J Sci Med Sport 2019; 22(12): 1355-60.
[http://dx.doi.org/10.1016/j.jsams.2019.07.006] [PMID: 31445948]

[37]   Gonçalves B, Coutinho D, Exel J, Travassos B, Lago C, Sampaio J. Extracting spatial-temporal features that describe a team match demands when considering the effects of the quality of opposition in elite football. PLoS One 2019; 14(8): e0221368.
[http://dx.doi.org/10.1371/journal.pone.0221368] [PMID: 31437220]

[38]   Khajavi F. Principal component analysis in drug excipient interactions. ECTA Scientific Pharmacol 2022; 3(4): 47-52.

[39]   Zhang Z, Castelló A. Principal components analysis in clinical studies. Ann Transl Med 2017; 5(17): 351.

[http://dx.doi.org/10.21037/atm.2017.07.12] [PMID: 28936445]

[40]   Giuliani A. The application of principal component analysis to drug discovery and biomedical data. Drug Discov Today 2017; 22(7): 1069-76.
[http://dx.doi.org/10.1016/j.drudis.2017.01.005] [PMID: 28111329]

[41]   Jolliffe IT, Cadima J. Principal component analysis: A review and recent developments. Philos Trans-Royal Soc, Math Phys Eng Sci 2016; 374(2065): 20150202.
[http://dx.doi.org/10.1098/rsta.2015.0202] [PMID: 26953178]

[42]   Shang HL. A survey of functional principal component analysis. AStA Adv Stat Anal 2014; 98(2): 121-42.
[http://dx.doi.org/10.1007/s10182-013-0213-1]

[43]   Gamble D, Bradley J, McCarren A, Moyna NM. Team performance indicators which differentiate between winning and losing in elite Gaelic football. Int J Perform Anal Sport 2019; 19(4): 478-90.
[http://dx.doi.org/10.1080/24748668.2019.1621674]

[44]   Pino-Ortega J, Gómez-Carmona CD, Nakamura FY, Rojas-Valverde D. Setting kinematic parameters that explain youth basketball behavior: Influence of relative age effect according to playing position. J Strength Cond Res 2022; 36(3): 820-6.
[http://dx.doi.org/10.1519/JSC.0000000000003543] [PMID: 32084109]

[45]   Rojas-Valverde D, Gómez-Carmona CD, Oliva-Lozano JM, Ibáñez SJ, Pino-Ortega J. Quarter's external workload demands of basketball referees during a European youth congested-fixture tournament. Int J Perform Anal Sport 2020; 20(3): 432-44.
[http://dx.doi.org/10.1080/24748668.2020.1759299]

[46]   Clark NR, Ma'ayan A. Introduction to statistical methods to analyze large data sets: Principal components analysis. Sci Signal 2011; 4(190): tr3.
[http://dx.doi.org/10.1126/scisignal.2001967] [PMID: 21917717]

[47]   Vidal NP, Manful CF, Pham TH, Stewart P, Keough D, Thomas R. The use of XLSTAT in conducting principal component analysis (PCA) when evaluating the relationships between sensory and quality attributes in grilled foods. MethodsX 2020; 7: 100835.
[http://dx.doi.org/10.1016/j.mex.2020.100835] [PMID: 32195148]

[48]   Rodopoulou MA, Tananaki C, Kanelis D, Liolios V, Dimou M, Thrasyvoulou A. A chemometric approach for the differentiation of 15 monofloral honeys based on physicochemical parameters. J Sci Food Agric 2022; 102(1): 139-46.
[http://dx.doi.org/10.1002/jsfa.11340] [PMID: 34056719]

[49]   Ballabio D. A MATLAB toolbox for principal component analysis and unsupervised exploration of data structure. Chemom Intell Lab Syst 2015; 149: 1-9.
[http://dx.doi.org/10.1016/j.chemolab.2015.10.003]

[50]   Guo ZW, Song MM, Zhang J, *et al.* Changes and prediction on metabolic function of intestinal microflora in severe burn patients at early stage by 16S ribosomal RNA sequencing. Zhonghua Shaoshang Zazhi Chinese Journal of Burns 2021; 37: 1-9.

[51]   Smith PF. On the application of multivariate statistical and data mining analyses to data in neuroscience. J Undergrad Neurosci Educ 2018; 16(2): R20-32.
[PMID: 30057506]

[52]   Salih Hasan BM, Abdulazeez AM. A review of principal component analysis algorithm for dimensionality reduction. Journal of Soft Computing and Data Mining 2021; 2(1): 20-30.
[http://dx.doi.org/10.30880/jscdm.2021.02.01.003]

[53]   Ming-Yuan Shieh. Juing- Shian Chiou, Yu- Chia Hu, Kuo- Yang Wang. Applications of PCA and SVM-PSO based real time face recognition system. Math Probl Eng 2014; 14.

[54]   Yoo C, Shahlaei M. The applications of PCA in QSAR studies: A case study on CCR5 antagonists. Chem Biol Drug Des 2018; 91(1): 137-52.

[http://dx.doi.org/10.1111/cbdd.13064] [PMID: 28656625]

[55] Qureshi NA, Suthar V, Magsi H, Sheikh MJ, Pathan M, Qureshi B. Application of principal component analysis (PCA) to medical data. Indian J Sci Technol 2017; 10(20): 1-9.
[http://dx.doi.org/10.17485/ijst/2017/v10i20/91294]

[56] Ghosh D, Chattopadhyay P. Application of principal component analysis (PCA) as a sensory assessment tool for fermented food products. J Food Sci Technol 2012; 49(3): 328-34.
[http://dx.doi.org/10.1007/s13197-011-0280-9] [PMID: 23729852]

[57] Shenai Prathamesh M, Xu Z, Yang Z. Principal component analysis: Engineering applications. InTech 2012; pp. 25-40.

[58] Martis RJ, Acharya UR, Mandana KM, Ray AK, Chakraborty C. Application of principal component analysis to ECG signals for automated diagnosis of cardiac health. Expert Syst Appl 2012; 39(14): 11792-800.
[http://dx.doi.org/10.1016/j.eswa.2012.04.072]

Quality by Design in Pharmaceutical Development: Current Advances and Future Prospects

Popat Mohite[1,*], Amol Gholap[1], Sagar Pardeshi[1], Abhijeet Puri[1] and **Tanavirsing Rajput[1]**

[1] *AET's St. John Institute of Pharmacy and Research, Palghar-401404, Maharashtra, India*

Abstract: QbD, or Quality by Design, is a cutting-edge methodology adopted extensively in the pharmaceutical industry. It is defined objects, such as the product's safety and effectiveness. QbD's primary focus in the pharmaceutical industry is ensuring the product's security and usefulness. Quality by Design (QbD) seeks to instill high standards of excellence in the blueprinting process. The International Council for Harmonization (ICH) has developed guidelines and elements that must be adhered to guarantee the consistent, high-quality development of pharmaceuticals. This chapter provides updated guidelines and elements, including quality risk management, pharmaceutical quality systems, QbD in analytical methods and pharmaceutical manufacturing, process control, vaccine development, pharmacogenomic, green synthesis, *etc*. QbD was briefly defined, and several design tools, regulatory-industry perspectives, and QbD grounded on science were discussed. It was portrayed that significant effort was put into developing drug ingredients, excipients, and manufacturing processes. Quality by design (QbD) is included in the manufacturing process's development, and the result is steadily improving product quality. Quality target product profiles, critical quality attributes, analytical process techniques, critical process parameters control strategy and design space are elements of many pharmaceutical advancements. Some of the topics covered included the application of QbD to herbal products, food processing, and biotherapeutics through analytical process techniques. We are still exploring and compiling all the data and metrics required to link and show the benefits of QbD to all stakeholders. Nevertheless, the pharmaceutical sector is quickly using the QbD process to create products that are reliable, efficient, and of high quality. Soon, a more profound comprehension of the dosage form parameters supported by the notion of QbD will benefit Risk management and process and product design, optimizing complex drug delivery systems.

Keywords: Drug delivery system, Process and product design, Quality by design, Risk management.

---

* **Corresponding author Popat Mohite:** AET's St. John Institute of Pharmacy and Research, Palghar-401404, Maharashtra, India; E-mail: mohitepb@gmail.com

Dilpreet Singh and Prashant Tiwari (Eds.)

1. INTRODUCTION

The fundamental pillars for human safety following drug ingestion in all dose forms are the safety and efficacy of the product. The actual analysis of the parameters for the pharmaceutical product's stability, quality and potency is performed in the drug development phase, especially during the pilot phase of the formulation, to provide the utmost care to the patients or customers [1]. The drug product evolved from the drug development phase, which includes several parameters like drug discovery and clinical trials. It also includes animal studies, the impact of laboratory testing, and a thorough analysis of the formations. The drug development of many formulations must comply with the stringent guidelines given by regulatory authorities [2]. Almost all regulatory agencies worldwide are asking for testing to establish and confirm product identity, purity, and quality, along with strength and stability determination, prior to its market launch for the customers. So, pharmaceutical validation, along with process control, is one of the main controlling factors for the fate of the finished pharmaceutical product. The compromise in the quality of the pharmaceutical products induced harmful effects on the patients; and also decreased the overall therapeutic effects for the anticipated actions for the given medicine. So, maintenance of the quality of the medicine is essential for the customers, and for the past two decades, many industries have been doing this with the help of conventional methods for quality maintenance.

1.1. Conventional Approach *vs* Design Approach

The conventional approach contains a series of works assigned with sequential testing and analysis of the raw materials, product manufacturing type of the process, and the testing of the finished products. Such an approach is called a quality by testing method and has some significant gaps like experiment cost, time management and alignment with the Food and drug administration (FDA) guidelines for the same. To address these issues, one can redesign the experiments as per needs in the repetitive mode to achieve the required specification, further putting an extra cost burden on the formulators and customers. So, these issues are well countered by implementing the quality by design (QbD) approach. This type of method ensures the quality of the finished pharmaceutical product with the help of process control and quality risk management (QrM) [3]. The detailed principles of QbD have been explained in the International Conference of Harmonization (ICH) annex Q8 (R2), while the implementation of the QbD guidelines was given in the ICHQ9 and ICHQ10. Such principles are advised to implement during the pharmacy product development to ensure the finished products' quality, safety and efficacy. In 1970, J. Juran described the concept of QbD, while the actual implementation of the same was initiated in 1980 in Toyota through the six-sigma approach. Several other industrial sectors, like automobiles and aeronautics, have started implementing QbD for better quality products. Most industries use different technologies along with QbD, like lean Six sigma. The design six sigma was also popularized during the same years and delivered better product quality. The FDA has seen this type of exploration and successful implementation of the QbD concept and realized to implement the same for better quality control of the finished pharmaceutical products on the significant levels for biologics and pharmaceutical drug markets. FDA has implemented QbD and given several guidelines for further updates for effective and smooth conduction into today's 21st century of pharma globally [4 - 8].

1.2. QbD Paradigm and Regulatory Authorities

As part of the QbD paradigm, different regulatory bodies working in pharma sectors have launched some QrM along with tools for control of the pharmaceutical product quality through the manufacturing process. This was a significant shift in understanding that the quality was based on manufacturing, not something we could alter or change at the end of the production process [9]. Usually, we have implemented different tests to care for the quality measures of the finished product by passing them into these tests, but the quality of the finished products lies in the process by which it is manufactured for the same. So, rather than working on product quality checks, the proper control of the manufacturing process was highly appreciated and consciously accepted by the FDA for the better future of pharmaceutical industries during those days. This concept will help reduce product formulation defects and ensure product quality. The conventional manufacturing process differs from QbD, majorly due to the less post-approval process burden in the latter. Regulatory authorities across the globe are always demanding some more amount of additional research into the post-approval phase. Some significant agencies like the United States (U.S.) FDA and European Medicines Agency (EMA) have enforced specific regulatory changes as a part of the post-approval studies, which further involve clinical trials. The clinical trials can be interventional and observational and treated as non-interventional studies for the post-approval phase. The design options and related data sources are vital for the study design selection.

The safety and effectiveness of the approved drug's data are required for post-approval studies, and these studies are tracked to completion with the help of regulatory authorities and registries across the globe. Independent variables impact the process of formulation of pharmaceutical products. These aspects are further explored and analyzed in depth using tools such as the Ishikawa diagram (I.D.) and relative risk matrix analysis (RRMA). The process also incorporates failure mode effect analysis (FMEA). All of these tools are connected to the critical quality attributes (CQA). Key Performance Indicators (KPIs) and Critical Process Parameters (CPPs) serve as the primary foundations of CQA [10, 11].

1.3. Contribution of Ishikawa Diagram

The Ishikawa diagram (ID) resembles a side view of a fish and was initially introduced by Ishikawa and the team from Japan. The ID continues to be utilized across various industries for accurate problem diagnosis and root cause analysis within organizations. The improvement in the quality through leadership can be achieved more significantly from that knowledge of I.D. The diagram is designed so that each large bone of the fish gives some branches into the smaller one, and the diagram is worked from the right to the left side, considering all the possible reasons or causes which can induce the problems [12]. A thorough analysis of the I.D. can be achieved with the help of four primary tools, including identification of the problem, a factor vital for the impact on the process, identification of the possible causes and further analysis of the diagram. The visual presentation of data into the I.D. will give more explicit pictures for the all-different facet of the factors that can interact in the given case study [13].

1.4. Impact of FEMA on Quality Improvement

The FEMA is also one of the crucial factors in quality management, also called potential failure modes and effect analysis or failure modes, effects, and criticality analysis (FMECA). The FEMA was started in 1940 by the U.S. military and involved a step-by-step approach to identifying all possible design, manufacturing, and assembly failures. Moreover, it is generally treated as a process analysis tool. The failure modes are the routes, or modes leading for the process to fail, significantly affecting the customer. These modes may be potential or actual and require a thorough analysis to evaluate their contribution to the total failures of the process. One of the significant purposes of implementing the FMEA is to reduce or decrease failures by focusing on the factors that are the most vital or priority to cause the failures [14]. The ideal condition to implement the FEMA is the conceptual earlier stage of the design, and it will also help give its valuable contribution to quality improvement during the product or service. So, there are many ways to implement the FEMA for processes, products, or services. One way is we can apply the same for designing the sectors mentioned above, or the second way is redesigning the same sectors after the quality function deployment. FEMS is also implemented if the service, process, or product is applied differently. Before developing control plans regarding the new or modified process, one can use FEMA, too [15].

1.5. RRMA into QbD and Process Failures

The RRMA is implemented as a risk management scheme during the project. Risk is one of the potentially unavoidable factors in any complex type of programme. Risk management contains all related activities for possible hazard identification and assessment [16]. The selection of the proper responses, along with risk monitoring, can be done as part of risk management. The International risk management standard ISO 31000 contains several vital stages, including organizational context definition, risk analysis, evaluation, and treatment. The monitoring, review, and treatment of the risk are also involved. Risk analysis in any field is vital for understanding the gravity of the risk and its damage. In the healthcare sector, the same is related to patient safety, and organizations are involved in the same for subsequent evaluation and the treatment of the risk [17].

1.6. Role of CQA, KPI and CPP within QbD

The CQA focuses on physical, chemical, and biological aspects or traits that affect product quality. It also covers the microbiological aspect of the product too, and it ensures the selection of limit, range, and distribution impact on the quality aspects of the product. It is somewhat difficult to measure the CQA directly in production. The KPI and CPP are related to the quality attributes and the manufacturing process involving upstream and downstream portions for the same. The CPP in the pharmaceutical product deals with the process variables impacting the CQA; hence, they are essential to control to obtain desired quality of the product. The CPP is covered under ICH Q8 (R2) and is related to the Process Analytical technologies (PAT) framework [18]. The thorough monitoring of the CPP will give a clear picture of the process knowledge from the foundation to the results, process control parameters, risk analysis, and quality control. The control strategies are also involved in the same. The CPP plays a significant role in risk analysis, quality risk management, and risk assessment, which helps foster the QbD. The process engineers often use the CPP to transform raw materials into the desired quality product with time management, less burden for retrospection and consistency through the production with the design of experiments (DoE) studies. The overall goal of these tasks is to get safer and more efficacious pharmaceutical products. The KPIs are used for the indication that every production step should follow the route as per our expectations to meet quality product and consistently maintain it throughout the process [19 - 21].

2. ICH Q8 PHARMACEUTICAL DEVELOPMENT

2.1. Overview of ICH Q8 (R2) Pharmaceutical Development

This section is significant for pharmaceutical product development for submitting documents to the regulatory authorities through new drug applications (NDA) or marketing authorization applications (MAA). It is presented in module 3 of the standard technical document (CTD) under section 3.2.P.2 and shares essential data required to understand the scientific background concept of design space along with manufacturing controls. It also deals with the specifications and QbD, which focus on the new version of the principle for pharmaceutical product quality. This section treats the quality from the build-up process rather than mere testing of products; to build the quality of the product, the proper design is mandatory. This section also deals with the commercial manufacturing section and pharmaceutical product development and demands a systematic and scientific approach from pharmaceutical industries for the drug development phase. Some of the unique points of ICHQ8 are described in the ICHQ11 related to the drug substance [22].

2.2. Quality by Testing Approach *vs* QbD Approach

The Pharmaceutical industry's primary goal is to provide the best quality product to the patients as required without any side effects or other issues. So, all the products are designed to meet this criterion and increase market sales of the products for the overall business growth of the pharmaceutical organization. Many factors from the last few decades are explored, like compound properties, target product profile (TPP), and previous experience with the same product. The available manufacturing facility and equipment also played a significant role. The quality-by-testing approach believed in the product's quality check during the manufacturing and end product testing rather than designing from the initial phase of product formulation. Medicines are treated as one of the goods required for time-to-time innovation into manufacturing and strategies to cope with increased health burdens along with critical issues of resistance and tolerance. In quality, testing the material meets all specifications as per the FDA or other regulatory standards are only allowed for market release. If such compounds fail to achieve this, they are subjected to reprocessing of products. One of the significant factors contributing to failures in such systems is a proper understanding of processes and factors which impact the quality of the products. The manufacturer, in such cases, has to restart the process till they address the root cause of failures to make products get accepted as per the FDA standards for failed batches. Such a process produces lower cost efficiency and product variation, resulting in limited drug safety. On the other hand, with the help of QbD, there is significant modulation in the pharmaceutical industry's approach to pharmaceutical quality regulation with the help of predefined objectives, risk assessment analysis, and proactive methods for product development. The QbD is a more scientific method for study, including the study of prior knowledge, DOE, PAT, and QRM tools. So, it will help the manufacturer understand product development from the regulatory perspective and give a rational choice for the manufacturer to select the perfect development strategy [23].

2.3. QbD Elements in Pharmaceutical Development

Many factors affect QbD for pharmaceutical developments, including quality target product profile (QTPP), identification of the critical quality attributes (CQA), determination of the critical material attributes (CMA) for the excipients and that of the active pharmaceutical ingredients [24]. It also includes selecting the proper manufacturing process for the drug's physicochemical properties and thoroughly understanding the scale-up principles [25]. Process refinement deals with continual improvement with the help of a check-up process on the elements. Quality by Design (QbD) is a method of managing quality risks and defining desired product qualities from the outset of a project. QbD applications are governed by the International Conference on Harmonization (ICH) guidelines Q8-Q11, the most important of which is ICH Q8 (R2), which was approved in 2009 and described the fundamentals of QbD in great depth. QbD has been in use for more than a decade, and in that time, companies have been asked to include additional details regarding their use in the development of pharmaceutical products in regulatory dossiers [26]. The elements of pharmaceutical product development are shown in Fig. (**1**).

2.4. Formulation Development

The formulation development and the product design deal with developing robust products to take care of the QTPP and patients' needs. It correlated to the investigation of interrelationships for the critical attributes and parameters. These studies are essential for formulation development as without these studies, any suboptimal changes in the formulation may have the capability to change the properties of the same, which will impact the drug product quality and adverse effects induced by the same. So, this study includes establishing the relationship between CQA and the input type of variable like CMA for excipients and drug substances [28]. It also takes care of the controlled strategy based on the scientific approach. The implementation of CMA is done to ensure future robustness to that of the formulation. The definition of the experiments and risk assessment is also presented.

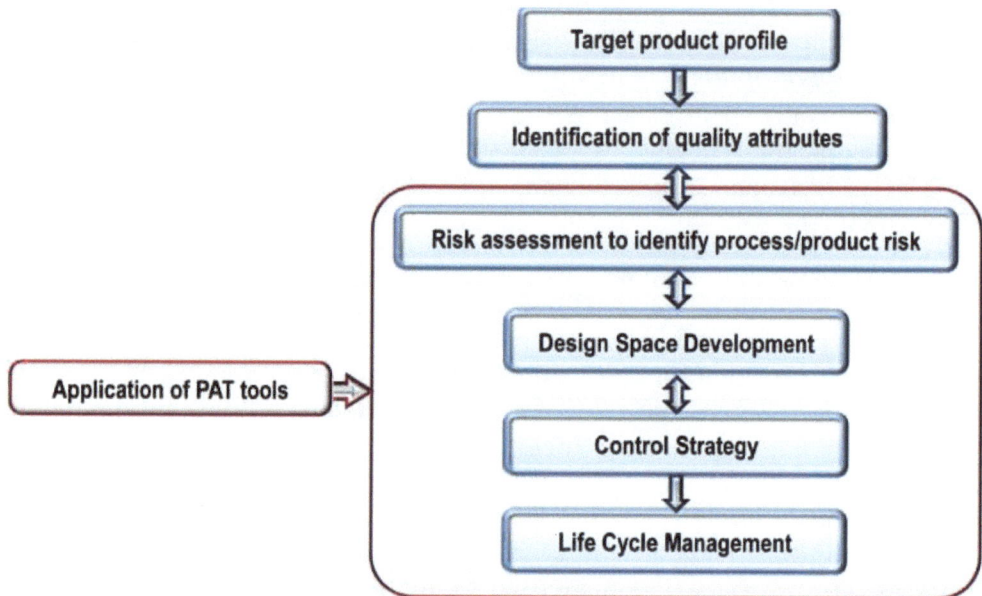

Fig. (1). Elements of quality by design [27].

3. QUALITY BY DESIGN (QBD) TOOLS: APPLICATION IN PRODUCT DEVELOPMENT

Quality by design (QbD) is a method of choice in formulation development to get robust and quality products followed by continuous improvement [29]. Quality by Design (QbD) is a methodical strategy with well-defined goals that place a premium on knowing one's stuff and, using that info, keeping one's processes in check [30]. QbD is particularly helpful in designing robust processes with well-understood operational limits and their significance. The pharmaceutical industry finds it challenging and time-consuming to implement Quality by Design (QbD). Using QbD in formulation development minimizes the trial cost of validation batches. Therefore, the current need is to implement QbD principles, develop a better understanding, and possibly design simple templates to develop complex formulations [31].

For understanding, we can take the example of the freeze-drying process. Freeze-drying, from a business perspective, is a time-consuming and costly procedure. As a result, the process must be brief, repeatable, and robust. To design a freeze-drying method, one must first define and control the essential formulation parameters in the process design. Drying chamber pressure, eating fluid temperature, and timeframe of both drying stages throughout primary and secondary drying are some of the freeze-drying process parameters that significantly impact the quality of finished products [32]. If such variables are not controlled, they can cause the dried cake's product to denaturate, melt, or collapse. So, systematic development of the freeze-dried product by implementing QbD is most required to obtain a quality product [33]. According to ICH Q8 (R2), different practical QbD tools can be used to develop robust quality formulations (Fig. **2**).

Fig. (2). Different quality by design tools.

3.1. Quality Product Quality Profile (QTPP)

The QTPP defines targeted product characteristics to be incorporated into the product and development process [34]. The QTPP embodies the overarching goals of the pharma product development program's quality. Tactics, past knowledge, experience, or the availability of equipment and resources could influence the selection of a QTPP. QTPP is a knowledge-based system, so it should be utilized to identify drug product characteristics of freeze-dried products [35]. Some probable QTPP elements of freeze-dried products are represented in Fig. (**3**).

3.2. Critical Material Attributes (CMA's), Critical Material Parameters (CPPs), and Critical Quality Attributes (CQA's)

Considering the material's intended purpose and knowledge and comprehension of its physicochemical, biological, and microbiological features, which can impact product development, the required quality for freeze-drying must be established. As a result, before beginning the development cycle, CMAs should be identified [36, 37]. A vital process parameter should be examined or managed to guarantee that the system produces the highest quality since its variability influences a crucial quality attribute. Several unit operations make up the freeze-drying process [38].

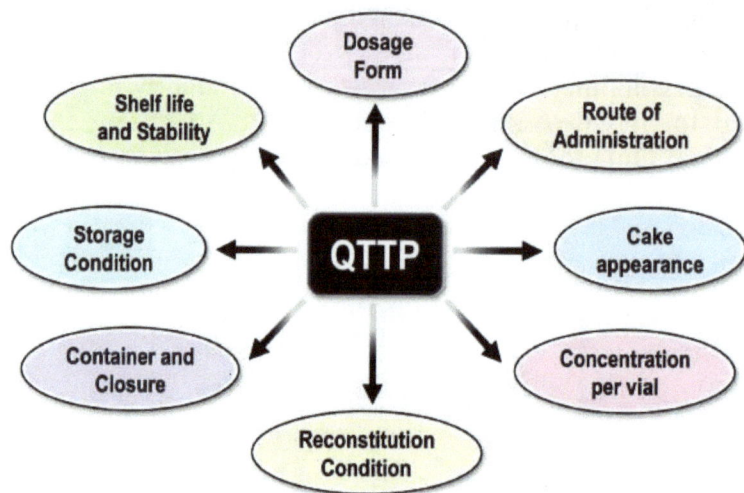

Fig. (3). QTPP elements of freeze-dried products.

CQAs are qualities or features with physicochemical, biological, or microbiological origins that must lie within a specific range [39]. The freeze-dried product's CQAs usually impact product quality, purity, stability, efficacy, strength, drug release, *etc*. CQAs usually are linked with the active substance, additives, intermediates (in-process steps), and drug products in freeze-drying. The possible CMAs, CPPs and CQAs for the freeze-dried product are depicted in Table **1**.

Table 1. Probable CMAs, CPPs and CQAs of freeze-dried product development [40].

Critical Material Attributes (CMA's)	Critical Process Parameters(CPPs)	Critical Quality Attributes (CQAs)
Buffer	Shelf temperature	Cake appearance
Surfactant	Freezing ramp rate	Reconstitution time
Fill volume	Freezing temperature	pH
Total solid content	Hold time for freezing	Residual moisture content
Ionic strength	Chamber pressure at primary drying	Potency
Configuration of vial	Ramp rate to primary drying hold	Concentration of solution
Stopper	Temperature for primary drying	Particulate matter
	Time required for primary drying	Content uniformity
	Ramp rate for secondary drying	Drug impurity
-	Temperature for secondary drying	Sterility
	Time required for secondary drying	-

3.3. Risk Assessment

By evaluating the potential impacts of Critical Material Attributes (CMA) and Critical Process Parameters (CPP) on the drug product's Critical Quality Attributes (CQAs), risk assessment serves as a substantial piece of scientific evidence and a tool utilized for Quality Risk Management. Previous models used for risk assessment include risk ranking, fishbone diagram, and Failure Mode Effects Analysis (FMEA) [41, 42]. The risk assessment of freeze-dried products involves linking input and process variables to CQAs. The relative risk of each variable was ranked as high, medium, or low based on criticality. We will delve deeper into factors that can significantly affect the CQAs of freeze-dried goods. Characteristics with low impact do not require further attention [43].

3.4. Design of Experiment (DOE) and Design Space

In DOE, inputs (variables) and output (response) form a complicated matrix in which the input variables may be interlinked or independent [44]. It is challenging to understand all potential variations and their influence on the standard of a pharmaceutical. DOE tools help scientifically collate most variations in a minimum number of experiments. Some possible input variables are the solution concentration, amount of excipients, freezing rate, annealing temperature, and secondary drying temperature [45]. The goal of the DOE in the freeze-dried product is to fine-tune essential values identified in risk assessment and to achieve the desired quality product. Two types of designs can be used in DOE, *i.e.*, screening design and optimization design. The screening design aims to test pairs of combinations in only a few experimental runs, thereby saving time and resources. Some commonly used screening designs are fractional factorial, Plackett-Burman, and Taguchi [46]. Optimization design will further develop the critical variables identified in the screening design. Standard optimization designs are Box-Behnken, central composite, and full factorial designs.

The design space laid out the connection between the input, procedure, and the product. Knowledge gained through risk assessment and the design of trials for its impacts on CQAs can be used to define the operational ranges of CMAs and CPPs [47]. Each parameter's impact on freeze-dried goods was calculated, and the optimal operating ranges were chosen as the design space for a stable formulation. Design space can be determined using ANOVA, response surface diagram (RSD), interaction plot, contour plot, main effect, and other statistical methods. It was assumed that any procedure within the design space would yield a result of the specified quality.

3.5. Control Strategy

Process monitoring, reasonable controls, testing for lot release, in-process controls, comparability testing, characterization testing, and stability testing are all part of the strategy's planned set of controls. The Control Strategy development requires an organized process involving a team of experts, linkage of the development of a product to the manufacturing process, and process equipment controls. An example of a control approach for a freeze-dried product is based on the findings of in-depth product and process analyses [48].

3.5.1. Process Analytical Technology (PAT)

Incorporating Process Analytical Technology (PAT) into the management strategy is a viable option. ICH Q8 (R2) defines PAT as the practice of maintaining processes within an established tolerance range [49]. PAT enables real-time monitoring of Critical Process Parameters (CPPs), Critical Material Attributes (CMAs), or Critical Quality Attributes (CQAs), indicating whether the process is being maintained within the defined design space and providing insights into decision-makers on whether to proceed [50]. The outcomes of PAT, CMAs, and CQAs can be analyzed online. Both PAT techniques outperform traditional Quality Assurance (QA) testing in identifying product defects. If any changes in the environment or input materials are identified that could impact the pharmaceutical product's quality, PAT facilitates active control of CMAs and CPPs, allowing for swift adjustment of operational parameters.

PAT uses four essential components: process analytical chemistry instruments, process tracking & control, continuous process optimization, and information management. Building a scientific comprehension of the method is necessary for multivariate data analysis, as is identifying key characteristics of materials and process variables that affect final product quality and incorporating this information into process management, which is the same as the process comprehending in the context of quality by design [51]. Tools for process analytical chemistry offer in-situ, instantaneous information about the process's state. The original data from PAT tools is connected to CQAs through multivariate data analysis. Process controls modify crucial variables according to data analysis to satisfy CQAs. The information acquired from the process serves as a springboard for further process enhancement. Studies performed in FDA research facilities revealed the potential of various PAT techniques and chemometric methodologies.

4. QUALITY-BY-DESIGN: CURRENT TRENDS IN PHARMACEUTICAL DEVELOPMENT

Lack of proper justifications for the conclusions that were drawn, inadequate visual representations of the factor interactions, unclear design space limits, shortages of details on the statistical reliability of models, a lack of framework in the information provided, *etc.*, are a few of the problems regulatory bodies run into when evaluating a QbD-based registration dossier [46]. Collaboration between researchers from academia, industry, and regulatory organisations is necessary to solve the abovementioned issues. Numerous scientific projects focus on modeling products and processes, in-line method evaluation, and design space planning. The scientifically supported Quality by Design (QbD) concept should be implemented using this information as a foundation. The following are a few examples of peer-reviewed instances of QbD module development [52].

Using the QbD method, we demonstrated a connection between the drug's CPPs, CQAs, and its therapeutic efficacy. As shown by our investigation of extended-release theophylline tablets, not all compendial studies effectively convey the therapeutic benefits of product variants. Inputs to the design space were broken down into critical and noncritical characteristics based on a quantitative assessment of the risk of failure and toxicity. In another study, the effects of composition parameters and process variables on mixture quality and mixing endpoint were assessed using a combined Quality by Design (QbD) and Discrete Element Model (DEM) simulation approach to characterize a mixing unit operation. To outline steps for process characterization, QbD was utilized to define content uniformity as a Critical Quality Attribute (CQA), establish its connection to blend uniformity, identify critical variables that could impact the effectiveness of blending operations, and rank the risks associated with these variables. The findings were used to establish a design space, create a three-dimensional knowledge space, and formulate an appropriate control strategy.

Similarly, the spray drying of insulin designed for respiratory administration was examined using QbD principles [53]. The impacts of method and formulation parameters on insulin stability and particulate characteristics were estimated. Using a well-planned experiment and multivariate data analysis, we identified key elements in the manufacturing process and established causal links between particle properties. The principal component method was employed to find connections between the two sets of variables.

5. STRATEGIES FOR EVALUATING RISK IN PHARMACEUTICAL PRODUCTION PROCEDURES

Throughout the many stages of a drug's product lifecycle, Quality Risk Management assesses, mitigates, disseminates, and reviews any potential threats to the drug's quality [54]. Risk has two components: how likely it is that something terrible will happen and how bad it could be if it did. The Quality Risk Management (QRM) system can't be created without a thorough risk assessment. Assessing risks entails systematically arranging relevant data for use in making choices. Risk assessment relies primarily on three factors:

5.1. Risk Identification

Theoretical analysis, historical data, stakeholder concerns, *etc.*, will all be systematically utilised to pinpoint the origins of any potential risks.

5.2. Risk Analysis

Risk assessment for the identified threats; and

5.3. Risk Evaluation

Evaluation of the relative importance of the estimated hazards by qualitative and quantitative comparison [55].

Drug compounds' physicochemical properties are studied early on in the development process. This includes the substance's appearance, polymorphism, solubility, particle size, hygroscopicity, pH solubility/temperature, salt screening, pKa, stability profiling, *etc.* Such studies help discover a drug's biopharmaceutical and stability qualities and comprehend the impact of different variables on such properties (Table **2**).

Table 2. Common methods for managing risk in pharmaceutical development [56].

Risk Management Tool	Attributes	Potential Applications
Basic Tools		
• Analysis of Diagrams • Flowcharts • To-Do Lists • Analyzing Procedures • Linking causes and effects with causal diagrams	• They commonly used methods for data collection, organisation, and decision-making within the realm of RM.	• Statistical or other empirical evidence is gathered to back up less complicated cases of deviation, complaint, default, *etc.*

(Table 2) cont.....

Risk Management Tool	Attributes	Potential Applications
• Risk ranking and filtering	• Technique for evaluating and ranking risks • Analysis of many quantitative and qualitative aspects associated with each risk and the application of weights and risk ratings is a common practice.	• Audit and evaluate active regions or places with the highest priority. • It's helpful when comparing risks and repercussions, especially when complex and varied.
Advanced Tools		
Fault tree analysis (FTA)	• A technique used to investigate the origins of a problem or failure. • Can analyse many causes of a failure at once by tracking down causal chains, typically used for single-failure evaluations of systems or subsystems. • Detailed knowledge of the process is crucial for isolating the causes of any observed effects.	• Look into customer complaints. • Analyze discrepancies.
Hazard operability analysis (HAZOP)	• The application presupposes unsafe situations arise when plans and procedures deviate from their intended state. • Makes use of a methodical approach to spot potential discrepancies between the actual use and the intended purposes of the product.	• Use of tools, machinery, and infrastructure for production • Commonly used in risk assessments for process safety.
Hazards analysis and critical control points (HACCP)	• Find and place process controls that reliably and effectively stop hazardous situations from happening. • The bottom-up method focuses on limiting the impact of potential dangers. • Highlights the importance of robust preventative controls in the detection. • Counts on having identified critical process parameters (CPPs) and having a thorough familiarity with the process beforehand. This device guarantees that CPPs will be followed.	• Preventative measures are more effective than reactive ones. • More useful for proactive than reactive uses. • Evaluation of CPPs' usefulness and the extent to which they can be consistently applied to any process.
Failure modes effects analysis (FMEA)	• Analyzes the causes of process failures and how they could affect results and product performance. • Removing, decreasing, or controlling risks is possible after failure modes have been identified. • Each failure mode is assigned a relative "risk score" as the final result.	• Determine critical processes and parameters in a manufacturing process; evaluate supporting facilities.

6. QBD IN PHARMACEUTICAL DEVELOPMENT

Currently, the QbD tools are used in different ways in pharmaceutical development to control the quality of finished products. The tools are recently used by different researchers in Pharmaceutics, Pharmaceutical analysis and Pharmacognosy, depending on their research area, and they became successful tools for them, as listed below.

6.1. QbD Approach in Process Control

The steps required for creating a medication are commonly divided into smaller, more achievable tasks. Both batch processing and continuous production lines can be employed to execute these unit tasks. Mixing, milling, granulating, drying, compressing, and coating are all instances of unit operations, as they lead to distinct physical or molecular transformations. A process is deemed well-understood when (1) all significant sources of variability are identified and understood; (2) variability is controlled throughout the process; and (3) product quality attributes can be predicted with precision and consistency. The input operating parameters (such as speed and flow rate) or process state variables (such as temperature and pressure) of a process step or unit action are referred to as process parameters. To achieve the desired quality, it is necessary to monitor or control a process parameter whose variability affects a crucial quality attribute [57].

6.2. QbD in the Development of Analytical Methods and Pharmaceutical Manufacturing

The QbD approach has been extended to the analytical process, where the method is defined using QbD, then evaluated and thoroughly studied to achieve the best method performance. The primary target of applying QbD to analytical method development is to build quality into the analytical method during the development stage, which has a structural approach. The chosen approach is tested to ensure it is robust and rugged by being evaluated for risks using risk assessment tools. These studies help to understand how well a method works, how to enhance it, and how to provide the risk management control approach so that the method works as intended after it has been validated [58].

AQbD is a methodical strategy for method development that manages the entire analytical process life cycle. The analytical process is now incorporated into the QbD procedures. The primary use of quality by design (QbD) concepts for developing analytical methods is concentrated on incorporating quality into the analytical technique while it is being developed. QbD defines the methodology employed to assess and compare various approaches in order to achieve optimal method performance. The established method is then subjected to testing, followed by a risk assessment, to identify potential risks and assess its robustness to withstand them. These studies contribute to the analysis of method performance, refinement, and the development of a risk management strategy. This guarantees that the method functions as anticipated once it has been validated [59].

Method validation requires defining the analytical target profile (ATP), deciding on the most critical method features and responses, and pinpointing the most crucial method parameters and variables. Critical parameters can be identified and optimised using screening and response-surface experimental designs, both made possible by statistical analysis. A mathematical model that provides a forecast elaborates the connection between the factors and the result. The Method Operable Design Region (MODR), where dependable method performance is guaranteed, is described, consisting of multivariate combinations of elements matching the CMA requirements [60].

The complete life cycle of an analytical technique must be shown to be appropriate for the task at hand. To achieve this goal, a QbD approach was proposed for pharmaceutical analysis, which contains 3 steps Method Development and Design, Method Producing High - quality, and Procedure Performance Verification following process validation in the predicted maker. Part of identifying ATP includes deciding on method requirements, such as the types of analytes of interest (products and contaminants), the type of analytical technique to be used, and the exact nature of the product being tested. An initial risk assessment will be performed to foresee analytical criticality and technique requirements. The standard analytical procedure ATP includes the following procedures: First, choose your APIs and impurities; second, pick your methods (HPLC, HPTLC, UV, *etc.*); and third, pick your method specifications (test profile, impurities, solvent residue).

6.2.1. QbD Approach in Chromatographic Techniques

A systematic methodology should be used for the analytical quality by design (QbD) technique in the actual method development process. Compliance with established objectives is the goal of the QbD method development. You can use HPLC as an example to show the goal of the QbD technique development. High-performance liquid chromatography (HPLC) is commonly used for analyzing active pharmaceutical ingredients (API) to isolate and quantify the active molecule and any critical quality attributes (CQA0) that may affect API quality. Specificity, linearity, accuracy, precision, robustness, and ruggedness are characteristics the specifications must meet to pass regulatory muster [60].

6.2.2. Strategy for HPLC Method Development

Based on the Quality by Design principles, researchers have presented a new technique for producing high reversed-phase liquid chromatography technology. Vamsi Krishna and colleagues investigated eberconazole nitrate's oxidative and photolytic breakdown kinetics using an HPLC method and a Quality by Design (QbD) approach. The tests were conducted utilising a complete factorial design (*i.e.*, 3 levels and 3 components) to identify and rationally analyse the impacts of TBAH quantity, buffer pH, and organic phase concentration on the capacity factor of EBZ. The tests were conducted utilising a complete factorial design (*i.e.*, 3 levels and 3 components) to identify and rationally analyse the impacts of TBAH quantity, buffer pH, and organic phase concentration on the capacity factor of EBZ. Previous univariate investigations and chromatographic intuition guided the selection of the components and range for consideration. Statistics were performed with Statistica on the collected data (Version 6.0). The results indicate the effect of each factor on the capacity factor. TBAH concentration and organic phase concentration were found to be significant and linear. Also, they reported that as TBAH concentration increases, capacity factor increases. The contour plot shows no effect of pH on the capacity factor. So, the developed methods and stability study indicates that QbD has helped the researchers select various parameters [61].

In another study, Piera Iuliani and colleagues employed a fractional factorial design to examine the impact of altering the pH of the mobile phase, in-column temperature (T), and the organic modifier CH3CN on the HPLC response of the analyzed mixture of NSAIDs. Additionally, they utilized a central composite design (CCD) involving pH and varying concentrations while maintaining the temperature at a constant 350 degrees Celsius. Using a response surface methodology (RSM) strategy, the optimal circumstances for analysis were determined throughout the development of the technology. Experiments were conducted, a response model was fitted to the experimental data, and the estimated response model was optimized. RSM has many benefits, including smaller sample sizes, shorter run times, and the power to provide helpful information for statistical analysis, such as data showing relationships between experimental variables [62].

The work shows that a UHPLC procedure for determining the impurities character of the putative CPL409116 substance (JAK/ROCK blocker) can be developed systematically, step-by-steeply, using the AQbD with DOE strategy, both in the preclinical and clinical stages of drug discovery research. The full factorial design model was utilised to examine the effect of varying amounts of critical method parameters (CMPs) on critical method attributes (CMAs) during the screening process. The Ph, mobile phase composition, and other columns have all been examined for the CMP method's ability to resolve peaks. The overall AQbD method, which includes screening, optimization, and validation processes to create a new technique for quantifying the complete profile of nine contaminants of a unique medicinal drug, highlighted the novelty of this work [63].

Hitesh Patel and colleagues conducted Validation experiments according to ICH criteria and established a chromatographic method based on UPLC-MS. This investigation selected the percent drug release (D.R.) at 60 minutes as the most important quality measure. One of the statistical methods utilised to dig deeper into the design's results was an analysis of variance (ANOVA). In the best-case scenario, dissolution was performed in 500 mL of pH 7.4 buffer using a USP apparatus of I (Basket) spinning at 120 rpm. Through QbD-guided research into dissolving conditions, a sensitive UPLC-MS technique for the quantification of digoxin was developed. Low-dose digoxin pills can benefit from this method for routine quality control tests, including dissolving and quantification [64].

6.2.3. Strategy for HPTLC Method Optimization and Development

In one of the studies based on HPTLC, Amol Patil and co-workers used a Quality design approach was used in HPTLC to measure anagliptin tablet dosage form. The required methods parameters are adjusted using central composite design, the retention factors, peak height, the mobile phase ratio, and the time required for saturation. Response surface analysis was used to study the interactions, and statistical methods were used. Graphical and numerical optimization methods are used to define the said design space. The study showed that the developed HPTLC method using QbD produces improved performance [65].

In another study, a quality-by-design methodology was used by Amruta Khurd. They have developed an innovative HPTLC technique to estimate rivaroxaban in the tablet dosage form. According to studies, fewer experimental runs are required if the quality-by-design (QbD) idea can be used [66].

Monika L. Jadhav and co-workers developed and applied the Quality by Design (QbD) for Propafenone hydrochloride tablet dosage form estimation using a chromatographic and spectrophotometric methodology according to ICH Q8(R2) criteria. Several factors were varied in the QbD technique, and these variables were built into the Ishikawa diagram. Principal component analysis was used to identify the crucial parameters, along with direct observation. The following critical parameters for the HPTLC method are estimated to be necessary: methanol, absorbance as a detection method, a 10 cm precoated, aluminium-backed TLC plate, 250 nm light, 20 minutes of saturation, 8 millimetres of band width, 70 millimetres of the solvent front, 5 millilitres of mobile phase, 10 minutes of scanning, and a 10-millilitre scanning chamber (10 cm x 10 cm x 10 mm). The critical parameters for the zero-order spectrophotometric method were solvent, sample, a wavelength of 247.4 nm, a slit width of 1.0, a medium scan speed, and a sampling interval of 0.2. The critical parameters for the first-order derivative spectrophotometric method were a scaling factor of 5 and a delta lambda of 4. ICH Q2(R1) standards approved the techniques. They were discovered to be precise and reliable; the suggested procedures can be utilised for routinely estimating propafenone hydrochloride in the tablet dosage form [67].

Lidia Gurba-Brykiewicz and colleagues used the AQbD with the DOE technique to determine quantitatively the impurity profile of the recently generated CPL409116 substance (JAK/ROCK inhibitor) in the preclinical and clinical stage of drug development research. Critical method parameters (CMPs) such as stationary phase type (8 distinct columns), aqueous mobile phase pH (2.6, 3.2, 4.0, 6.8), organic mobile phase concentration (starting at 20% and ending at 85%) have been carefully evaluated (ACN). Any trustworthy methodology must satisfy the CMAs for peak symmetry (0.8 and 1.8) and resolution between peaks (2.0). As part of the screening process, they used a complete fractional design 2^2 to compare how various concentrations of CMPs affected the CMAs. The screening step's knowledge space was used to inform the robustness tests, which were implemented using a $2^{(4-1)}$ implementation of the fractional factorial design. The method operable design region (MODR) has been produced. Monte Carlo simulations were used to calculate the probability of achieving the CMAs' performance goals. The screening, optimization, and validation processes of the AQbD approach to developing a new method for quantitatively determining the complete profile of nine contaminants of a unique pharmaceutical substance using the structure-based pre-development stood out compared to previous work in the field. These were the final terms of employment: liquid mobile phase water, Zorbax Eclipse Plus C18 column Aqueous solution of HCOOH (10 mM 1 mM), pH 2.6, initial concentration of 20% 1% ACN, the final concentration of 85% 1% ACN, and column temperature of 30 °C ±2 °C. The procedure was validated by Q2 of the ICH guidelines (R1). The optimal procedure is well-defined, linear, accurate, and dependable. All impurities have LOQs and LODs at or near the reporting thresholds of 0.05% and 0.02%, respectively [68].

6.2.4. Method Development/Optimization Strategy for U.V.

Different researchers presented the application of QbD methods for spectrophotometric methods. They used the Analytical Quality by Design (AQbD) technique following ICH Q8(R2)., Sk. Mastanamma designed and verified the spectrophotometric method for quantifying tenofovir alafenamide in bulk and its synthetic mixture. The sample preparation method, solvent, wavelength, and experimental parameters, including scan speed, slit width, sampling interval, *etc.*, were all mapped out using Ishikawa diagrams. The sample preparation method, solvent, wavelength, and experimental parameters, including scan speed, slit width, sampling interval, *etc.*, were all mapped out using Ishikawa diagrams. Critical parameters were identified using a combination of direct observation and principal component analysis. Sample drug concentrations were determined in a straightforward spectrophotometric procedure employing distilled water, methanol, and acetonitrile at 258 nm. With distilled water, Beer's rule was followed at concentrations between 2.5 and 25 g/ml (r2=0.998) and 5 and 30 g/ml (r2=0.996). It was found that methanol and acetonitrile at concentrations of 10-50 g/ml (r2=0.997) gave the best results. The proposed drug analysis method was reliable, reproducible, and cost-effective. Quality-by-Design (QbD) implementation led to the development of more secure procedures that save time and money and generate consistent, dependable, and high-quality data at every stage of the process [69].

6.3. Quality-based Design for Novel Drug Delivery Systems

Many regulatory bodies worldwide recommend QbD-based product development. So the statistical design of experiments (DoE) is used to optimise nanocarriers like polymeric NPs, solid lipid NPs, micelles, niosomes, liposomes, nanostructured lipid carriers, liquid crystalline NPs, and lipid polymer hybrid NPs. DoE tools are the statistical methods that aid in comprehending how input factors (both separately and collectively) impact the final output responses. The development of nanocarriers usually uses Design of Experiments (DoE) methods such as screening design (Fractional Factorial Design, Full Factorial Design, and Plackett-Burman) and Response Surface Methodology. (RSM). Two standard RSM methods in the pharmaceutical sector are the Central Composite Design (CCD) and the Box-Behnken Design (BBD) [70].

6.3.1. Polymeric nanoparticles

Numerous forms of DoE have been employed to examine the effects of the polymer and the surfactant, the drug/polymer ratio, the solvent/nonsolvent ratio, the homogenization rate, and the homogenization time on nanocarrier production to optimise the process.To comprehend how these variables influence the responses of the polymeric NPs, screening models like FFD and PBD are used [70]. Quantities of polymer, oil, and surfactants, drug concentration, drug/polymer ratio, volumes of aqueous and organic phases, and homogenization rate are just some of the factors that need to be considered when designing polymeric nanoparticles for a specific therapeutic application [70].

Particle size, entrapment efficiency, zeta potential, polydispersity index, pH and drug loading capacity are only a few of the dependent factors of polymers-based nanocarriers that have been studied. In addition to the previously stated parameters, others related to the manufacturing of polymeric nanoparticles, such as their surface area, and uniformity, can be evaluated, and by using the ionic gelation technique, Pandey and colleagues [70] created chitosan NPs that contained erlotinib and assessed the independent variables using CCD. The findings indicated that the drug released at a rate of 39.78% after 24 hours in 0.1 N hydrochloride acid and phosphate buffer pH 6.6, with a maximum entrapment efficiency into NPs measuring 91.57%. The researchers concluded that the CCD was effectively used in the tests and that the polymeric nanoparticles produced could be a starting point for further analysis. The marketed product had a faster drug release rate, whereas the improved formulation had a delayed release rate and a more significant cytotoxic effect on cancer cells.

6.3.2. Polymeric micelles

Polymeric micelles are nanocarriers with a core-and-shell structure, ranging in size from 10 to 200 nm; however, the addition of a drug to the system can increase its size. Factors such as surfactants, combinations of surfactants, drug amount, concentration, rehydration buffer quantity, and ultrasonication time and amplitude can all be investigated for polymeric micelles. Ye and colleagues [70] utilized a Box-Behnken design (BBD) to analyze the relationship between operational factors and the apparent solubility of curcumin. Using this method, they generated 12 factorial points and 5 center points. The resulting data were predictive. The spherical micelles exhibited a diameter of 247 nm, an observed solubility of 4.44 g/mL, and a zeta potential of -13.3 mV. The study revealed that the optimized formulation did not require the use of surfactant and solvent.

6.3.3. Liposomes

Liposomes are spherical lipid nanostructures with an aqueous inner core surrounded by a bilayer framework composed primarily of phospholipids and cholesterol. Standard liposomes can have certain components modified to create structures with distinct characteristics [71]. It is possible to create pH-sensitive or thermosensitive liposomes using a different lipid. These liposomes are primarily used in cancer therapy to prevent lysosomal breakdown and encourage hyperthermia. Some studies have investigated how adjusting variables such as lipid/drug/cholesterol concentration and ratio affect liposome properties. Shah and Jobanputra investigated the effects of sonication time, amplitude, and drug/lipid ratio on BBD [72]. These factors also impacted the responses' size and entrapment's effectiveness. Results showing an inverse relationship between sonication duration and amplitude Research by Shah and Jobanputra on the effects of sonication time, amplitude, and drug/lipid ratio on BBD revealed that lowering these parameters would reduce particle size and encapsulation efficiency.

6.3.4. Microemulsions and nanoemulsions

Nanosystems comprising an oil phase combined with an aqueous phase with the proper amount of surfactant are known as microemulsions and nanoemulsions. These phases can spontaneously or create a colloidal dispersion with an external energy supply [73]. DoE has been used to better comprehend important variables' roles and produce or apply micro and nanoemulsions. Models for screening, including FFD and PBD, and RSM, including CCD and BBD, have been used. As previously stated, these systems combine aqueous and oily phases while a surfactant or combination of surfactants is involved. Therefore, it is common to practise studying factors such as formulation components, equipment, preparation conditions, and temperature. The oil/surfactant/water ratio and the surfactant blend are the primary formulation components that affect micro- and nanoemulsions. The preparation conditions are linked to temperature, homogeni-zation/sonication time, and intensity. Ahmad and colleagues [51] created nanoemulsions to enhance this nanosystem and used CCD. The oil/surfactant mixture ratio and sonication duration were evaluated in the earlier study, whereas the sonication's intensity and temperature were also assessed in the later study. Size, transmittance %, and polydispersity index were all the same between the two papers. Both works were optimised, and the authors felt that the resulting nanoemulsions were safe for topical application to the skin to speed wound healing.

6.3.5. Solid lipid nanoparticles

The unique properties of SLN are the presence of a solid lipid (between 0.1 and 30.0 percent) (w/w) and the presence of a surfactant (between 0.5 and 5.0 percent) in an aqueous solution at room temperature. The solid lipid mixture can comprise various formulations, surfactants, and co-surfactants [74].

In DoE models of SLNs, parameters such as entrapment efficiency and drug loading are typically added as responses because specific molecules can be difficult to encapsulate or because high encapsulation rates may occasionally be required. For instance, rosuvastatin calcium is encapsulated in SLN, primarily made of stearic acid and a lubricant (Poloxamer 407/Tween 80, 1:1% w/v). According to initial studies by the authors, drug encapsulation, drug loading, and size were significantly impacted by factors like stirring and component ratio. Drug encapsulation and loading were studied using a CCD, and the authors discovered that the number of lipids used made a difference. After that, they optimised the process and found that the actual values for drug loading and encapsulation efficiency matched their predictions [74].

6.4. QbD Approach in the Extraction of Phytochemicals & Polyherbal Formulation

Healing with plants predates recorded history. Even in ancient times, people utilized herbal remedies to address various conditions. Despite the remarkable advancements in modern medicine over the last few decades [75], herbal supplements remain crucial for maintaining global health. However, the application of Quality by Design (QbD) in herbal medication production is sporadic, despite the widespread use of herbal remedies. Manufacturing should be tailored to align with the goals of QbD, as this approach aims to shift formulation away from empirical trial-and-error methods. The manufacturing process must be developed within the framework of QbD because most herbal medicinal products are made using experience-based methods, and Due to the intricate chemical composition and multiple process parameters, such methods are unable to improve our understanding of the underlying process [76]. Because QbD attempts to reorient the transition from empirical trial-and-error methods in formulating to more predictable and present patterns, it is crucial to build the manufacturing process within its framework [77]. If commercial herbal products are ever going to reach a substantial degree of safety, efficacy, and quality, a QbD model must be adopted during development [78].

The complicated chemical constitution and various process factors mean that the prevalent use of experience-based methodologies during the production of herbal medicine items does nothing to advance our understanding of the manufacturing process. Consequently, it is crucial to establish a production procedure within the QbD framework; it aims to shift the emphasis from empirical trial-and-error methods to more tightly controlled environments to ensure consistent quality throughout a product's lifespan. For this reason, the commercialization of herbal products necessitates using a QbD model for their development to guarantee an adequate degree of safety, efficacy, and quality. Extraction of herbal medicine products is a relatively complicated procedure from a manufacturing standpoint since it involves a wide range of distinct components and unit operations that vary greatly. Most of the time, these procedures entail several components or variables. Due to the complicated chemical makeup of herbal medications, it is challenging to establish correlations between input critical sample attributes (CSAs) or critical process variables (CPVs) and critical quality parameters (CQPs) of end products. Because the extraction of herbal medications primarily depends on experience, the effect(s) of diverse input materials on the product is poorly understood, resulting in a depletion of final product quality [79]. Therefore, to enhance the extraction performance of botanical medication, it is necessary to have a thorough grasp of the interactions between the CSAs/CPVs and product CPVs. The extraction of bioactive components from herbal medicine products has been emphasized in several papers as an example of how such QbD paradigms might be used.

Variations in input quality, or sources, include:

• Crude drug source, processing, plant components, and impurities.

• Temperature, frequency, energy, extraction method.

• The development of apparatus such as microwaves, ultrasound generators, supercritical fluid extractors, and accelerated solvent extractors.

• Methods of measurement, such as those used to get a representative sample, a scale for preparation, and an analytical scale, a scale for weighing finished products.

• Concerns about the environment include relative humidity, temperature, and moisture concentration.

As a result, variance is often utilized to measure the overall variation in the process. Determine whether inputs, variables, or factors will affect the process. This is a challenge that is faced by every industry that deals with herbs. QbD (ICH Q9) is a set of tools that allows for the rigorous evaluation of all the available inputs to a process. This evaluation aims to identify the relatively few parts of the process that have the potential to have the most significant impact on the process. Screening designs that are typically employed in the process of botanical extraction are as follows:

• Complete design by factorial (2^k).

• Fractional factorial design (2^{k-p}).

• Plackett-Burman design.

• Taguchi design.

Mykhailenko *et al.* applied a Quality-by-Design approach to creating a process for extracting anticancer constituents from the pericarp of *Crocus Sativus* (*C. sativus*) Perianth. An approach was developed to collect extracts using a variety of solvents, and the best extractant was identified by chromatographic analysis to guarantee conformity with the QTPP standards. Crocus was defended by Q-markers utilizing the Herbal Chemical Marker Ranking System (HerbMars), which evaluates bioavailability, pharmacological efficacy, and the selected benchmark. A series of tests examined how changing factors in the water/ethanol extraction of compounds from raw materials affected the results (DoE). Using high-performance liquid chromatography (HPLC), they could identify 16 chemicals in the perianths of crocuses and determine their relative abundances. The chemical analysis of the crocus flower's perianth is a reliable indicator of its beauty (rutin, mangiferin, ferulic acid, isoquercitrin and crocin). The ISO 3632 criteria for spice saffron was also satisfied by Volyn stigmas (category I). Cytotoxicity tests using IGR39 and MDA-MB-231 melanoma and the hydroethanolic extract of *C. sativus* perianth were more effective than the water extract against triple-negative breast cancer cell lines. Therefore, evidence suggests that Quality by Design is a practical process development approach for producing high-quality herbal medicines [80].

Pan *et al.* studied the extraction, concentration, water precipitation, and chromatography methods needed to create a total of *Panax notoginseng (P. notoginseng)* saponin using various quality samples. A conclusive screening layout examined ten process parameters simultaneously. CPPs and CMAs were discovered with process CQAs. Models included CMAs, CPPs, and process CQAs. Monte Carlo simulations determined the design space with 0.90 confidence. Quality assurance should account for production influences to assure product quality. Ginsenoside and total saponin in the eluate were critical for quality assurance. Extraction, elution, and ethanol concentration were CPPs. Dry extractable *P. notoginseng* was important morphologically. Other CMAs included extractable notoginsenoside R1, Rg1, Rb1, and Rd. Premium and economy-grade *P. notoginseng* were differentiated using the NTS standard of Xuesaitong injection. NTS should not be made from low-quality *P. notoginseng*. High-quality extracts with optimized parameters are achievable. Excellent scientific testing confirms two superior *P. notoginseng* batches. This study's quality control technique considered the procedure's similarity to the herbal constituents [81].

Koli *et al.* formulated a tablet containing multiple herbs and determined the best way to make it. Using the Risk Priority Number, it was first evaluated the potential dangers posed by critical ingredients. The authors then developed a completely random factorial design in which the proportions of microcrystalline cellulose (MCC) and croscarmellose sodium (CCS) may be varied to reduce these risks. MCC 5–15% and CCS 0.5–5% were used as independent variables in a Design of Experiments (DoE) to create four trial formulations, and the friability and disintegration time of these samples ranged from 0.35–0.91% and 11.41–15.50 minutes, respectively. After testing four different formulations, three were chosen for short-term stability research because of their superior disintegration and friability profiles: F-1, F-2, and F-4. MCC and CCS were the free variables in Formulation F-1. Disintegration time (min) and friability (%) were both shown to be strongly influenced by the formulation components X1: MCC and X2: CCS (%). Thus, it demonstrated that improved outcomes were achieved by analyzing quality factors and optimizing the formulation of polyherbal tablets using quality by design [82].

Vijayaraj *et al.* applied the QbD approach to creating and verifying the RP-HPLC method for nanoparticle rutin analysis. RSM's experimental design explains the quadratic model's chromatographic response. A Central Composite Design (CCD) assessed each design variable at three levels to establish the quadratic regression model's coefficients. The CCD design determined the ideal flow rate, mobile phase pH, strength, and wavelength for optimal retention time, Peak area, and asymmetry factor chromatographic responses. Chromatographic levels and factors optimized experiment settings. Thirty experiments had four independent variables (A, B, C, and D) and three dependent outcomes. Using optimized independent variables, rutin's retention time, peak area, and asymmetry factor were 3.75 minutes, 1014.79 millivolts, and 1.26. Quantitation was 0.15 g/mL, and detection was 0.005 g/mL. The optimized assay condition was tested for linearity, accuracy, precision, and robustness per ICH guidelines. QbD optimization of the recommended rutin analysis method in chitosan-sodium alginate nanoparticles showed encouraging results [83].

6.5. QbD Approach in Green Synthesis

The goal of green synthesis, also known as green chemistry, is to develop a wide variety of scientific products and processes that are extremely low in harmful by-products, resource- and energy-efficient, and waste-free [84]. Green nanotechnology is a subset of green technology that incorporates green chemistry and green engineering ideas, with the term "green" referring to the usage of plant-based materials. Green nanotechnology enables the manufacture of more eco-friendly nanomaterials and nanoproducts without the use of harmful components and at low temperatures, while consuming less energy and utilizing renewable resources whenever feasible [85]. Plants act as stabilising and reducing agents, making green source-mediated synthesis a potent tool for fine-tuning the nanoparticles' size and shape. Furthermore, green nanoparticles' therapeutic effect is more significant than chemically produced nanoparticles [86]. QbD has been used in the green synthesis of nanotechnology-based drug delivery of nanoparticle compositions to improve the efficacy of the drug's release profile and targeting and to boost the system's pharmacodynamic, pharmacokinetic, and toxicity profiles. Multiple statistical models have shown that DoE improves the consistency of the synthesized product while decreasing the number of necessary tests. Several DoE design methods could aid in optimizing the green production of nanoparticulate systems [87]. QbD can be used in green source-mediated synthesis to check the desired quality profile of the final product by providing insight into how the synthesis's improved variable effects and resilient quality were achieved. The BBD is a powerful, effective, and systematic instrument for surface response approach that speeds up green synthesis and improves research output [75].

Green synthesis from manjistha extract was used to create a formulation for ZnO-NPs by Kaur *et al*. Particle size, polydispersity index and entrapment efficiency were among the many variables considered by the Design-Expert software to achieve the optimal formulation (PDI). Also included as covariates were reagent concentration (percent w/v), stirring rate (rpm), and duration of ultrasonic treatment (minutes). The BBD was used to enhance the performance of the ZnO-MJE-NPs. The software uncovered thirteen different formulations, and the data for each response was then fitted into several experimental design models to evaluate which model produced the best fit.

A study showed how independent factors affect the particle sizes of different formulations. The model's statistical significance was demonstrated by its F-value of 23.30. (p 0.05). ANOVA test results revealed that the quadratic model offered a proper fit. There was no statistically significant difference, according to the F-value. R^2 was higher for the quadratic model than for the other models. A good indicator was if the precession was more than 4. $ZnSO_4$ had a beneficial impact on particle size. The increased dispersion viscosity, with increased particle size with increases $ZnSO_4$ concentration. As sonication time (B) increased, particle size dropped. Nanoparticle size decreased with longer sonication times due to particle breaking during sonication. However, there was evidence that stirring speed (C) hurt particle size, this effect was not statistically significant. Larger increases in stirring speed led to decreased particle sizes due to particle breaking. Particle size graphs that are dimensional, contoured, as well as measured and calculated. When a quadratic model was applied, formulation optimization produced acceptable findings compared to earlier studies [77].

Kheshtzar *et al*. researched to determine the optimal reaction conditions for the environmentally friendly production of zero-valent iron nanoparticles (INPs) using an extract from pine tree leaves (*Pinus eldarica*). During the green synthesis process, the ideal reaction conditions were determined using a central composite face design, and the effective parameters were screened using a fractional factorial design [88]. The most fruitful factors for INP generation were the amounts of leaf extract and iron precursors. Using a two-stage statistical design of the experiment, we could fine-tune the synthesis parameters and, therefore, increase INP production.

The first step, which involved determining the critical factors that significantly affected the output of nanoparticles, was finished. The authors used a fractional factorial design to examine the correlation between plant iron precursor concentration, extract amount, total reaction time and reaction temperature. Next, an empirical model was developed to determine the optimum values for the variables of interest. To maximize the generation of INP, the optimal ratio of iron precursor to leaf extract was calculated using the central composite face (CCF) efficient factors. The influential variables identified in the screening phase were optimized using the CCF design and RSM. As determined by a plot analysis, the optimal final concentrations of leaf extract and iron precursor are 20–50 mM and 8–9 mL, respectively, for maximum INP production. Fig. (**1**) demonstrates that nanoparticle formation is inhibited when ferric chloride concentration is increased to more than 50 mM, and leaf extract volume is decreased to less than 8 mL. The model predicted a maximum productivity of 0.015 g per reaction (1.5 mg per mL of the reaction mixture) while using leaf extract and ferric chloride at 25 mM of each (9 mL). As a result, 0.015 g mL^{-1} of INPs were produced under ideal circumstances for the validation experiment. It was determined how different reaction parameters affected the manufacture of INPs to maximize the number of particles produced in each reaction volume. The reaction circumstances significantly impact the physicochemical features of the formed environmentally friendly nanoparticles. Therefore, examining the impact of the synthesis conditions and associated components on the reactivity, form, and other properties of the produced nanoparticles is necessary. A statistically planned experiment showed that the two most crucial factors in the green synthesis reaction were the iron precursor concentration and the amount of leaf extract used [78, 89].

Selvam *et al.* presented a green, quick, and simple approach to manufacturing silver nanoparticles (AgNPs) by employing *Tinospora cordifolia* (*T. cordifolia*) as a constraining and decreasing factor. Using a response surface methodology-based bioreactor design, the effects of silver nitrate ($AgNO_3$) concentration, fresh weight of *T. cordifolia* leaf, incubation period, and pH on silver reduction were studied (BBD). The optimal conditions were 1.25 mM $AgNO_3$, an incubation duration of 15 hours, a temperature of 45 degrees Celsius, and a pH of 5.8. (4.5). Once the reactants are heated (60 °C), the extract of *T. cordifolia* leaves can reduce silver ions into AgNPs within 30 minutes, as the acquired reddish brown hue shows. The best conditions for synthesizing the AgNPs were determined with the help of the BBD. Using Fisher's F test, the quadratic regression model was statistically significant; thus, the ANOVA results confirmed this. The model term was statistically significant if the value of Prob > F' (0.0001). With an F-value of 26.06, the model was statistically significant. There was only a 0.01% possibility of observing a model F-value due to random noise. The value of R2 (0.9631), which is closer to 1.0 anticipated R2 (0.7872) or adjusted R2 (0.9261), showed that the model suited the data better than expected than the value of R2 (0.7872). In this investigation, we tested the model for the rhizome-mediated synthesis of AgNPs using three distinct methodologies: sequential model sum of squares, model summary statistics and lack of fit tests. AgNPs production utilizing *T. cordifolia* leaf extract was optimized by investigating the effects of varying concentrations of $AgNO_3$, incubation duration (h), temperature (°C), and pH. Using BBD, we were able to determine the ideal ranges for all of the variables [90].

6.6. QbD on the Processing of Biotherapeutics

The importance of products with a biotechnological origin has grown over the past few years, mainly when dealing with long-term illnesses like cancer and arthritis. Multiple samples are taken from multiple batch processing unit operations to ensure quality in biotechnological products and control processes. In process development and commercial-scale manufacturing, advanced analytical technologies can identify and address quality variables in materials and processes [91]. The nature of process development in biotherapeutics is complex and heterogeneous. Due to this, these processes usually face many challenges, as their complete analytical characterization is impossible [92]. Fig. (**4**) depicts the critical steps involved in the development and fabrication of biotherapeutics.

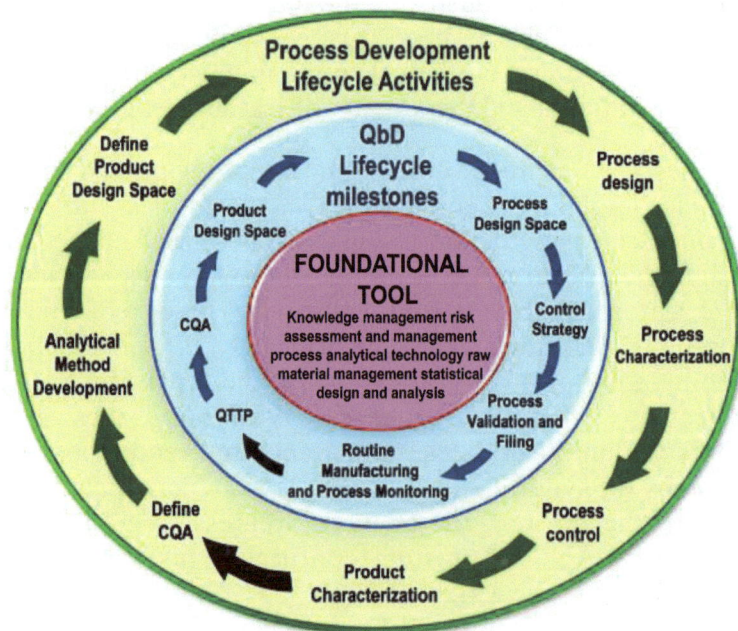

Fig. (4). Critical steps in the QbD approach for developing and manufacturing biotherapeutics.

These include:

• The process of determining the most critical aspects of quality -CQA ensures that the desired quality product with defined product limits is formulated.

• Method development for providing these elements.

• Reliable methods of process regulation to guarantee repeatable results.

• The procedure must be validated and documented to show that the control technique is effective.

• Constant monitoring throughout the product's lifespan guarantees reliable process performance [93].

Deepak Kumar *et al.* have researched the various media components' role in producing an optimum medium based on QbD principles to assure high cellular development of *E. coli* and high productivity of the Fab products. As claimed by QbD, the screened medium components resulted in a five-fold increase in the target protein titer (from 5 mg/L to 25 mg/L) compared to the basal medium [94].

6.7. QbD Approach in Immunoassays

Immunoassays quantify biomolecules of interest based on the specificity and selectivity of antibody reagents generated. Although analytical QbD is a relatively new invention, the ideas behind it, such as statistical quality assurance and goal-oriented planning, have been used as "design control" for diagnostic tests for a long time. Very few instances reflecting the progress of QbD-based immunoassays have been described beyond partial applications and particular instruments. During the process of developing a new technique, the Design of Experiments is being used to optimise the immunoassay parameters that influence Critical Material Attributes (CMAs) (DOEs) [95].

Using QbD principles, Yorovoi H. *et al.* created four immunoassays to back up vaccine release testing. They implemented systematic techniques to collect up-front objectives (ATP), determined probable method inputs (Fishbone diagram, cause-and-effects matrix), explored the design space (DOEs), and managed remaining risks to streamline the development process (FMEA). QbD-based method development was a more efficient and organised means of managing assay transfers and product life cycles, according to the study's findings [96].

Food-and-mouth disease diagnostic blood assay was created by Goris *et al.*, who also established international reference standards and quality control methodologies using a variety of quality instruments. The first assay was isolated from matching or outperforming the performance of two "gold standard" methods: a comparable ELISA and a cell-based viral neutralization test [97]. Control charts and statistical variance component analysis were employed in an interesting new way by Bak *et al*. Instead of relying on isolated positive controls that can only provide an accurate representation of a subset of the wells on an assay plate, they developed a system to track the overall plate's performance [98].

The authors of one of the studies initially used the QbD principle to create a technique for quantifying drugs in human plasma. The chemical of interest was elvitegravir. The studies were conducted to meet the standards set forth by the European Medicines Agency (EMA) for validating bioanalytical methods in a timely fashion and at a reasonable cost. Analytical performance and features of the HPLC-MS/MS approach adopted in the present investigation were equivalent to the limited number of methods previously published for elvitegravir. To establish the most appropriate method for isolating and quantifying elvitegravir in plasma samples, the scientists considered the chemical properties of the molecules and their specific application to standard clinical scenarios. The author concludes that the rigorous optimization procedure yielded a method that is both under statistical control (as suggested by the control charts) and strong enough to meet the needs of its intended use now [99].

7. REGULATORY AND INDUSTRY VIEW ON QBD

Early adoption of new technologies and risk-based methodologies in pharmaceutical product development has been encouraged since the FDA's 21st-century initiative (A Risk-Based Approach) was established in 2004. (FDA, 2004). An FDA official's (Woodcock, 2004) definition of QbD describes the characteristics of process and product performance that are statistically tailored to meet specific objectives rather than being generated just from experimental results from test batches. QbD can potentially increase the likelihood of a first-cycle approval, boost production efficiency, cut costs and product rejects, reduce or eliminate the need for conformity actions, streamline post-approval alterations and regulatory processes, facilitate more targeted inspections, and present opportunities for ongoing improvement. The FDA has provided case studies of ANDAs for immediate-release and extended-release drugs that illustrate how QbD principles have been implemented. Section 3.2.P.2 of Module 3 of the CTD file on the FDA website includes some instances (Quality). This article discusses the pharmaceutical advancement of acetriptan quick-release and provides an example of a modified-release tablet [74].

Organizations, including the EMA, FDA, and ICH, formed the ICH quality implementation working group (Q-IWG). In response to the ICH Q8(R2), ICH

Q9, and ICH Q10 guidelines, the Q-IWG developed many blueprints, workshop training materials, inquiries and responses, and a points-to-consider paper (published in 2011). This paper provides an interesting summary of how different modelling approaches can be used in QbD. The ICH Q-IWG manual divides models based on their respective contributions to ensuring product quality or regulatory submissions [74].

Pharmaceutical manufacturers should constantly remember that regulatory applications should contain sufficient details about research and manufacturing. Yet, sound risk management principles must be the foundation for all regulatory choices. Pfizer was among the first businesses to use the QbD and PAT ideas. These ideas gave the business better process knowledge, greater process capacity, higher quality products, and more flexibility in implementing changes for continual enhancement. Additionally, a significant portion of the QbD investment goes towards process development, with the advantage seen in commercial manufacturing [100]. Table **3** provides critical elements in the regulatory field of analytical QbD *vs* product QbD.

Table 3. Quality-by-Design for regulation: Analytical QbD *vs* product QbD.

Sr. No.	Stage	Product QbD	Analytical QbD
1	Stage 1	Specify the ideal product profile in terms of quality (QTTP)	Define analytical target profile (ATP)
2	Stage 2	Critical quality attributes	Critical quality attributes
3	Stage 3	Risk assessment	Risk assessment
4	Stage 4	Design Space	Method operable Design region
5	Stage 5	Methods of Regulation	Methods of Regulation
6	Stage 6	Managing the lifespan	Management of the life cycle

CONCLUSION

The rapid increase in the popularity of QbD and related tools suggests that the methodologies are not fads but solutions to contemporary production processes' requirements. QbD is an efficient design and production method because it uses DoE, risk assessment, and PAT tools to learn more about the materials and processes. This makes QbD viable and practical in the pharmaceutical industry. Applying AQbD offers a powerful analytical technique crucial to developing pharmaceutical products. The AQbD strategy makes comprehending risk management practises and management in depth easier. Optimization methods that adhere to AQbD standards must consider the potential for prediction errors and the propagation of those errors. Improved knowledge of how inputs like raw material characteristics and processing parameters might affect product quality is fostered by the QbD approach.

Additionally, it infuses organized, risk-based, holistic thinking into manufacturing processes. The QbD approach, which uses a risk-based approach, has the advantage of being more efficient in terms of time and money from the research stage to regulatory approval and large-scale manufacture. It can encompass the early stages of pharmaceutical research operations. To implement QbD and AQbD, it is necessary to harmonize terminology and concepts, train and educate human resources for industry and regulatory organizations and provide instructions on recording information developed when developing pharmaceutical methods. We advocate for more discussion between industry and regulatory bodies to help the QbD adoption.

REFERENCES

[1]   Allen LV Jr. Basics of compounding: Potency and stability testing. Int J Pharm Compd 2013; 17(3): 220-4.
        [PMID: 24046938]

[2]   Sangshetti JN, Deshpande M, Zaheer Z, Shinde DB, Arote R. Quality by design approach: Regulatory need. Arab J Chem 2017; 10 (Suppl. 2): S3412-25.
        [http://dx.doi.org/10.1016/j.arabjc.2014.01.025]

[3]   Mv L. Quality risk management (QRM): A review. J Drug Deliv Ther 2013; 149.

[4]   Deepak Kumar D, Ancheria R, Shrivastava S, Soni SL, Sharma M. Review on pharmaceutical quality by design (QbD). Asian Journal of Pharmaceutical Research and Development 2019; 7(2): 78-82.
        [http://dx.doi.org/10.22270/ajprd.v7i2.460]

[5]   Savitha S, Devi K. Quality by design: A review. J Drug Deliv Ther 2022; 12(2-S): 234-9.
        [http://dx.doi.org/10.22270/jddt.v12i2-S.5451]

[6]   Yu LX, Amidon G, Khan MA, *et al.* Understanding pharmaceutical quality by design. AAPS J 2014; 16(4): 771-83.
        [http://dx.doi.org/10.1208/s12248-014-9598-3] [PMID: 24854893]

[7]   Kalyane D, Raval N, Polaka S, Tekade RK. Quality by design as an emerging concept in the development of pharmaceuticals. The Future of Pharmaceutical Product Development and Research. Elsevier 2020; pp. 1-25.
        [http://dx.doi.org/10.1016/B978-0-12-814455-8.00001-3]

[8]   International conference on harmonisation. ICH Q8(R2) pharmaceutical development. Available from: https://database.ich.org/sites/default/files/Q8_R2_Guideline.pdf

[9]   Sangshetti JN, Deshpande M, Zaheer Z, Shinde DB, Arote R. Quality by design approach: Regulatory need. Arabian Journal of Chemistry. 2017; 10: S3412-S25.
        [http://dx.doi.org/10.1007/s43441-020-00254-9] [PMID: 33439461]

[10]   Werther W, Loughlin AM. Post-approval regulatory requirements. Principles and Practice of Clinical Trials. Springer International Publishing 2021; pp. 1-28.

[11]   EASE. Available from: https://www.ease.io/root-cause-analysis-how-to-use-a-fishbone-diagram/#:~:text=A%20fishbone%20diagram%2C%20also%20called,At%20the%20University%20of%20Tokyo

[12]   Wong KC. Using an Ishikawa diagram as a tool to assist memory and retrieval of relevant medical cases from the medical literature. J Med Case Reports 2011; 5(1): 120.

[http://dx.doi.org/10.1186/1752-1947-5-120] [PMID: 21447163]

[13]   Liliana L. A new model of Ishikawa diagram for quality assessment. IOP Conf Ser Mater Sci Eng 2016.
[http://dx.doi.org/10.1088/1757-899X/161/1/012099]

[14]   Chen H, Tao Z, Zhou C, Zhao S, Xing Y, Lu M. The effect of comprehensive use of PDCA and FMEA management tools on the work efficiency, teamwork, and self-identity of medical staff: A cohort study with Zhongda Hospital in China as an example. Contrast Media Mol Imaging 2022; 1-8.
[http://dx.doi.org/10.1155/2022/5286062] [PMID: 35685656]

[15]   ASQ (American Society for Quality). Available from: https://asq.org/quality-resources/fmea

[16]   Butreddy A, Bandari S, Repka MA. Quality-by-design in hot melt extrusion based amorphous solid dispersions: An industrial perspective on product development. Eur J Pharm Sci 2021; 158: 105655.
[http://dx.doi.org/10.1016/j.ejps.2020.105655] [PMID: 33253883]

[17]   Pascarella G, Rossi M, Montella E, *et al.* Risk analysis in healthcare organizations: Methodological framework and critical variables. Risk Manag Healthc Policy 2021; 14: 2897-911.
[http://dx.doi.org/10.2147/RMHP.S309098] [PMID: 34267567]

[18]   Demmon S, Bhargava S, Ciolek D, *et al.* A cross-industry forum on benchmarking critical quality attribute identification and linkage to process characterization studies. Biologicals 2020; 67: 9-20.
[http://dx.doi.org/10.1016/j.biologicals.2020.06.008] [PMID: 32665104]

[19]   Hamilton Company. Available from: https://www.hamiltoncompany.com/process-analytics/proces-
-analytical-technology/critical-quality-attributes#:~:text=A%20CQA%20is%20a%20physical,To%20
measure%20directly%20in%20production

[20]   Hamilton Company. Manufacturing KPIs in bioprocessing Available from: https://www.hamilton-
company.com/process-analytics/process-analytical-technology/manufacturing-kpis-in-bioprocessing

[21]   Hamilton Company. Critical process parameters Available from: https://www.hamiltoncompany.com/
process-analytics/process-analytical-technology/critical-process-parameters

[22]   Holm P, Allesø M, Bryder MC, Holm R. Q8(R2). ICH Quality Guidelines. John Wiley & Sons, Inc. 2017; pp. 535-77.
[http://dx.doi.org/10.1002/9781118971147.ch20]

[23]   Pielenhofer J, Meiser SL, Gogoll K, *et al.* Quality by design (QbD) approach for a nanoparticulate imiquimod formulation as an investigational medicinal product. Pharmaceutics 2023; 15(2): 514.
[http://dx.doi.org/10.3390/pharmaceutics15020514] [PMID: 36839835]

[24]   Ahirwar K, Shukla R. Preformulation Studies: A Versatile Tool in Formulation Design [Internet]. Drug Formulation Design. IntechOpen; 2023. Available from: http://dx.doi.org/10.5772/intechopen.110346.

[25]   Ahirwar K, Shukla R. Preformulation Studies: A Versatile Tool in Formulation Design [Internet]. Drug Formulation Design. IntechOpen; 2023.
[http://dx.doi.org/10.5772/intechopen.110346]

[26]   ter Horst JP, Turimella SL, Metsers F, Zwiers A. Implementation of Quality by Design (QbD) principles in regulatory dossiers of medicinal products in the European Union (EU) between 2014 and 2019. Ther Innov Regul Sci 2021; 55(3): 583-90.
[http://dx.doi.org/10.1007/s43441-020-00254-9] [PMID: 33439461]

[27]   Patil AS, Pethe AM. Quality by Design (QbD): A new concept for development of quality pharmaceuticals. Vol. 4.

[28]   Namjoshi S, Dabbaghi M, Roberts MS, Grice JE, Mohammed Y. Quality by Design: Development of the quality target product profile (QTPP) for semisolid topical products. Pharmaceutics 2020; 12(3): 287.
[http://dx.doi.org/10.3390/pharmaceutics12030287] [PMID: 32210126]

[29]   Mahajan R, Gupta K. Food and drug administration's critical path initiative and innovations in drug development paradigm: Challenges, progress, and controversies. J Pharm Bioallied Sci 2010; 2(4): 307-13.
[http://dx.doi.org/10.4103/0975-7406.72130] [PMID: 21180462]

[30]   Q8(R2) Pharmaceutical Development. Available from: https://www.fda.gov/regulatory-information/search-fda-guidance-documents/q8r2-pharmaceutical-development

[31]   Pardeshi SR, More MP, Patil PB, Mujumdar A, Naik JB. Statistical optimization of voriconazole nanoparticles loaded carboxymethyl chitosan-poloxamer based *in situ* gel for ocular delivery: *In vitro, ex vivo*, and toxicity assessment. Drug Deliv Transl Res 2022; 12(12): 3063-82.
[http://dx.doi.org/10.1007/s13346-022-01171-0] [PMID: 35525868]

[32]   Cheng HP, Tsai SM, Cheng CC. Analysis of heat transfer mechanism for shelf vacuum freeze-drying equipment. Adv Mater Sci Eng 2014; 2014: 1-7.
[http://dx.doi.org/10.1155/2014/515180]

[33]   Pardeshi S, More M, Patil P, *et al.* A meticulous overview on drying-based (spray-, freeze-, and spray-freeze) particle engineering approaches for pharmaceutical technologies. Dry Technol 2021; 39(11): 1447-91.
[http://dx.doi.org/10.1080/07373937.2021.1893330]

[34]   Khan A, Naquvi KJ, Haider MF, Khan MA. Quality by design- newer technique for pharmaceutical product development. Intelligent Pharmacy. 2023.

[35]   Yu LX. Pharmaceutical quality by design: Product and process development, understanding, and control. Pharm Res 2008; 25(4): 781-91.
[http://dx.doi.org/10.1007/s11095-007-9511-1] [PMID: 18185986]

[36]   Azad MA, Capellades G, Wang AB, *et al.* Impact of critical material attributes (CMAs)-particle shape on miniature pharmaceutical unit operations. AAPS PharmSciTech 2021; 22(3): 98.
[http://dx.doi.org/10.1208/s12249-020-01915-6] [PMID: 33709195]

[37]   Pardeshi, S.R., Deshmukh, N.S., Telange, D.R. *et al.* Process development and quality attributes for the freeze-drying process in pharmaceuticals, biopharmaceuticals and nanomedicine delivery: a state-of-the-art review. Futur J Pharm Sci 2023; 9: 99.
[http://dx.doi.org/10.1186/s43094-023-00551-8]

[38]   Awotwe-Otoo D, Agarabi C, Wu GK, Casey E, Read E, Lute S, *et al.* Quality by design: Impact of formulation variables and their interactions on quality attributes of a lyophilized monoclonal antibody. Int J Pharm. 2012; 438 (1–2): 167–75.
[http://dx.doi.org/10.1016/j.ijpharm.2012.08.033.] [PMID: 24854893]

[39]   Namjoshi S, Dabbaghi M, Roberts MS, Grice JE, Mohammed Y. Quality by Design: Development of the Quality Target Product Profile (QTPP) for Semisolid Topical Products. Pharmaceutics. 2020 Mar 23; 12(3): 287.
[http://dx.doi.org/10.3390/pharmaceutics12030287]

[40]   Desai N, Purohit R. Design and development of clopidogrel bisulfate gastroretentive osmotic formulation using quality by design tools. AAPS PharmSciTech 2017; 18(7): 2626-38.
[http://dx.doi.org/10.1208/s12249-017-0731-3] [PMID: 28247292]

[41]   Desai N, Purohit R. Development of novel high density gastroretentive multiparticulate pulsatile tablet of clopidogrel bisulfate using quality by design approach. AAPS PharmSciTech 2017; 18(8): 3208-18.
[http://dx.doi.org/10.1208/s12249-017-0805-2] [PMID: 28550603]

[42]   Mockus LN, Paul TW, Pease NA, *et al.* Quality by design in formulation and process development for a freeze-dried, small molecule parenteral product: a case study. Pharm Dev Technol 2011; 16(6): 549-76.
[http://dx.doi.org/10.3109/10837450.2011.611138] [PMID: 21932931]

[43]   Jankovic A, Chaudhary G, Goia F. Designing the design of experiments (DOE) – An investigation on

the influence of different factorial designs on the characterization of complex systems. Energy Build 2021; 250: 111298.
[http://dx.doi.org/10.1016/j.enbuild.2021.111298]

[44]   Patel SM, Jameel F, Sane SU, Kamat M. Lyophilization process design and development using QbD principles. In: Jameel F, Hershenson S, Khan M, Martin-Moe S, Eds. Quality by design for biopharmaceutical drug product development. New York, NY: Springer 2015; pp. 303-29.
[http://dx.doi.org/10.1007/978-1-4939-2316-8_14]

[45]   Ferreira NN, Miranda RR, Moreno NS, Pincela Lins PM, Leite CM, Leite AET, *et al.* Using design of experiments (DoE) to optimize performance and stability of biomimetic cell membrane-coated nanostructures for cancer therapy. Frontiers in Bioengineering and Biotechnology. 2023;11.
[http://dx.doi.org/10.3389/fbioe.2023.1120179]

[46]   Rathore A. Quality RA. Design Space for Biotech Products Available from: https://www.biopharmin-ternational.com/view/quality-design-space-biotech-products

[47]   Radhakrishnan V, Davis P, Hiebert D. Scientific approaches for the application of QbD principles in lyophilization process development. In: Warne N, Mahler H, Eds. Challenges in Protein Product Development. Cham: Springer 2018; pp. 441-71.
[http://dx.doi.org/10.1007/978-3-319-90603-4_20]

[48]   Rößler M, Huth PU, Liauw MA. Process analytical technology (PAT) as a versatile tool for real-time monitoring and kinetic evaluation of photocatalytic reactions. React Chem Eng 2020; 5(10): 1992-2002.
[http://dx.doi.org/10.1039/D0RE00256A]

[49]   Kumar R, Mittal A, Kulkarni MP. Quality by design in pharmaceutical development. Computer Aided Pharmaceutics and Drug Delivery. Singapore: Springer Nature Singapore 2022; pp. 99-127.
[http://dx.doi.org/10.1007/978-981-16-5180-9_4]

[50]   Chanda A, Daly AM, Foley DA, *et al.* Industry perspectives on process analytical technology: Tools and applications in API development. Org Process Res Dev 2015; 19(1): 63-83.
[http://dx.doi.org/10.1021/op400358b]

[51]   Grangeia HB, Silva C, Simões SP, Reis MS. Quality by design in pharmaceutical manufacturing: A systematic review of current status, challenges and future perspectives. Eur J Pharm Biopharm 2020; 147: 19-37.
[http://dx.doi.org/10.1016/j.ejpb.2019.12.007] [PMID: 31862299]

[52]   Powar TA, Hajare AA. QbD based approach to enhance the *in-vivo* bioavailability of ethinyl estradiol in sprague-dawley rats. Acta Chim Slov 2020; 67(1): 283-303.
[http://dx.doi.org/10.17344/acsi.2019.5441] [PMID: 33558931]

[53]   Kumar N, Jha A. Quality risk management during pharmaceutical 'good distribution practices' – A plausible solution. Bull Fac Pharm Cairo Univ 2018; 56(1): 18-25.
[http://dx.doi.org/10.1016/j.bfopcu.2017.12.002]

[54]   Fukuda IM, Pinto CFF, Moreira C dos S. Saviano AM, Lourenço FR. Design of Experiments (DoE) applied to pharmaceutical and analytical quality by design (QbD). Braz J Pharm Sci 2018; 54.

[55]   Gupta NV, Reddy V, Gupta V, Raghunandan V, Kashyap N. Quality risk management in pharmaceutical industry: A review. Development of Nanocarriers Vol 6

[56]   Alemayehu D, Alvir J, Levenstein M, Nickerson D. A data-driven approach to quality risk management. Perspect Clin Res. 2013; 4(4): 221-226.
[http://dx.doi.org/10.4103/2229-3485.120171]

[57]   Katekar V, Sangule D, Bhurbhure O, Ingle P, Dhage S, Jadhav K. A review on quality by design approach in analytical methods. J Drug Deliv Ther 2022; 12(3-S): 255-61.
[http://dx.doi.org/10.22270/jddt.v12i3-S.5386]

[58]   HG S, Ruba Priya MG, Murugan V. Implementation of Quality by Design Approach for Method

Development and Validation. Res. J Pharm Technol, 2022; 436-40.

[59]   Tome T, Žigart N, Časar Z, Obreza A. Development and optimization of liquid chromatography analytical methods by using AQbD principles: Overview and recent advances. Org Process Res Dev 2019; 23(9): 1784-802.
[http://dx.doi.org/10.1021/acs.oprd.9b00238]

[60]   Krishna MV, Dash RN, Jalachandra Reddy B, Venugopal P, Sandeep P, Madhavi G. Quality by Design (QbD) approach to develop HPLC method for eberconazole nitrate: Application oxidative and photolytic degradation kinetics. J Saudi Chem Soc 2016; 20: S313-22.
[http://dx.doi.org/10.1016/j.jscs.2012.12.001]

[61]   Iuliani P, Carlucci G, Marrone A. Investigation of the HPLC response of NSAIDs by fractional experimental design and multivariate regression analysis. Response optimization and new retention parameters. J Pharm Biomed Anal 2010; 51(1): 46-55.
[http://dx.doi.org/10.1016/j.jpba.2009.08.001] [PMID: 19758780]

[62]   Veza I, Spraggon M, Fattah IMR, Idris M. Response surface methodology (RSM) for optimizing engine performance and emissions fueled with biofuel: Review of RSM for sustainability energy transition. Results in Engineering. 2023; 18: 101213.

[63]   P.R. HV. Beeraka NM, Kumar P, Patel HB, Gurupadayya BM. UPLC-MS-Based Method Development, Validation, and Optimization of Dissolution Using Quality by Design Approach for Low Dose Digoxin: A Novel Strategy. Curr Pharm Anal 2022; 18(8): 841-51.

[64]   Patil A, Shirkhedkar A. Application of quality by design in the development of hptlc method for estimation of anagliptin in bulk and in-house tablets. Eurasian Journal of Analytical Chemistry 2017; 12(5): 443-58.
[http://dx.doi.org/10.12973/ejac.2017.00181a]

[65]   Khurd AS, Doshi KV. Quality by design-based optimization and validation of a high-performance thin-layer chromatography method for the estimation of rivaroxaban in bulk and its pharmaceutical dosage form. J Planar Chromatogr Mod TLC 2019; 32(6): 505-10.
[http://dx.doi.org/10.1556/1006.2019.32.6.9]

[66]   Jadhav ML, Tambe SR. Implementation of QbD approach to the analytical method development and validation for the estimation of propafenone hydrochloride in tablet dosage form. Chromatogr Res Int 2013; 1-9.
[http://dx.doi.org/10.1155/2013/676501]

[67]   Gurba-Bryśkiewicz L, Dawid U, Smuga DA, *et al.* Implementation of QbD approach to the development of chromatographic methods for the determination of complete impurity profile of substance on the preclinical and clinical step of drug discovery studies. Int J Mol Sci 2022; 23(18): 10720.
[http://dx.doi.org/10.3390/ijms231810720] [PMID: 36142622]

[68]   Mastanamma S, Srilakshmi PSB, Ramadevi N, Prathyusha D, Rani MV, Rani MV. Analytical quality by design approach for the development of uv-spectophotometric method in the estimation of tenofovir alafenamide in bulk and its laboratory synthetic mixture. Research Journal of Pharmacy and Technology 2018; 11(2): 499.
[http://dx.doi.org/10.5958/0974-360X.2018.00091.4]

[69]   Tavares Luiz M, Santos Rosa Viegas J, Palma Abriata J, *et al.* Design of experiments (DoE) to develop and to optimize nanoparticles as drug delivery systems. Eur J Pharm Biopharm 2021; 165: 127-48.
[http://dx.doi.org/10.1016/j.ejpb.2021.05.011] [PMID: 33992754]

[70]   Guimarães D, Cavaco-Paulo A, Nogueira E. Design of liposomes as drug delivery system for therapeutic applications. Int J Pharm 2021; 601: 120571.
[http://dx.doi.org/10.1016/j.ijpharm.2021.120571] [PMID: 33812967]

[71]   Shah VH, Jobanputra A. Enhanced ungual permeation of terbinafine HCl delivered through liposome-

loaded nail lacquer formulation optimized by QbD approach. AAPS PharmSciTech 2018; 19(1): 213-24.
[http://dx.doi.org/10.1208/s12249-017-0831-0] [PMID: 28681334]

[72] Dasgupta N, Ranjan S, Gandhi M. Nanoemulsion ingredients and components. Environ Chem Lett 2019; 17(2): 917-28.
[http://dx.doi.org/10.1007/s10311-018-00849-7]

[73] Duan Y, Dhar A, Patel C, *et al.* A brief review on solid lipid nanoparticles: Part and parcel of contemporary drug delivery systems. RSC Advances 2020; 10(45): 26777-91.
[http://dx.doi.org/10.1039/D0RA03491F] [PMID: 35515778]

[74] Chintamani RB, Salunkhe KS, Kharat AR, Chintamani SR, Singh RP, Chavan MJ. Preparation, characterization and evaluation of green synthesis nanoparticle of hydro alcoholic floret extract of *Brassica oleracea var. italica plenck* (broccoli) using QbD approach for breast tumor cells T-47D treatment. Int J Sci Technol Res 2020; 9(2).

[75] Zhang L, Yan B, Gong X, Yu LX, Qu H. Application of quality by design to the process development of botanical drug products: A case study. AAPS PharmSciTech 2013; 14(1): 277-86.
[http://dx.doi.org/10.1208/s12249-012-9919-8] [PMID: 23297167]

[76] Ali J, Pramod K, Tahir MA, Charoo NA, Ansari SH. Pharmaceutical product development: A quality by design approach. Int J Pharm Investig 2016; 6(3): 129-38.
[http://dx.doi.org/10.4103/2230-973X.187350] [PMID: 27606256]

[77] Khan IA, Smillie T. Implementing a "quality by design" approach to assure the safety and integrity of botanical dietary supplements. J Nat Prod 2012; 75(9): 1665-73.
[http://dx.doi.org/10.1021/np300434j] [PMID: 22938174]

[78] Saroj Seema, Shah Priya, Jairaj Vinod and Rathod Rajeshwari, Green Analytical Chemistry and Quality by Design: A Combined approach towards Robust and Sustainable Modern Analysis, Current Analytical Chemistry 2018; 14(4).
[http://dx.doi.org/10.2174/1573411013666170615140836]

[79] Mykhailenko O, Ivanauskas L, Bezruk I, Petrikaitė V, Georgiyants V. Application of quality by design approach to the pharmaceutical development of anticancer crude extracts of *Crocus Sativus Perianth.* Sci Pharm 2022; 90(1): 19.
[http://dx.doi.org/10.3390/scipharm90010019]

[80] Pan J, He S, Zheng J, *et al.* The development of an herbal material quality control strategy considering the effects of manufacturing processes. Chin Med 2019; 14(1): 38.
[http://dx.doi.org/10.1186/s13020-019-0262-9] [PMID: 31572490]

[81] Mannur VS. Development of directly compressible polyherbal tablets by using QbD approach a novel immunomodulatory material. J med pharma and allied sci 2022; 11(6): 5476-84.
[http://dx.doi.org/10.55522/jmpas.V11I6.4520]

[82] Vijayaraj S, Palei NN, Archana D, Lathasri K, Rajavel P. Quality by design (QbD) approach to develop stability indicating HPLC method for estimation of rutin in chitosan-sodium alginate nanoparticles. Braz J Pharm Sci 2020; 56: e18793.
[http://dx.doi.org/10.1590/s2175-97902020000318793]

[83] Diallo MS, Fromer NA, Jhon MS. Nanotechnology for sustainable development: Retrospective and outlook. J Nanopart Res 2013; 15(11): 2044.
[http://dx.doi.org/10.1007/s11051-013-2044-0]

[84] Ahmed SF, Mofijur M, Rafa N, *et al.* Green approaches in synthesising nanomaterials for environmental nanobioremediation: Technological advancements, applications, benefits and challenges. Environ Res 2022; 204(Pt A): 111967.
[http://dx.doi.org/10.1016/j.envres.2021.111967] [PMID: 34450159]

[85] Yarovoi H, Frey T, Bouaraphan S, Retzlaff M, Verch T. Quality by design for a vaccine release

immunoassay: A case study. Bioanalysis 2013; 5(20): 2531-45.
[http://dx.doi.org/10.4155/bio.13.198] [PMID: 24138626]

[86]   Goris N, De Clercq K. Quality assurance/quality control of foot and mouth disease solid phase competition enzyme-linked immunosorbent assay-Part II. Quality control: comparison of two charting methods to monitor assay performance. Rev Sci Tech 2005; 24(3): 1005-16.
[http://dx.doi.org/10.20506/rst.24.3.1630] [PMID: 16642771]

[87]   Bak H, Ekeroth L, Houe H. Quality control using a multilevel logistic model for the Danish pig Salmonella surveillance antibody-ELISA programme. Prev Vet Med 2007; 78(2): 130-41.
[http://dx.doi.org/10.1016/j.prevetmed.2006.10.001] [PMID: 17175047]

[88]   Baldelli S, Marrubini G, Cattaneo D, Clementi E, Cerea M. Application of quality by design approach to bioanalysis: Development of a method for elvitegravir quantification in human plasma. Ther Drug Monit 2017; 39(5): 531-42.
[http://dx.doi.org/10.1097/FTD.0000000000000428] [PMID: 28650901]

[89]   Zhang L, Mao S. Application of quality by design in the current drug development. Asian J Pharma Sci 2017; 12(1): 1-8.
[http://dx.doi.org/10.1016/j.ajps.2016.07.006] [PMID: 32104308]

[90]   Peraman R, Bhadraya K, Padmanabha Reddy Y. Analytical quality by design: A tool for regulatory flexibility and robust analytics. Int J Anal Chem 2015; 2015: 1-9.
[http://dx.doi.org/10.1155/2015/868727] [PMID: 25722723]

[91]   Schad M, Gautam S, Grein TA, Käß F. Process Analytical Technologies (PAT) and Quality by Design (QbD) for bioprocessing of virus-based therapeutics. In: Gautam S, Chiramel AI, Pach R, Käß F, Eds. Bioprocess and Analytics Development for Virus-based Advanced Therapeutics and Medicinal Products (ATMPs). Cham: Springer 2023; pp. 295-328.
[http://dx.doi.org/10.1007/978-3-031-28489-2_13]

[92]   Singh J, Yadav P, Yadav S. A comprehensive review on regulatory quality-by-design (QbD) guidelines for pharmaceutical development. Pharmaceutics 2020; 12(8): 761.
[PMID: 32806507]

[93]   Bahl D, Bakhshi P, Rawat S. Quality by design: A systematic review. Int J Appl Pharm 2020; 12(3): 12-8.

[94]   Schad M, Gautam S, Grein TA, Käß F. Process Analytical Technologies (PAT) and Quality by Design (QbD) for Bioprocessing of Virus-Based Therapeutics. In: Gautam S, Chiramel AI, Pach R, editors. Bioprocess and Analytics Development for Virus-based Advanced Therapeutics and Medicinal Products (ATMPs). Cham: Springer International Publishing; 2023. p. 295-328.
[http://dx.doi.org/10.1007/978-3-031-28489-2_13]

[95]   Lohiya RT, Tambe A, Jadhav M, Kapadiya N. Quality by design (QbD) based pharmaceutical product development: A review. J Appl Pharm Sci 2021; 11(04): 156-66.

[96]   Suthar B, Patel R, Mehta T. Quality by design (QbD) and its application in pharmaceutical industry: A comprehensive review. Int J Pharm Sci Res 2020; 11(5): 2011-25.

[97]   Rathore AS. Quality by Design for Biopharmaceuticals: Principles and Case Studies. John Wiley & Sons 2009.
[http://dx.doi.org/10.1002/9780470466315]

[98]   Prajapati VD, Jani GK, Moradiya NG, Randeria NP. Quality by Design (QbD): An integrated approach of analytical and formulation development for optimization of the analytical procedure. J Pharm Biomed Anal 2012; 67-68: 22-30.

[99]   Singh S, Srinivasan KK, Gowthamarajan K. Quality by design approach for formulation development: A systematic review. J Young Pharm 2011; 3(2): 115-21.

[100]   Bansal AK, Dighe V, Gaikwad D, Gupta V, Jangam S, Tyagi R. Quality by design: A comprehensive review. Int J Pharm Sci Res 2013; 4(10): 3690-707.

CHAPTER 5

Virtual Tools and Screening Designs for Drug Discovery and New Drug Development

Sonal Dubey[1,*]

[1] *College of Pharmaceutical Sciences, Dayananda Sagar University, Kumaraswamy Layout, Bengaluru-560111, India*

Abstract: The synergy between virtual tools and screening designs has catalyzed a transformative shift in drug discovery and new drug development. Leveraging computational models, molecular simulations, and artificial intelligence, virtual tools empower researchers to predict molecular interactions, assess binding affinities, and optimize drug-target interactions. This predictive capacity expedites the identification and prioritization of promising drug candidates for further investigation. Simultaneously, screening designs facilitate systematic and high-throughput evaluation of vast compound libraries against target proteins, enabling the rapid identification of lead compounds with desired pharmacological activities. Advanced data analysis techniques, including machine learning, enhance the efficiency and accuracy of hit identification and optimization processes. The integration of virtual tools and screening designs presents a holistic approach that accelerates the drug discovery pipeline. By expounding on rational drug design, these tools guide the development of novel compounds with enhanced properties. Furthermore, this approach optimizes resource allocation by spotlighting high-potential candidates and minimizing costly experimental iterations. As an outcome of this convergence, drug discovery processes are becoming more precise, efficient, and cost-effective. The resulting drug candidates exhibit improved efficacy, specificity, and safety profiles. Thus, the amalgamation of virtual tools and screening designs serves as a potent catalyst for innovation in drug discovery and new drug development, ensuring the delivery of transformative therapies to address unmet medical challenges. In this chapter, we shall be discussing different tools in detail with actual examples leading to successful stories.

Keywords: Drug discovery, Drug design, Designing software, Molecular modelling, Molecular docking, QSAR, Virtual screening.

1. INTRODUCTION

The landscape of drug discovery and development has witnessed a remarkable transformation with the integration of virtual tools and screening designs. These

[*] **Corresponding author Sonal Dubey:** College of Pharmaceutical Sciences, Dayananda Sagar University, Kumaraswamy Layout, Bengaluru-560111, India; E-mail: drsonaldubey-pharmacy@dsu.edu.in

Dilpreet Singh and Prashant Tiwari (Eds.)

innovative technologies have emerged as game-changers, revolutionizing the traditional approach to drug development [1]. In this comprehensive review, we will explore the profound impact and significance of virtual tools and screening designs in reshaping the pharmaceutical industry. Historically, drug discovery has been a time-consuming and costly endeavor, with a high attrition rate due to the limitations of experimental approaches. However, the advent of virtual tools has introduced a paradigm shift. These tools encompass a wide range of computational techniques that enable the prediction and optimization of drug-target interactions, molecular properties, and ADMET (absorption, distribution, metabolism, excretion, and toxicity) profiles [2]. As a result, the drug discovery process has become more efficient and cost-effective, saving both time and resources.

Screening designs, both virtual and experimental, are pivotal in identifying promising drug candidates from vast chemical libraries. Virtual screening employs computational methods to prioritize compounds with potential binding affinities for target proteins. This approach not only expedites the selection of lead compounds but also minimizes the number of compounds that need to be synthesized and tested experimentally [3]. Experimental screening, on the other hand, involves high-throughput assays that rapidly evaluate the biological activity of compounds, further enhancing the efficiency of the drug discovery pipeline. The synergy between virtual tools and screening designs is evident in their ability to optimize hit-to-lead and lead optimization phases. Virtual tools facilitate rational drug design by providing insights into the structure-activity relationships (SAR) and helping researchers modify molecular structures to enhance potency, selectivity, and other desired properties. Coupled with screening techniques, this approach accelerates the identification of promising drug candidates, expediting the transition from bench to bedside [4].

While the benefits of virtual tools and screening designs are substantial, it is important to acknowledge their limitations. Virtual models are based on assumptions and approximations, and their accuracy heavily relies on the quality of available data. Furthermore, experimental validation is essential to confirm the predictions made by virtual tools. Screening designs also require careful selection of assays and validation strategies to ensure the reliability of results. In this chapter, we aim to provide a comprehensive understanding of the role that virtual tools and screening designs play in modern drug discovery and development. By delving into their applications, benefits, and limitations, we aspire to shed light on their transformative potential. As the pharmaceutical industry continues to evolve, virtual tools and screening designs stand as indispensable tools in the pursuit of novel and effective therapeutic agents.

2. CONCEPT OF DRUG DESIGN

Drug design is a dynamic and interdisciplinary field at the intersection of chemistry, biology, and computational science, focused on creating novel therapeutic agents to treat diseases. This process involves a systematic approach that integrates various scientific principles to develop molecules with desired properties for specific targets. The intricate role of molecules, proteins, and cellular pathways forms the foundation of drug design, aiming to achieve efficacy, safety, and specificity. Drug design is a complex process that involves identifying a target region, developing a lead compound that can interact with the target protein, and testing the safety and efficacy of that compound [5]. Virtual tools have increasingly become important in drug design, enabling researchers to expedite and optimize the drug discovery process [6]. Computational modeling has emerged as a crucial tool in drug discovery, allowing researchers to simulate and predict the behavior of molecules and their interactions with others. Various software programs are available on the market for this purpose, including molecular dynamics simulation packages, molecular docking software, and QSAR (quantitative structure-activity relationship) modeling tools [7].

2.1. Quantitative Structure-activity Relationship (QSAR)

QSAR is a method used in drug discovery to predict the activity of molecules based on their chemical structure. QSAR has a rich history dating back to the 1960s when it was first introduced as a way to analyse the relationship between the structure of organic compounds and their biological activity [7]. QSAR modelling tools are a type of computational modelling software that can be used to predict the activity of molecules based on their chemical structure. These tools use statistical methods to identify patterns in the relationships between the structure of a molecule and its activity. This information can be used to design new compounds with specific activity profiles [8, 9].

The development of QSAR was driven by the need to better understand the mechanism of drug action and to design more effective drugs. The earliest QSAR models were based on simple statistical methods and used only a small number of molecular descriptors to predict the biological activity of a compound. Over time, the field has evolved to include more complex models that incorporate a larger number of molecular descriptors and account for the three-dimensional structure of molecules [10 - 12]. Descriptors are quantitative measures that represent the physical, chemical, and biological properties of molecules. Descriptors are used to represent the structure and properties of molecules and are often used as inputs to QSAR models. There are several types of descriptors that help us in designing

QSAR modelling, each with its strengths and limitations. The main approaches used in QSAR modelling are:

2.1.1. Topological Approach

The topological approach is the simplest approach to QSAR modelling and involves the use of molecular graphs to represent chemical structures. This approach uses topological indices, which are numerical values calculated from the molecular graph, as descriptors to predict biological activity. Topological indices describe the molecular weight, the number of atoms, the number of ring sizes, the shape, and the branching of the molecular graph. They are often used to predict physicochemical properties such as logP (lipophilicity) and boiling point [13 - 16].

2.1.2. Physicochemical Approach

The physicochemical approach involves the use of physicochemical descriptors, which describe the physical and chemical properties of a molecule, to predict biological activity. These descriptors include properties such as molecular weight, polarizability, and refractivity. Physicochemical descriptors are often used to predict the solubility, stability, and transport of drugs in the body [13, 17, 18].

2.1.3. Quantum Chemical Approach

The quantum chemical approach involves the use of quantum mechanical calculations to predict the electronic structure and properties of molecules. This approach uses quantum chemical descriptors, which describe the electronic structure and reactivity of a molecule, to predict biological activity. Quantum chemical descriptors include properties such as ionization potential, electron affinity, and dipole moment [17, 19, 20].

2.1.4. Molecular Mechanics Approach

The molecular mechanics approach involves the use of classical mechanics to simulate the motion and interactions of molecules. This approach uses molecular mechanics descriptors, which describe the energy and geometry of a molecule, to predict biological activity. Molecular mechanics descriptors include properties such as total energy, bond length, and bond angle [21, 22].

2.1.5. Hybrid Approach

The hybrid approach combines two or more of the above approaches to improve the accuracy and robustness of QSAR models. For example, a hybrid approach might combine topological indices with physicochemical descriptors to predict the biological activity of a molecule. Hybrid approaches can be more powerful than individual approaches and are often used to predict complex biological activities [23].

There are two primary types of QSAR models: 2D-QSAR and 3D-QSAR. 2D-QSAR models employ two-dimensional molecular representations to predict biological activity, while 3D-QSAR models consider the three-dimensional structure of molecules and their interactions with targets. Although 3D-QSAR models are more intricate and computationally intensive, they offer the potential for more accurate predictions of biological activity [24 - 26].

2.2. 2D-QSAR

is a relatively simple and widely used approach that relies on the calculation of descriptors based on two-dimensional molecular structures. These descriptors include physicochemical properties, such as molecular weight, logP, and polarizability, as well as topological indices, such as the number of rings, the number of atoms, and the branching index. The calculated descriptors are then used to develop a mathematical model that relates the biological activity of the compounds to their structural features. The model can be used to predict the activity of new compounds that have similar structural features to those in the training set [27 - 30]. One of the main advantages of 2D-QSAR is that it is computationally efficient and can be used to analyze large data sets. However, the approach has limitations in that it does not take into account the three-dimensional shape of the molecules, which can be important for activity. For example, two compounds may have similar two-dimensional structures, but their three-dimensional shapes may be different, leading to differences in activity. This limitation can be addressed by using 3D-QSAR modelling [31 - 33].

2.3. 3D-QSAR

Comparative molecular field analysis (CoMFA), is a more advanced approach that takes into account the three-dimensional shape of the molecules. 3D-QSAR involves the alignment of a set of molecules in a common reference frame, which allows the calculation of the steric and electrostatic fields at each grid point. The fields are then used to derive three-dimensional descriptors, which are used to develop a mathematical model that relates the biological activity of the compounds to their structural features. A probe atom is usually used to map the 3D structure of the molecule by placing a 3D grid around it (like a mesh box), and XYZ coordinates are located and documented to graphically represent the structure (Fig. **1**). The model can be used to predict the activity of new compounds that have similar structural features to those in the training set [32]. One of the main advantages of 3D-QSAR is that it can account for the three-dimensional shape of the molecules, which can be important for activity. However, the approach has limitations in that it is computationally intensive and can be sensitive to the alignment of the molecules. In addition, the quality of the model depends on the quality of the three-dimensional structures of the molecules.

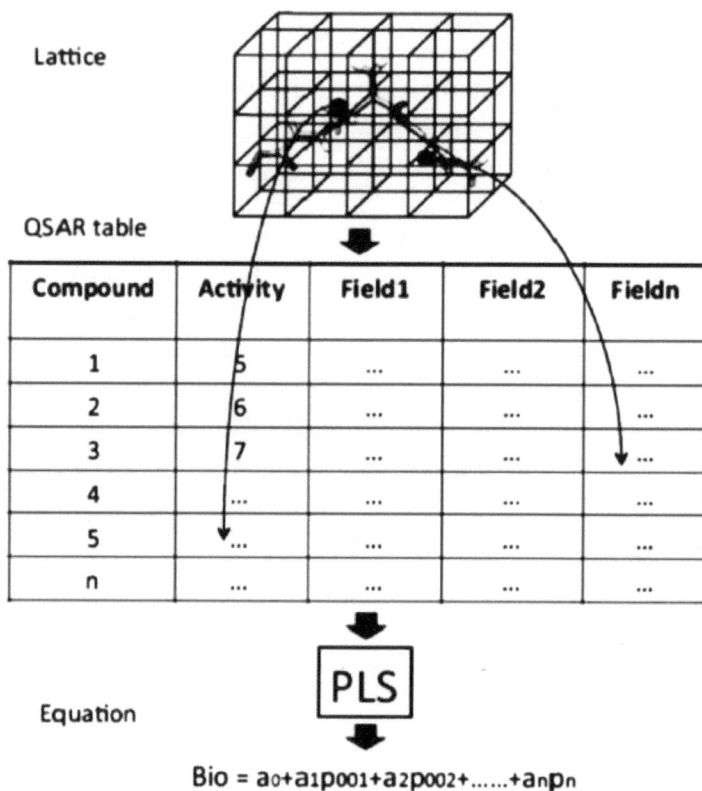

Fig. (1). Pictorial representation of the calculation of various descriptors using the probe in the grid to derive a mathematical model.

Fig. (2) depicts the various steps involved in CoMFA analysis, by which the bioactive conformation of the molecules can be generated and aligned over each other. The steric and electronic parameters are measured using the probe atom, and the final results are presented in terms of contour plots that indicate favorable and unfavorable regions in the 3D space.

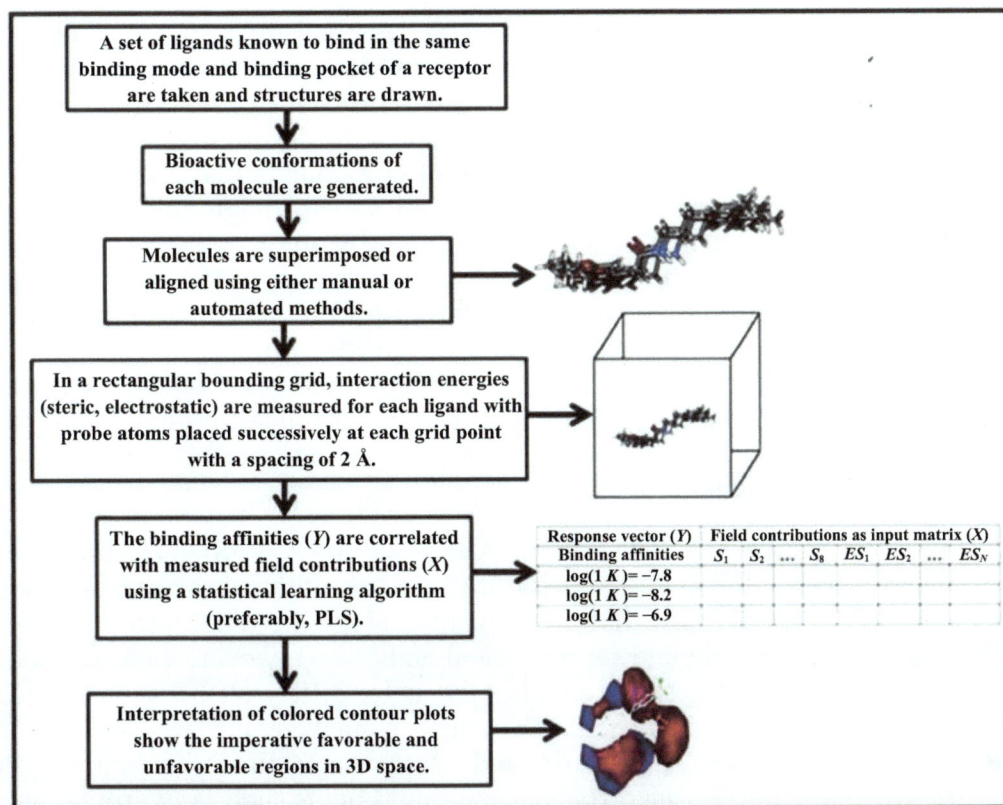

Fig. (2). Steps involved in CoMFA.

Another approach to 3D QSAR is the Comparative Molecular Similarity Indices Analysis (CoMSIA), which is a powerful quantitative structure-activity relationship (QSAR) method that is used to analyze the relationship between the structure of a molecule and its biological activity. CoMSIA is a modification of the traditional molecular similarity analysis (CoMFA), and it incorporates additional fields in addition to steric and electrostatic interactions to increase its predictive power [34]. CoMSIA uses a grid-based approach to analyze the 3D structure of molecules, and it calculates the similarity between different molecules based on their physicochemical properties, such as charge, hydrophobicity, and polarizability. In CoMSIA, five different similarity fields are calculated *i.e.* steric, electrostatic, hydrophobic, hydrogen bond donor and hydrogen bond acceptor. These fields were selected to cover the major contributions to ligand binding. Similarity indices are computed at regularly spaced grid points for the pre-aligned molecules. CoMSIA generates a 3D grid around the molecule and employs partial least squares (PLS) regression analysis to examine the connection between physicochemical properties and biological activity [35, 36]. CoMSIA has several advantages over other QSAR methods. It is less sensitive to the choice of molecular alignment, which can be a significant source of variability in other methods. It can also handle larger datasets than traditional CoMFA, which can be useful when analyzing large libraries of compounds. In addition, CoMSIA is capable of predicting the biological activity of compounds that are structurally distinct from the training set, making it a useful tool for lead optimization and scaffold hopping [24]. CoMSIA has been successfully applied in several areas of drug discovery, including the design of inhibitors for HIV-1 protease, the development of new antitumor agents, and the optimization of antifungal agents. One example of the successful application of CoMSIA is the design of novel antitumor agents based on the structure of pyrrolopyrimidines. CoMSIA was used to analyze the relationship between the physicochemical properties of the compounds and their antitumor activity, and it was able to predict the activity of new compounds with high accuracy [35, 37].

CoMFA (Comparative Molecular Field Analysis) and CoMSIA (Comparative Molecular Similarity Indices Analysis) are both quantitative structure-activity relationship (QSAR) methods that are used to analyze the relationship between the structure of a molecule and its biological activity. However, there are some key differences between the two methods. One major difference between CoMFA and CoMSIA is the types of fields they use to analyze the 3D structure of molecules. CoMFA uses only steric and electrostatic fields to analyze the structure, whereas CoMSIA incorporates additional fields such as hydrophobicity and polarizability to increase its predictive power. Another difference between the two methods is their sensitivity to molecular alignment. CoMFA is highly sensitive to the choice of molecular alignment, and small changes in the alignment can lead to significant changes in the calculated fields. CoMSIA, on the other hand, is less sensitive to molecular alignment and can produce more accurate results even when the molecules are not perfectly aligned. CoMSIA also has the advantage of being able to handle larger datasets than CoMFA. This can be useful when analyzing large libraries of compounds or when searching for new lead compounds [38]. Despite these differences, both CoMFA and CoMSIA have been widely used in drug discovery and have been shown to be effective in predicting the biological activity of compounds. The choice of which method to use depends on the specific goals of the study and the nature of the molecules being analyzed [39].

Numerous software packages are accessible for QSAR modeling, such as MOE, Schrödinger, and ChemAxon. These software options employ machine learning algorithms to detect patterns within extensive datasets containing chemical structures and activity information [40]. Subsequently, these algorithms can predict the activity of novel compounds based on their chemical composition. An illustrative achievement of QSAR in drug discovery is the creation of the medication Celecoxib, utilized for pain and inflammation management [41]. Celecoxib is a selective inhibitor of the enzyme cyclooxygenase-2 (COX-2), which is involved in the production of inflammatory mediators. Researchers used QSAR to design new compounds with improved selectivity for COX-2 over the related enzyme COX-1. This led to the development of Celecoxib, which is more selective for COX-2 and has fewer side effects than earlier non-selective COX inhibitors [42].

Another such example of the use of QSAR in drug discovery is the development of the drug Gefitinib, which is used to treat non-small cell lung cancer. Gefitinib is a selective inhibitor of the epidermal growth factor receptor (EGFR), which is overexpressed in many types of cancer [43]. Researchers used QSAR to design new compounds with improved selectivity for EGFR over other members of the same protein family. This led to the development of Gefitinib, which has shown promising results in clinical trials [44 - 46]. In addition to these examples, QSAR has also been used in the development of a variety of other drugs, including antivirals, antimicrobials, and anticancer agents. By using QSAR, researchers can accelerate the drug discovery process, identify new drug candidates, optimize drug properties, and improve the safety and efficacy of new treatments [47 - 50].

3. MOLECULAR MODELLING

Molecular modelling is another powerful virtual tool for drug design and discovery. It allows scientists to visualize the 3D structure of a drug molecule and its interactions with the target protein, and to optimize the drug's properties, such as its binding affinity, selectivity, and pharmacokinetics. In this article, we will explore the different types of molecular modelling techniques and their applications in drug discovery [51, 52].

3.1. Protein Modelling

If the structure of the protein is not known at all or is not fully established, protein modelling can be done using either homology modelling or *de novo* modelling. *Homology modelling* is a computational method that predicts the 3D structure of a protein based on its sequence similarity to a known protein structure. This method can be used to predict the 3D structure of the target protein and optimize the drug's interactions with it [53, 54]. *De novo* modelling is a computational method that predicts the 3D structure of a protein from scratch without relying on any known protein structures. This method can be used when the target protein has no known homologues, or when the homology models are inaccurate [55].

3.2. Lead Modelling

Molecular modelling of the lead compound can be done by various approaches.

3.2.1. Lead Optimization

Molecular modelling can be used to optimize the properties of a lead compound, such as its binding affinity, selectivity, and pharmacokinetics. This can help to improve the potency and safety of the drug candidate [56 - 58].

3.2.2. Scaffold Hopping

This refers to the search for compounds that have similar activity but contain different core structures. Besides activity, other molecular properties might also be considered. Molecular modelling can be used to identify new chemical scaffolds that have similar properties to a known drug molecule, but with improved potency or safety. This can help generate new drug candidates that have different chemical structures but similar biological activity [59, 60].

3.2.3. Protein Engineering

Molecular modelling can be used to design new proteins or modify existing ones to improve their stability, binding affinity, or other properties. This can be used to generate new targets for drug discovery [61 - 63].

3.2.4. Virtual Screening

Molecular modelling can be used to screen large libraries of compounds to identify potential drug candidates. This can help to prioritize compounds for further experimental testing and reduce the time and cost of drug development [64, 65].

3.2.5. Toxicity Prediction

Molecular modelling can be used to predict the potential toxicity of a drug candidate. This can help to identify potential safety issues early in the drug development process [66, 67].

3.3. Software for Molecular Modelling

There are several software packages available for molecular modelling, each with its own strengths and limitations [68 - 70]. Some popular software packages for molecular modelling include:

3.3.1. Schrödinger

Schrödinger offers a suite of software tools for molecular modelling, including homology modelling, docking, and molecular dynamics simulations.

3.3.2. MOE

MOE is a comprehensive software package for molecular modelling, including homology modelling, docking, molecular dynamics simulations, and virtual screening.

One of the most important virtual tools in drug design is computational modelling. This involves the use of software programs and algorithms to simulate the behaviour of molecules and predict their interactions with other molecules. Computational modelling can be used to study the three-dimensional structure of proteins, the binding of ligands to receptors, and the pharmacokinetics of drugs. By using computational modelling, researchers can design drugs with specific properties, such as improved binding affinity, reduced toxicity, or increased selectivity [71].

Computational modelling of the drug Tafamidis, which is used to treat a rare genetic disorder called transthyretin amyloidosis, is one such example. Transthyretin is a protein that can misfold and accumulate in the body, leading to tissue damage and organ failure. Researchers used molecular dynamics simulations to study the interactions between Tafamidis and transthyretin, and to design new derivatives of the drug with improved binding affinity. These derivatives were then tested in preclinical and clinical studies, and Tafamidis was eventually approved for use in patients [72]. Another example of the use of computational modelling in drug discovery is the development of the drug Raltegravir, which is used to treat HIV. Raltegravir is an integrase inhibitor that prevents the integration of the HIV genome into the host cell DNA. Researchers used molecular docking software to identify potential integrase inhibitors, and then used molecular dynamics simulations to study the interactions between the inhibitors and the integrase enzyme. Based on these simulations, Raltegravir was designed and tested in preclinical and clinical studies, and was eventually approved for use in patients [73]. In addition to these examples, computational modelling has also been used in the development of a variety of other drugs, including anticancer agents, antivirals, and antibiotics. By using these tools, researchers can accelerate the drug discovery process, identify new drug candidates, optimize drug properties, and improve the safety and efficacy of new treatments [71].

4. MOLECULAR DOCKING

is a computational method used in drug discovery to predict the binding mode and affinity of small molecules to a protein target. It has become an integral part of drug discovery, as it allows researchers to identify potential lead compounds and optimize their binding affinity and selectivity [74]. The history of molecular docking can be traced back to the 1980s, when researchers first began using computer simulations to study protein-ligand interactions. Over the years, the field has evolved, and various approaches to molecular docking have been developed [75]. There are two main types of molecular docking: rigid docking and flexible docking [76 - 78].

4.1. Rigid Docking

In rigid docking, the protein target is held in a fixed conformation, while the ligand is allowed to move freely. This approach is useful when the protein target has a well-defined binding site and a rigid structure.

4.2. Flexible Docking

In flexible docking, both the protein target and the ligand are allowed to move, which can take into account the flexibility of both molecules. This approach is useful when the protein target has a flexible structure or when the ligand needs to adapt to the binding site. There are various software programs available for molecular docking, including AutoDock, GOLD, Glide, and MOE. These programs use different algorithms and scoring functions to predict the binding mode and affinity of small molecules to a protein target. For example, AutoDock uses a genetic algorithm to search for the best binding pose, while GOLD uses a genetic algorithm combined with a scoring function based on protein-ligand interactions.

Molecular docking has been used in many drug discovery projects to identify and optimize lead compounds. For example, in the development of the anti-HIV drug maraviroc, molecular docking was used to identify potential inhibitors of the CCR5 receptor, which is involved in the entry of the Human Immunodeficiency Virus (HIV) into human cells. The lead compound was then optimized using molecular docking and other computational methods to improve its binding affinity and selectivity. Another example is the development of the anti-cancer drug imatinib, which targets the Bcr-Abl fusion protein that is responsible for chronic myeloid leukemia. Molecular docking was used to design imatinib and optimize its binding affinity to the Bcr-Abl protein [79, 80].

5. MOLECULAR DYNAMIC SIMULATION (MDS)

MDS is another important virtual tool in drug design. This involves the use of algorithms to simulate the motion of atoms and molecules over time. Molecular dynamics simulations can be used to study the interactions between a drug molecule and its target and to predict how changes in the drug molecule's structure or conformation might affect its activity. Molecular dynamics simulations can also be used to study the pharmacokinetics of drugs by predicting their absorption, distribution, metabolism, and excretion in the body. Molecular dynamics (MD) is a computational method that simulates the motion of atoms and molecules in a system over time. In drug design, MD simulations can provide insights into the behaviour of drug molecules and their interactions with target proteins. MD simulations can help researchers understand the structure-function relationships of molecules and optimize the design of drug candidates. The theory of MD simulations can be traced back to the 1950s, when scientists began using computers to study the behaviour of physical systems. In the 1970s, molecular dynamics simulations were first applied to study biomolecules, such as proteins and nucleic acids. Over the years, the field has evolved, and various approaches to MD simulations have been developed [80, 81].

There are several approaches to MD simulations, including explicit solvent MD, implicit solvent MD, and coarse-grained MD. In explicit solvent MD, water molecules and other solvent molecules are explicitly represented in the simulation. In implicit solvent MD, solvent effects are accounted for through an implicit solvent model. In coarse-grained MD, multiple atoms are represented by a single "bead," which allows for simulations of larger systems over longer timescales.

There are various software programs available for MD simulations, including GROMACS, AMBER, CHARMM and NAMD. These programs use different force fields and algorithms to simulate molecular systems. For example, GROMACS uses a combination of molecular mechanics and molecular dynamics to simulate the motion of molecules in a system. MD simulations have many applications in drug design. MD simulations can be used to study the binding mode of drug molecules to target proteins and predict the binding affinity of different compounds. MD simulations can also be used to study the stability and conformational changes of proteins in the presence of ligands. Additionally, MD simulations can be used to identify potential binding sites on proteins and design novel ligands to target those sites. In recent years, machine learning approaches have been applied to MD simulations to accelerate the drug discovery process. For example, researchers have developed methods to predict protein-ligand binding affinity using MD simulations and machine learning algorithms [82].

6. VIRTUAL SCREENING

An essential virtual tool in drug design is virtual screening, which employs computational techniques to pinpoint compounds likely to interact with a target molecule. Virtual screening enables the exploration of vast chemical compound databases, leading to the discovery of potential drug candidates that might have been overlooked otherwise. Additionally, this approach facilitates the enhancement of existing compound properties by recognizing modifications that could bolster their effectiveness or diminish their toxicity. Virtual tools have also found applications in drug repurposing, wherein new applications for existing drugs are identified. By utilizing computational methods to sift through extensive drug databases, researchers can unearth compounds with potential activity against novel targets. This approach proves cost-effective for developing new treatments, given that existing drugs have already undergone comprehensive testing for safety and efficacy [83]. Fig. (**3**) shows the step-by-step approach to virtual screening. It begins with the preparation of the protein of interest and ligands, followed by choosing the screening method to be employed. By doing so, Hits are generated, which should be validated, and the most promising candidates are selected.

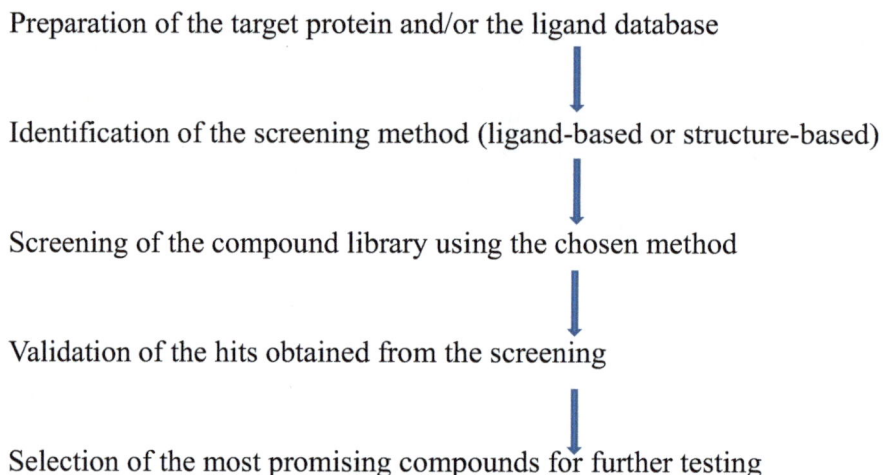

Preparation of the target protein and/or the ligand database

Identification of the screening method (ligand-based or structure-based)

Screening of the compound library using the chosen method

Validation of the hits obtained from the screening

Selection of the most promising compounds for further testing

Fig. (3). The general flowchart of virtual screening.

6.1. Types of Virtual Screening

Virtual screening can be divided into two main types: ligand-based screening and structure-based screening.

6.1.1. Ligand-Based Screening

Ligand-based screening relies on the similarity of the properties of a known active molecule and the compounds in the library. This method is based on the assumption that similar compounds will have similar biological activities. The most commonly used techniques for ligand-based screening are pharmacophore-based screening, similarity-based screening, and machine learning-based screening.

6.1.1.1. Similarity-based Screening

This approach is an integrated ligand-based virtual screening approach that identifies potential drug candidates based on their structural similarity to known active compounds. The basic assumption behind this approach is that compounds with similar structures are likely to have similar biological activities. This is because similar structural features are likely to interact with the same binding site or biological target [84].

The process of similarity-based screening can be divided into four main steps:

6.1.1.1.1. Selection of the Reference Compounds

A set of known active compounds is selected as the reference set. The reference set can be obtained from a database of known bioactive compounds, or it can be selected based on experimental data from previous drug discovery projects.

6.1.1.1.2. Calculation of Molecular Descriptors

A set of molecular descriptors is calculated for each compound in the reference set. These descriptors represent the structural and physicochemical properties of the molecules, such as molecular weight, lipophilicity, and hydrogen bond donor/acceptor properties.

6.1.1.1.3. Calculation of Similarity Scores

The similarity between each compound in the reference set and the compounds in the screening library is calculated using a similarity scoring algorithm. The most commonly used algorithm is the Tanimoto coefficient, which is based on the number of common structural features between the two molecules.

6.1.1.1.4. Ranking and Validation of the Hits

The compounds in the screening library are ranked based on their similarity scores. The top-ranked compounds are then validated using experimental methods to confirm their binding affinity and selectivity.

Several software packages are available for similarity-based screening, including ChemMine, ChemBridge, and MolSoft. These software packages can be used to calculate molecular descriptors, calculate similarity scores, and analyze the results. Similarity-based screening has several advantages over other virtual screening approaches. It is a ligand-based approach, which means that it does not require structural information of the target protein. This makes it particularly useful for targets where the 3D structure is unknown or difficult to obtain. Additionally, similarity-based screening can be used to identify compounds that have novel scaffolds, which can lead to the discovery of new drug classes [84]. Similarity-based screening has been successfully used in several drug discovery projects. For example, it was used to identify new inhibitors of the protein kinase enzyme, which is a promising target for the treatment of cancer. A set of known active compounds was selected as the reference set, and a compound library was screened using a similarity scoring algorithm. Several hits were identified, and their binding affinity was confirmed using molecular docking. One of the hits was found to have potent inhibitory activity against protein kinase and was further optimized to improve its potency and selectivity [85].

6.1.1.2. Pharmacophore-based Screening

This approach is an integrated ligand-based virtual screening approach that identifies potential drug candidates based on their ability to fit into a pharmacophore model. A pharmacophore is a spatial arrangement of functional groups that are necessary for a molecule to bind to a target protein and produce a biological response. The pharmacophore model can be derived from the 3D structure of a known ligand that binds to the target protein or from the active site of the protein itself [86]. The pharmacophore model is used to screen a library of compounds to identify those that match the pharmacophore features. The compounds that match the pharmacophore model are then ranked based on their ability to fit the model and their predicted binding affinity for the target protein. Pharmacophore-based screening has several advantages over other virtual screening approaches. It is a ligand-based approach, which means that it does not require structural information of the target protein. This makes it particularly useful for targets where the 3D structure is unknown or difficult to obtain. Additionally, pharmacophore-based screening can be used to identify compounds that have novel scaffolds, which can lead to the discovery of new drug classes.

The process of pharmacophore-based screening can be divided into four main steps:

6.1.1.2.1. Development of the Pharmacophore Model

The pharmacophore model can be derived from the 3D structure of a known ligand or from the active site of the target protein. The model consists of a set of pharmacophore features, such as hydrogen bond acceptors, hydrogen bond donors, hydrophobic groups, and aromatic rings.

6.1.1.2.2. Preparation of the Compound Library

A diverse library of compounds is selected for screening. The compounds are pre-processed to remove any duplicates and to standardize their structures.

6.1.1.2.3. Screening of the Compound Library

The pharmacophore model is used to screen the compound library. The compounds that match the pharmacophore features are selected as potential hits.

6.1.1.2.4. Validation of the Hits

The potential hits are validated using molecular docking or other experimental methods to confirm their binding affinity and selectivity.

Several software packages are available for pharmacophore-based screening, including Ligand Scout, MOE, and Discovery Studio. These software packages can be used to develop pharmacophore models, screen compound libraries, and analyze the results.

Pharmacophore-based screening has been successfully used in several drug discovery projects [87, 88]. For example, it was used to identify new inhibitors of the histone deacetylase enzyme, which is a promising target for the treatment of cancer. A pharmacophore model was developed based on the structure of a known inhibitor, and a compound library was screened using the model. Several hits were identified, and their binding affinity was confirmed using molecular docking [89]. One of the hits was found to have potent inhibitory activity against histone deacetylase and was further optimized to improve its potency and selectivity.

6.1.1.3. Machine learning-based Screening

It is a computational drug discovery approach that uses artificial intelligence algorithms to predict the activity of potential drug candidates. Machine learning models are trained on a large set of known active and inactive compounds, and the resulting model is used to predict the activity of new compounds [90].

The process of machine learning-based screening can be divided into four main steps:

6.1.1.3.1. Data Preparation

A large set of known active and inactive compounds is assembled. Each compound is characterized by a set of molecular descriptors, which represent its structural and physicochemical properties [91].

6.1.1.3.2. Feature Selection

A subset of the molecular descriptors is selected for use in the machine learning model. This is done to eliminate redundant or irrelevant descriptors that can negatively affect the performance of the model.

6.1.1.3.3. Model Training

A machine learning algorithm is trained on the selected molecular descriptors and the corresponding activity data. The most commonly used algorithms include random forests, support vector machines, and neural networks. The resulting model is optimized to achieve the highest possible accuracy and predictive power.

6.1.1.3.4. Virtual Screening

The trained machine learning model is used to predict the activity of potential drug candidates. A compound library is screened using the model, and the top-ranked compounds are selected for further experimental validation [92].

Several software packages are available for machine learning-based screening, including KNIME, ChemML, and RDKit. These software packages can be used to pre-process the molecular data, train the machine learning models, and analyze the results [93].

Machine learning-based screening has several advantages over other virtual screening approaches. It can analyze large and complex datasets, and it can handle multiple targets simultaneously. Additionally, machine learning-based screening can identify new chemical scaffolds and predict the activity of compounds that have not been previously tested. Machine learning-based screening has been successfully used in several drug discovery projects. For example, it was used to identify potential inhibitors of the SARS-CoV-2 virus that causes COVID-19 [94]. A machine learning model was trained on a set of known antiviral compounds, and the resulting model was used to predict the activity of a large compound library. Several hits were identified, and their antiviral activity was confirmed using cell-based assays [61, 62, 95, 96].

6.1.2. Structure-Based Screening

Structure-based screening is based on the structural information of the target protein. This method involves the use of molecular docking to predict the binding affinity of a compound for the protein. The compounds are docked into the binding site of the protein, and their binding energies are calculated. The compounds with the lowest binding energies are considered potential hits [97]. Several databases are available for virtual screening, like PubChem is a public database of chemical compounds and their biological activities. ZINC is a database of commercially available compounds that can be used for virtual screening. ChemBank is a database of chemical compounds and their biological activities, *etc.*

Virtual Screening has helped in the development of various active molecules and drugs. One such example is the discovery of Falcipain-2 inhibitors. Virtual screening was used to identify potential inhibitors of the falcipain-2 enzyme, which is a promising target for the development of new drugs for malaria. A database of over 2.5 million compounds was screened using a structure-based approach, and 30 hits were identified. The compounds were further tested and two compounds were found to have potent inhibitory activity against falcipain-2 [98]. Another example is the identification of potential inhibitors of the HIV protease enzyme, which is a key target for the development of new HIV drugs using Virtual screening. A database of over 2 million compounds was screened using a ligand-based approach, and 100 hits were identified. The compounds were further tested, and several compounds were found to have potent inhibitory activity against HIV protease [99].

One of the most important advantages of virtual tools in drug design is the ability to reduce the time and cost of drug discovery. By using computational methods to design and screen compounds, researchers can identify potential drug candidates more quickly and efficiently than by using traditional methods. This can help to accelerate the drug discovery process, and to bring new treatments to patients more quickly. Another advantage of virtual tools is the ability to optimize drug properties. By using computational modelling to study the interactions between a drug molecule and its target, researchers can design compounds with improved binding affinity, reduced toxicity, or increased selectivity. This can help to improve the safety and efficacy of drugs, and to reduce the risk of side effects [100 - 102]. Virtual screening is a valuable technique in drug discovery as it can significantly reduce the time and cost of drug development. It allows researchers to quickly and efficiently screen large compound libraries to identify potential drug candidates. Virtual screening can also be used to identify novel chemical scaffolds and optimize existing molecules.

CONCLUSION

The development of new drugs is a long and expensive process involving multiple stages of testing and evaluation. Virtual tools and screening designs have become an essential part of drug discovery and development, as they enable researchers to identify potential drug candidates faster and more efficiently. Virtual screening involves the use of computational methods to analyze large databases of compounds to identify those that are likely to be active against a specific target. There are several types of virtual screening, including pharmacophore-based screening, similarity-based screening, and machine-learning-based screening. These techniques have become increasingly popular due to their ability to reduce the time and cost required to identify potential drug candidates. In addition to virtual screening, molecular modelling tools have been developed to design and optimize drug candidates. For example, molecular docking can be used to simulate the interaction between a drug candidate and its target protein, while molecular dynamics can be used to predict the behaviour of a drug candidate in a biological environment. Quantitative structure-activity relationship (QSAR) models have also been developed to predict the biological activity of compounds based on their chemical structure. Virtual tools and screening designs have revolutionized the process of drug discovery and development. They enable researchers to analyze large datasets quickly and efficiently, increasing the likelihood of identifying promising drug candidates. In addition, they reduce the time and cost required to develop new drugs, making the process more accessible to smaller research groups and pharmaceutical companies. Overall, virtual tools and screening designs are essential for modern drug discovery and development. They have the potential to accelerate the pace of drug discovery, leading to the development of more effective treatments for a wide range of diseases.

REFERENCES

[1]   Zhong WZ, Zhou SF. Molecular science for drug development and biomedicine. Int J Mol Sci 2014; 15(11): 20072-8.
        [http://dx.doi.org/10.3390/ijms151120072] [PMID: 25375190]

[2]   Paul D, Sanap G, Shenoy S, Kalyane D, Kalia K, Tekade RK. Artificial intelligence in drug discovery and development. Drug Discov Today 2021; 26(1): 80-93.
        [http://dx.doi.org/10.1016/j.drudis.2020.10.010] [PMID: 33099022]

[3]   Nag S, *et al.* Deep learning tools for advancing drug discovery and development. 3 Biotech 2022; 12(5).

[4]   Mak KK, Pichika MR. Artificial intelligence in drug development: Present status and future prospects.

Drug Discov Today 2019; 24(3): 773-80.
[http://dx.doi.org/10.1016/j.drudis.2018.11.014] [PMID: 30472429]

[5]   Mohs RC, Greig NH. Drug discovery and development: Role of basic biological research. Alzheimers Dement (N Y) 2017; 3(4): 651-7.
[http://dx.doi.org/10.1016/j.trci.2017.10.005] [PMID: 29255791]

[6]   Deore AB, Dhumane JR, Wagh R, Sonawane R. The stages of drug discovery and development process. Asian Journal of Pharmaceutical Research and Development 2019; 7(6): 62-7.
[http://dx.doi.org/10.22270/ajprd.v7i6.616]

[7]   Katsila T, Spyroulias GA, Patrinos GP, Matsoukas MT. Computational approaches in target identification and drug discovery. Comput Struct Biotechnol J 2016; 14: 177-84.
[http://dx.doi.org/10.1016/j.csbj.2016.04.004] [PMID: 27293534]

[8]   Zhang L, Tan J, Han D, Zhu H. From machine learning to deep learning: Progress in machine intelligence for rational drug discovery. Drug Discov Today 2017; 22(11): 1680-5.
[http://dx.doi.org/10.1016/j.drudis.2017.08.010] [PMID: 28881183]

[9]   D. Segall M. Multi-parameter optimization: Identifying high quality compounds with a balance of properties. Current pharmaceutical design. 2012 Mar 1; 18(9): 1292-310.

[10]   Adamson,G.W., and Bawden, D.A. Substructural analysis methods for structure-activity correlation of heterocyclic compounds using Wiswesser linenotation. J. Chem. Inf. Comput. Sci. 1977; 17: 164-171.

[11]   Zhang R, Zhao J, Yang Y, Lu Z, Shi W. Understanding electronic and optical properties of La and Mn co-doped anatase TiO_2. Computational Condensed Matter 2016; 6: 5-17.
[http://dx.doi.org/10.1016/j.cocom.2016.03.001]

[12]   Firth NC, Atrash B, Brown N, Blagg J. MOARF, An integrated workflow for multiobjective optimization: Implementation, synthesis, and biological evaluation. J Chem Inf Model 2015; 55(6): 1169-80.
[http://dx.doi.org/10.1021/acs.jcim.5b00073] [PMID: 26054755]

[13]   Franke R, Huebel S, Streich WJ. Substructural QSAR approaches and topological pharmacophores. Environ Health Perspect 1985; 61: 239-55.
[http://dx.doi.org/10.1289/ehp.8561239] [PMID: 3905376]

[14]   Hemmateenejad B, Mehdipour AR, Popelier PLA. Quantum topological QSAR models based on the MOLMAP approach. Chem Biol Drug Des 2008; 72(6): 551-63.
[http://dx.doi.org/10.1111/j.1747-0285.2008.00731.x] [PMID: 19090922]

[15]   Gozalbes R, Doucet J, Derouin F. Application of topological descriptors in QSAR and drug design: history and new trends. Curr Drug Targets Infect Disord 2002; 2(1): 93-102.
[http://dx.doi.org/10.2174/1568005024605909] [PMID: 12462157]

[16]   Palyulin VA, Radchenko EV, Zefirov NS. Molecular field Topology analysis method in QSAR studies of organic compounds. J Chem Inf Comput Sci 2000; 40(3): 659-67.
[http://dx.doi.org/10.1021/ci980114i] [PMID: 10850771]

[17]   Tropsha A. Predictive quantitative structure-activity relationship modeling. Compr Med Chem II 2006; 4: 149-65.

[18]   Hansch C. The physicochemical approach to drug design and discovery (QSAR). Drug Dev Res 1981; 1(4): 267-309.
[http://dx.doi.org/10.1002/ddr.430010403]

[19]   De Benedetti PG, Fanelli F. Multiscale quantum chemical approaches to QSAR modeling and drug design. Drug Discov Today 2014; 19(12): 1921-7.
[http://dx.doi.org/10.1016/j.drudis.2014.09.024] [PMID: 25281852]

[20]   Oluwaseye A, Uzairu A, Shallangwa GA, Abechi SE. Quantum chemical descriptors in the QSAR studies of compounds active in maxima electroshock seizure test. J King Saud Univ Sci 2020; 32(1):

75-83.
[http://dx.doi.org/10.1016/j.jksus.2018.02.009]

[21]   Sizochenko N, Majumdar D, Roszak S, Leszczynski J. Application of quantum mechanics and molecular mechanics in chemoinformatics. Handb Comput Chem 2017; 2041-63.
[http://dx.doi.org/10.1007/978-3-319-27282-5_52]

[22]   Braga RC, Andrade CH. QSAR and QM/MM approaches applied to drug metabolism prediction. Mini Rev Med Chem 2012; 12(6): 573-82.
[http://dx.doi.org/10.2174/138955712800493807] [PMID: 22587770]

[23]   Haghshenas H, Kaviani B, Firouzeh M, Tavakol H. Developing a variation of 3D-QSAR/MD method in drug design. J Comput Chem 2021; 42(13): 917-29.
[http://dx.doi.org/10.1002/jcc.26514] [PMID: 33719136]

[24]   Doweyko AM. Three-dimensional quantitative structure-activity relationship: The state of the art. Compr Med Chem II 2006; 4: 575-95.

[25]   Cruciani G, Carosati E, Clementi S. Three-dimensional quantitative structure-property relationships. Pract Med Chem Second Ed 2003; 405-16.
[http://dx.doi.org/10.1016/B978-012744481-9/50029-5]

[26]   Silakari O, Singh PK. QSAR: Descriptor calculations, model generation, validation and their application. Concepts Exp Protoc Model Informatics Drug Des 2021; pp. 29-63.

[27]   Roy K, Das R. A review on principles, theory and practices of 2D-QSAR. Curr Drug Metab 2014; 15(4): 346-79.
[http://dx.doi.org/10.2174/1389200215666140908102230] [PMID: 25204823]

[28]   Lewis RA, Wood D. Modern 2D QSAR for drug discovery. Wiley Interdiscip Rev Comput Mol Sci 2014; 4(6): 505-22.
[http://dx.doi.org/10.1002/wcms.1187]

[29]   Neves BJ, Braga RC, Melo-Filho CC, Moreira-Filho JT, Muratov EN, Andrade CH. QSAR-based virtual screening: Advances and applications in drug discovery. Front Pharmacol 2018; 9: 1275.
[http://dx.doi.org/10.3389/fphar.2018.01275] [PMID: 30524275]

[30]   Ugbe FA, Shallangwa GA, Uzairu A, Abdulkadir I. A combined 2-D and 3-D QSAR modeling, molecular docking study, design, and pharmacokinetic profiling of some arylimidamide-azole hybrids as superior *L. donovani* inhibitors. Bull Natl Res Cent 2022; 46(1): 189.
[http://dx.doi.org/10.1186/s42269-022-00874-1]

[31]   Kwon S, Bae H, Jo J, Yoon S. Comprehensive ensemble in QSAR prediction for drug discovery. BMC Bioinformatics 2019; 20(1): 521.
[http://dx.doi.org/10.1186/s12859-019-3135-4] [PMID: 31655545]

[32]   Verma J, Khedkar V, Coutinho E. 3D-QSAR in drug design: A review. Curr Top Med Chem 2010; 10(1): 95-115.
[http://dx.doi.org/10.2174/156802610790232260] [PMID: 19929826]

[33]   Akamatsu M. Current state and perspectives of 3D-QSAR. Curr Top Med Chem 2002; 2(12): 1381-94.
[http://dx.doi.org/10.2174/1568026023392887] [PMID: 12470286]

[34]   Klebe G, Abraham U, Mietzner T. Molecular similarity indices in a comparative analysis (CoMSIA) of drug molecules to correlate and predict their biological activity. J Med Chem 1994; 37(24): 4130-46.
[http://dx.doi.org/10.1021/jm00050a010] [PMID: 7990113]

[35]   Klebe G, Abraham U. Comparative molecular similarity index analysis (CoMSIA) to study hydrogen-bonding properties and to score combinatorial libraries. J Comput Aided Mol Des 1999; 13(1): 1-10.
[http://dx.doi.org/10.1023/A:1008047919606] [PMID: 10087495]

[36]   Roy K, Kar S. Introduction to 3D-QSAR. Underst Basics QSAR Appl Pharm. Sci Risk Assess 2015;

291-317.

[37]   Tsakovska I, Pajeva I, Alov P, Worth A. Recent advances in the molecular modeling of estrogen receptor-mediated toxicity. Adv Protein Chem Struct Biol 2011; 85: 217-51.
[http://dx.doi.org/10.1016/B978-0-12-386485-7.00006-5] [PMID: 21920325]

[38]   Ul-Haq Z, Wadood A, Uddin R. CoMFA and CoMSIA 3D-QSAR analysis on hydroxamic acid derivatives as urease inhibitors. J Enzyme Inhib Med Chem 2009; 24(1): 272-8.
[http://dx.doi.org/10.1080/14756360802166665] [PMID: 18608766]

[39]   Sharma R, Dhingra N, Patil S. CoMFA, CoMSIA, HQSAR and molecular docking analysis of ionone-based chalcone derivatives as antiprostate cancer activity. Indian J Pharm Sci 2016; 78(1): 54-64.
[http://dx.doi.org/10.4103/0250-474X.180251] [PMID: 27168682]

[40]   Doytchinova IA, Flower DR. Toward the quantitative prediction of T-cell epitopes: CoMFA and coMSIA studies of peptides with affinity for the class I MHC molecule HLA-A*0201. J Med Chem 2001; 44(22): 3572-81.
[http://dx.doi.org/10.1021/jm010021j] [PMID: 11606121]

[41]   Xu Y, He Z, Liu H, *et al.* 3D-QSAR, molecular docking, and molecular dynamics simulation study of thieno[3,2- *b*]pyrrole-5-carboxamide derivatives as LSD1 inhibitors. RSC Advances 2020; 10(12): 6927-43.
[http://dx.doi.org/10.1039/C9RA10085G] [PMID: 35493862]

[42]   Ashraf S, Ranaghan KE, Woods CJ, Mulholland AJ, Ul-Haq Z. Exploration of the structural requirements of Aurora Kinase B inhibitors by a combined QSAR, modelling and molecular simulation approach. Sci Rep 2021; 11(1): 18707.
[http://dx.doi.org/10.1038/s41598-021-97368-3] [PMID: 34548506]

[43]   Araújo PHF, Ramos RS, da Cruz JN, *et al.* Identification of potential COX-2 inhibitors for the treatment of inflammatory diseases using molecular modeling approaches. Molecules 2020; 25(18): 4183. Polanski J.
[http://dx.doi.org/10.3390/molecules25184183] [PMID: 32932669]

[44]   Chemoinformatics PJ. Compr Chemom 2009; 4: 459-506.

[45]   Elrayess R, Abdel Aziz YM, Elgawish MS, Elewa M, Elshihawy HA, Said MM. Pharmacophore modeling, 3D-QSAR, synthesis, and anti-lung cancer evaluation of novel thieno[2,3- *d*][1,2,3]triazines targeting EGFR. Arch Pharm (Weinheim) 2020; 353(2): 1900108.
[http://dx.doi.org/10.1002/ardp.201900108] [PMID: 31894866]

[46]   Wu X, Li M, Qu Y, *et al.* Design and synthesis of novel Gefitinib analogues with improved anti-tumor activity. Bioorg Med Chem 2010; 18(11): 3812-22.
[http://dx.doi.org/10.1016/j.bmc.2010.04.046] [PMID: 20466555]

[47]   Gonçalves RB, Ferraz WR, Calil RL, Scotti MT, Trossini GHG. Convergent QSAR Models for the Prediction of Cruzain Inhibitors. ACS Omega. 2023 Oct 13; 8(42): 38961-38982.
[http://dx.doi.org/10.1021/acsomega.3c03376] [PMID: 37901514] [PMCID: PMC10601054]

[48]   Ugbe FA, *et al.* Computational design, molecular properties, ADME, and toxicological analysis of substituted 2,6-diarylidene cyclohexanone analogs as potent pyridoxal kinase inhibitors. Silico Pharmacol 2023; 11(1).

[49]   Garro Martinez JC, Vega-Hissi EG, Andrada MF, Estrada MR. QSAR and 3D-QSAR studies applied to compounds with anticonvulsant activity. Expert Opin Drug Discov 2015; 10(1): 37-51.
[http://dx.doi.org/10.1517/17460441.2015.968123] [PMID: 25297377]

[50]   Yang GF, Huang X. Development of quantitative structure-activity relationships and its application in rational drug design. Curr Pharm Des 2006; 12(35): 4601-11.
[http://dx.doi.org/10.2174/138161206779010431] [PMID: 17168765]

[51]   Gu Y, Li M. Molecular modeling. Handb. Benzoxazine Resins 2011; pp. 103-10.

[52]   Barbosa NSV, Lima ERA, Tavares FW. Molecular Modeling in Chemical Engineering. Ref Modul Chem Mol Sci Chem Eng. 2017.
[http://dx.doi.org/10.1016/B978-0-12-409547-2.13915-0]

[53]   Rodriguez R, Chinea G, Lopez N, Pons T, Vriend G. Homology modeling, model and software evaluation: Three related resources. Bioinformatics 1998; 14(6): 523-8.
[http://dx.doi.org/10.1093/bioinformatics/14.6.523] [PMID: 9694991]

[54]   Krieger E, Nabuurs SB, Vriend G. Homology Modeling. Struct Bioinforma 2005; pp. 509-23.

[55]   Greener JG, Kandathil SM, Jones DT. Deep learning extends *de novo* protein modelling coverage of genomes using iteratively predicted structural constraints. Nat Commun 2019; 10(1): 3977.
[http://dx.doi.org/10.1038/s41467-019-11994-0] [PMID: 31484923]

[56]   Kubota K, Funabashi M, Ogura Y. Target deconvolution from phenotype-based drug discovery by using chemical proteomics approaches. Biochim Biophys Acta Proteins Proteomics 2019; 1867(1): 22-7.
[http://dx.doi.org/10.1016/j.bbapap.2018.08.002] [PMID: 30392561]

[57]   Chan JNY, Nislow C, Emili A. Recent advances and method development for drug target identification. Trends Pharmacol Sci 2010; 31(2): 82-8.
[http://dx.doi.org/10.1016/j.tips.2009.11.002] [PMID: 20004028]

[58]   Jenkins JL, Bender A, Davies JW. *In silico* target fishing: Predicting biological targets from chemical structure. Drug Discov Today Technol 2006; 3(4): 413-21.
[http://dx.doi.org/10.1016/j.ddtec.2006.12.008]

[59]   Bajorath J. Computational scaffold hopping: Cornerstone for the future of drug design? Future Med Chem 2017; 9(7): 629-31.
[http://dx.doi.org/10.4155/fmc-2017-0043] [PMID: 28485634]

[60]   Sun H, Tawa G, Wallqvist A. Classification of scaffold-hopping approaches. Drug Discov Today 2012; 17(7-8): 310-24.
[http://dx.doi.org/10.1016/j.drudis.2011.10.024] [PMID: 22056715]

[61]   Lavecchia A. Machine-learning approaches in drug discovery: Methods and applications. Drug Discov Today 2015; 20(3): 318-31.
[http://dx.doi.org/10.1016/j.drudis.2014.10.012] [PMID: 25448759]

[62]   Kennedy T. Managing the drug discovery/development interface. Drug Discov Today 1997; 2(10): 436-44.
[http://dx.doi.org/10.1016/S1359-6446(97)01099-4]

[63]   Venkatesh S, Lipper RA. Role of the development scientist in compound lead selection and optimization. J Pharm Sci 2000; 89(2): 145-54.
[http://dx.doi.org/10.1002/(SICI)1520-6017(200002)89:2<145:AID-JPS2>3.0.CO;2-6]   [PMID: 10688744]

[64]   Weng G. Exploring protein-protein interactions by peptide docking protocols. Methods Enzymol 2002; 344: 577-86.
[http://dx.doi.org/10.1016/S0076-6879(02)44741-6] [PMID: 11771411]

[65]   Altschul SF, Gish W, Miller W, Myers EW, Lipman DJ. Basic local alignment search tool. J Mol Biol 1990; 215(3): 403-10.
[http://dx.doi.org/10.1016/S0022-2836(05)80360-2] [PMID: 2231712]

[66]   Liu J, Lei X, Zhang Y, Pan Y. The prediction of molecular toxicity based on BiGRU and GraphSAGE. Comput Biol Med 2023; 153: 106524.
[http://dx.doi.org/10.1016/j.compbiomed.2022.106524] [PMID: 36623439]

[67]   Zang Q, Mansouri K, Williams AJ, *et al. In silico* prediction of physicochemical properties of environmental chemicals using molecular fingerprints and machine learning. J Chem Inf Model 2017;

57(1): 36-49.
[http://dx.doi.org/10.1021/acs.jcim.6b00625] [PMID: 28006899]

[68] Roy A, Kucukural A, Zhang Y. I-TASSER: A unified platform for automated protein structure and function prediction. Nat Protoc 2010; 5(4): 725-38.
[http://dx.doi.org/10.1038/nprot.2010.5] [PMID: 20360767]

[69] Waterhouse A, Bertoni M, Bienert S, *et al.* SWISS-MODEL: Homology modelling of protein structures and complexes. Nucleic Acids Res 2018; 46(W1): W296-303.
[http://dx.doi.org/10.1093/nar/gky427] [PMID: 29788355]

[70] Yuan X, Shao Y, Bystroff C. Ab initio protein structure prediction using pathway models. Comp Funct Genomics 2003; 4(4): 397-401.
[http://dx.doi.org/10.1002/cfg.305] [PMID: 18629080]

[71] Aminpour M, Montemagno C, Tuszynski JA. An overview of molecular modeling for drug discovery with specific illustrative examples of applications. Molecules 2019; 24(9): 1693.
[http://dx.doi.org/10.3390/molecules24091693] [PMID: 31052253]

[72] Zhou S, Ge S, Zhang W, *et al.* Conventional molecular dynamics and metadynamics simulation studies of the binding and unbinding mechanism of TTR stabilizers AG10 and tafamidis. ACS Chem Neurosci 2020; 11(19): 3025-35.
[http://dx.doi.org/10.1021/acschemneuro.0c00338] [PMID: 32915538]

[73] Clarke DF, Mirochnick M, Acosta EP, *et al.* Use of modeling and simulations to determine raltegravir dosing in neonates: A model for safely and efficiently determining appropriate neonatal dosing regimens: Impaact P1110. J Acquir Immune Defic Syndr 2019; 82(4): 392-8.
[http://dx.doi.org/10.1097/QAI.0000000000002149] [PMID: 31658182]

[74] Morris GM, Lim-Wilby M. Molecular docking. Methods Mol Biol 2008; 443: 365-82.
[http://dx.doi.org/10.1007/978-1-59745-177-2_19] [PMID: 18446297]

[75] Stanzione F, Giangreco I, Cole JC. Use of molecular docking computational tools in drug discovery. Prog Med Chem 2021; 60: 273-343.
[http://dx.doi.org/10.1016/bs.pmch.2021.01.004] [PMID: 34147204]

[76] Berry M, Fielding B, Gamieldien J. Practical considerations in virtual screening and molecular docking. emerg trends comput biol bioinformatics. Syst Biol Algorithms Softw Tools 2015; 487-502.

[77] Tiwari A, Singh S. Computational approaches in drug designing. Bioinforma Methods Appl 2021; 207-17.

[78] Lamb ML, Jorgensen WL. Computational approaches to molecular recognition. Curr Opin Chem Biol 1997; 1(4): 449-57.
[http://dx.doi.org/10.1016/S1367-5931(97)80038-5] [PMID: 9667895]

[79] Dar AM, Mir S. Molecular docking: Approaches, types, applications and basic challenges. J Anal Bioanal Tech 2017; 8(2): 1-3.
[http://dx.doi.org/10.4172/2155-9872.1000356]

[80] Shoichet BK, McGovern SL, Wei B, Irwin JJ. Lead discovery using molecular docking. Curr Opin Chem Biol 2002; 6(4): 439-46.
[http://dx.doi.org/10.1016/S1367-5931(02)00339-3] [PMID: 12133718]

[81] Cho AE, Rinaldo D. Extension of QM/MM docking and its applications to metalloproteins. J Comput Chem 2009; 30(16): 2609-16.
[http://dx.doi.org/10.1002/jcc.21270] [PMID: 19373896]

[82] Chan HCS, Shan H, Dahoun T, Vogel H, Yuan S. Advancing drug discovery *via* artificial intelligence. Trends pharmacol sci 2019; 40(8): 592-604.
[http://dx.doi.org/10.1016/j.tips.2019.06.004] [PMID: 31320117]

[83] Hamza A, Wei NN, Zhan CG. Ligand-based virtual screening approach using a new scoring function.

J Chem Inf Model 2012; 52(4): 963-74.
[http://dx.doi.org/10.1021/ci200617d] [PMID: 22486340]

[84]   Kristensen TG, Nielsen J, Pedersen CNS. Methods for Similarity-based Virtual Screening. Comput Struct Biotechnol J 2013; 5(6): e201302009.
[http://dx.doi.org/10.5936/csbj.201302009] [PMID: 24688702]

[85]   Bajorath J. Machine learning and similarity-based virtual screening techniques. Silico Drug Discov Des 2013; 134-46.
[http://dx.doi.org/10.4155/ebo.12.419]

[86]   Seidel T, Ibis G, Bendix F, Wolber G. Strategies for 3D pharmacophore-based virtual screening. Drug Discov Today Technol 2010; 7(4): e221-8.
[http://dx.doi.org/10.1016/j.ddtec.2010.11.004] [PMID: 24103798]

[87]   Sanachai K, Mahalapbutr P, Hengphasatporn K, *et al.* Pharmacophore-based virtual screening and experimental validation of pyrazolone-derived inhibitors toward janus kinases. ACS Omega 2022; 7(37): 33548-59.
[http://dx.doi.org/10.1021/acsomega.2c04535] [PMID: 36157769]

[88]   Liu C, Yin J, Yao J, Xu Z, Tao Y, Zhang H. Pharmacophore-based virtual screening toward the discovery of novel anti-echinococcal compounds. Front Cell Infect Microbiol 2020; 10: 118.
[http://dx.doi.org/10.3389/fcimb.2020.00118] [PMID: 32266168]

[89]   Kumar BK, Faheem , Sekhar KVGC, *et al.* Pharmacophore based virtual screening, molecular docking, molecular dynamics and MM-GBSA approach for identification of prospective SARS-CoV-2 inhibitor from natural product databases. J Biomol Struct Dyn 2022; 40(3): 1363-86.
[http://dx.doi.org/10.1080/07391102.2020.1824814] [PMID: 32981461]

[90]   Carpenter KA, Cohen DS, Jarrell JT, Huang X. Deep learning and virtual drug screening. Future Med Chem 2018; 10(21): 2557-67.
[http://dx.doi.org/10.4155/fmc-2018-0314] [PMID: 30288997]

[91]   Hnatyshyn S, Thayasivam U, Hnatyshin V, White C. Machine learning algorithms for metabolomics applications. Identif Data Process Methods Metabolomics 2015; 96-110.
[http://dx.doi.org/10.4155/fseb2013.14.163]

[92]   Carpenter KA, Huang X. Machine learning-based virtual screening and its applications to alzheimer's drug discovery: A review. Curr Pharm Des 2018; 24(28): 3347-58.
[http://dx.doi.org/10.2174/1381612824666180607124038] [PMID: 29879881]

[93]   Carracedo-Reboredo P, Liñares-Blanco J, Rodríguez-Fernández N, *et al.* A review on machine learning approaches and trends in drug discovery. Comput Struct Biotechnol J 2021; 19: 4538-58.
[http://dx.doi.org/10.1016/j.csbj.2021.08.011] [PMID: 34471498]

[94]   Sliwoski G, Lowe EW. Computational fragment-based drug design. Silico Drug Discov Des 2013; 22-32.
[http://dx.doi.org/10.4155/ebo.13.335]

[95]   Lounkine E, Keiser MJ, Whitebread S, *et al.* Large-scale prediction and testing of drug activity on side-effect targets. Nature 2012; 486(7403): 361-7.
[http://dx.doi.org/10.1038/nature11159] [PMID: 22722194]

[96]   Xiao X, Min JL, Lin WZ, Liu Z, Cheng X, Chou KC. iDrug-Target: Predicting the interactions between drug compounds and target proteins in cellular networking *via* benchmark dataset optimization approach. J Biomol Struct Dyn 2015; 33(10): 2221-33.
[http://dx.doi.org/10.1080/07391102.2014.998710] [PMID: 25513722]

[97]   Maia EHB, Assis LC, de Oliveira TA, da Silva AM, Taranto AG. Structure-based virtual screening: From classical to artificial intelligence. Front Chem 2020; 8: 343.
[http://dx.doi.org/10.3389/fchem.2020.00343] [PMID: 32411671]

[98]   Alberca LN, Chuguransky SR, Álvarez CL, Talevi A, Salas-Sarduy E. *In silico* guided drug

repurposing: Discovery of new competitive and non-competitive inhibitors of falcipain-2. Front Chem 2019; 7(Aug): 534.
[http://dx.doi.org/10.3389/fchem.2019.00534] [PMID: 31448257]

[99]   Okafor SN, Angsantikul P, Ahmed H. Discovery of novel HIV protease inhibitors using modern computational techniques. Int J Mol Sci 2022; 23(20): 12149.
[http://dx.doi.org/10.3390/ijms232012149] [PMID: 36293006]

[100]   Lavecchia A, Giovanni C. Virtual screening strategies in drug discovery: A critical review. Curr Med Chem 2013; 20(23): 2839-60.
[http://dx.doi.org/10.2174/09298673113209990001] [PMID: 23651302]

[101]   Wu KJ, Lei PM, Liu H, Wu C, Leung CH, Ma DL. Mimicking strategy for protein-protein interaction inhibitor discovery by virtual screening. Molecules 2019; 24(24): 4428.
[http://dx.doi.org/10.3390/molecules24244428] [PMID: 31817099]

[102]   Wermuth CG, *et al.* Strategies in the search for new lead compounds or original working hypotheses. Pract Med Chem Fourth Ed 2015; 73-99.
[http://dx.doi.org/10.1016/B978-0-12-417205-0.00004-3]

Predicting Drug Properties: Computational Strategies for Solubility and Permeability Rates

Anshita Gupta Soni[1,*], **Renjil Joshi**[1], **Deependra Soni**[2], **Chanchal Deep Kaur**[3], **Swarnlata Saraf**[4] and **Pankaj Kumar Singh**[5]

[1] *Shri Rawatpura Sarkar Institute of Pharmacy, Kumhari, Durg, Chhattisgarh, India*

[2] *Faculty of Pharmacy, MATS University Campus, Aarang, Raipur, Chhattisgarh, India*

[3] *Rungta Institute of Pharmaceutical Sciences and Research, Raipur, Chhattisgarh, India*

[4] *University Institute of Pharmacy, Pt.Ravishankar Shukla University, Raipur, Chhattisgarh, India*

[5] *Department of Pharmaceutics, National Institute of Pharmaceutical Education and Research NIPER), Hyderabad, Telangana-500037, India*

Abstract: The oral bioavailability of a medicine can be considerably influenced by its water solubility, which can also have an impact on how the drug is dispersed through the body. To decrease the likelihood of failures in the late phases of drug development, aqueous solubility must be taken into account early in the drug research and development process. By using computer models to predict solubility, combinatorial libraries might be screened to identify potentially problematic chemicals and exclude those with insufficient solubility. In addition to predicting solubility from chemical structure, the explanation of such models can provide insight into correlations between structure and solubility and can direct structural improvement to improve solubility while preserving the effectiveness of the medications under study. Such model development is a difficult procedure that calls for taking into account a wide range of variables that may affect how well the model performs in the end. In this article, various solubility modeling techniques are presented. Despite many studies on model creation, predicting the solubility of various medications remains difficult. One of the primary reasons for the poor trustworthiness of many of the suggested models is the quality of the experimental data that may be used to simulate solubility, which is becoming more widely acknowledged. Consequently, increased availability of trustworthy data produced using the same experimental technique is necessary to fully realize the potential of the established modeling tools.

Keywords: Computational tools, Caco-2, Ligand-based computer-aided drug discovery, PAMPA, Permeability, Solubility.

---

* **Corresponding author Anshita Gupta Soni:** Shri Rawatpura Sarkar Institute of Pharmacy, Kumhari, Durg, Chhattisgarh, India; E-mail: anshita1912@gmail.com

Dilpreet Singh and Prashant Tiwari (Eds.)

1. INTRODUCTION

The pharmaceutical research pipeline, from drug discovery through production, uses the solubility of therapeutic agents in a specific solvent system as a crucial characteristic [1]. Reliable solubility forecasting is crucial in this industry for guiding experimental work, accelerating time to market, and lowering material costs [1]. Furthermore, it is essential for predictive models to encompass a wide range of drug/drug-like solutes and organic solvents, as various solvents are utilized throughout different stages and operations within unit activities. In this context, we propose the utilization of a combined, cross-solvent structure for solubility prediction, grounded in data. API Absorption is altered by various factors, including aqueous drug solubility and intestinal drug permeability [2]. Computational tools have been developed to predict these factors, aiding drug discovery and formulation. Among these tools, one model stands out for its high predictive power using only chemical compositions as features, achieving an accuracy of around 92%. *In vitro*, methods such as Parallel Artificial Membrane Permeability Assay, Caco-2, and rat intestinal canals are also used to screen compounds for permeability and absorption [3]. These methods are valuable research tools for studying solute-membrane interactions and cell culture techniques for nasal drug permeability [4].

Important parameters for medication absorption include drug solubility and permeability, and a variety of computational methods have been established to forecast these [5]. The "ligand-based computer-aided drug discovery" (LB-CADD) method is one model that stands out for its excellent predictive potential. This method includes analyzing ligands that interact with a target of interest without needing to know the target's structure. LB-CADD methods use reference structures to represent compounds with physicochemical properties relevant to the desired interactions [6]. In addition, *in vitro* methods such as "Parallel Artificial Membrane Permeability Assay", Caco-2, and rat intestinal canals are used to study solute-membrane interactions, whereas cell culture techniques are used for nasal drug permeability [7]. Moreover, molecular surface areas have been referred to as permeability and solubility descriptors. These tools have the potential to aid drug discovery and formulation by providing precise medication permeability and solubility predictions. However, it is important to note that computational tools for solubility-permeability prediction have their limitations. They rely on the availability of accurate and diverse training datasets, as well as the quality of input data and descriptors used for modeling. The performance of these tools can vary depending on the chemical space and applicability domain of the models.

Hence, computational tools for solubility-permeability prediction have emerged as valuable resources in the drug discovery process. They offer rapid and cost-effective means of assessing the solubility and permeability characteristics of compounds, aiding in the selection and optimization of drug candidates. With ongoing advancements in computational techniques and the accumulation of experimental data, these tools will continue to evolve and contribute to the development of safe and effective drugs.

2. COMPUTATIONAL MODEL FOR PREDICTING PERMEABILITY AND SOLUBILITY OF DRUG

For the *in silico* prediction of drug membrane permeability, the "immobilized artificial membrane" (IAM) and "immobilized liposome chromatography" (ILC) procedures are frequently utilized [8]. By examining ligands that interact with a target of interest, the ligand-based computer-aided drug discovery (LB-CADD) strategy may also be used to forecast solubility [9]. However, it is important to note that LB-CADD neglects the dynamic nature of the ligands, which can be considered in alternative approaches [10]. *In silico* screening, "hit-to-lead and lead-to-drug optimization, and DMPK/ADMET" property optimization can all benefit from these strategies [11].

2.1. Computational Model for Predicting Permeability of Drug

To reach their objective, most medications must cross at least one cellular membrane. Low membrane permeability commonly leads to mediocre or nonexistent *in vivo* effectiveness, even though the potency of a drug depends on how strongly it binds to its target. Therefore, a thorough understanding of how a particular species is divided in the membrane is essential from the perspectives of pharmacokinetics and logical drug design. In eukaryotic systems, a molecule can travel through a membrane either actively or passively. A transport protein moves a membrane-crossing molecule by active transport, which utilizes energy (such as ATP hydrolysis). In contrast, passive transport includes a molecule diffusing through the membrane without the need for outside aid or energy input. This is the most prevalent method of medication transportation through membranes. The rate at which a compound passively diffuses across a membrane depends on several factors, including the partition coefficient, diffusion coefficient, and the concentration that traverses the membrane [12]. The processes of membrane binding and diffusion are contingent upon the chemical properties of small molecules, such as lipophilicity, molecular weight, and measures of molecular polarity [13]. Consequently, creating effective medications necessitates achieving a delicate equilibrium among all these characteristics within a molecular scaffold, which is a formidable task.

A pivotal parameter in drug design is the drug's permeation across membranes. A drug intended for intracellular targets but with poor permeability would exhibit subpar performance. Numerous *in-vitro* and *in-silico* permeability prediction models have been developed. Among the most popular and relatively straightforward *in-vitro* techniques are the "parallel artificial membrane permeability assay (PAMPA)," the "immobilized artificial membrane (IAM)" technique, and "immobilized liposome chromatography." These methods offer high throughput and play a crucial role in drug discovery. The PAMPA *in-vitro* method was initially introduced by Kansy *et al.* in 1998 [14]. Subsequently, it has been adapted for predicting blood-brain barrier (BBB) permeability, even though its original application was for rapid forecasting of passive permeability across the gastrointestinal tract [15]. It has demonstrated its reliability as a predictor of BBB penetration.

Quantitative structure permeability relationship (QSPR) models are commonly applicable to predict the permeability of the drug [16]. These models take into consideration the lipophilicity, molecular size, and polarity of molecules to predict their permeability across biological membranes [17]. The accuracy of these models depends on their domain of applicability and the availability of reliable experimental data [18]. In addition to QSPR, other techniques such as "immobilized artificial membrane (IAM) and immobilized liposome chromatography" (ILC) are also predicting drug membrane permeability [19]. The Caco-2 model is another method for predicting solubility and permeability but has some inconsistencies [20]. There have been suggestions made to increase the precision of these models [21]. Furthermore, a new method based on molecular dynamics simulations has been developed to determine a number that closely coincides with the Caco-2 cell line assay's measurement of chemical permeability [22]. These techniques can be useful for *in silico* screening, "hit-to-lead and lead-to-drug optimization, and DMPK/ADMET" property optimization in drug discovery and development [11].

2.2. Parallel Artificial Membrane Permeability Assay (PAMPA)

PAMPA was utilized to assess compounds for passive diffusion in this study, employing the mildest precoated PAMPA plate technology. Essentially, this system involves a 96-well plate with two compartments, each filled with fluid—one acting as the donor well and the other as the acceptor well. A DOPC phospholipid-coated polyvinylidene fluoride filter plate separates these compartments. Substances were dissolved in Hanks Balanced Salt Solution at a concentration of 100 μM [23]. A volume of 0.3 mL of the solution was added to the donor wells, while the acceptor well was positioned over the donor well and filled with 0.2 mL of blank Hanks Balanced Salt Solution. Incubation at 25±0.5°C for 5 hours followed. Upon completion of the incubation period, solutions from both wells were extracted. Compounds were then reserved for UPLC analysis.

Some chemicals exhibited extremely minimal traversal to the acceptor well, rendering the identification of the substances through UPLC challenging. For such compounds, extraction into methanol was performed, followed by vacuum concentration and resuspension in a minimal suitable quantity. Compounds present in the solution were subsequently quantified using the UPLC apparatus [24].

2.3. Immobilized Artificial Membrane (IAM) Method

Widespread usage of the "immobilized artificial membrane" (IAM) technology in drug discovery and development to simulate biological cell membranes and predict drug membrane permeability [23]. IAM chromatography is an ambivalent tool that can be used for multiple integrations in the pharmaceutical field because it is on the dividing line between passive diffusion and binding [24]. The IAM column is very helpful for high throughput drug membrane permeability prediction [25]. The results can help researchers determine how well a drug will be absorbed by the body and how it will interact with cell membranes.

An excellent tool for predicting drug membrane permeability at high throughput is the Immobilized Artificial Membrane (IAM) Column. The outcomes from this column are quicker to get and less expensive than those from more established *in-vitro* techniques like intestinal tissue and Caco-2 Cells. The most prevalent phospholipid found in cell membranes is phosphatidylcholine (PC) [26]. IAM chromatography phases closely match the surface of a genuine cell membrane when created from PC analogues. IAM phases are therefore advantageous for the separation of membrane proteins and the study of drug-membrane interactions because they have a high affinity for membrane proteins [27].

The membrane-forming phospholipids are covalently bonded to silica to create the IAM surface. The IAM column produces findings that are quicker to get and more closely resemble those obtained using more conventional *in vitro* techniques, such as intestinal tissue and Caco-2 cells. Some phases, like ODS silica, for instance, only retain analytes by their hydrophobicity [27]. IAM more closely resembles how analytes interact with biological membranes, which are capable of a variety of interactions including hydrophobic, ion pairing, and hydrogen bonding. Phospholipophilicity is a group of interactions that the IAM column can measure. These developments have resulted in the creation of several additional IAM phases, including the "IAM.PC.DD2 and the IAM Fast-screen Mini Column", which are used to forecast drug membrane permeability.

2.4. Immobilized Liposome Chromatography (ILC) Technique

Drug research and development have embraced Immobilized Liposome Chromatography (ILC) technology for conducting permeability assessments [28]. This method revolves around utilizing a stationary phase comprising permeable silica microspheres where liposomes are immobilized [29]. These liposomes serve as models of biological cell membranes, and the ILC technique is employed to forecast drug membrane permeability [30]. The ILC column proves to be valuable in gauging the extent of a drug's absorption within the body and its interactions with cell membranes [31].

2.5. Caco-2 Model for Predicting Permeability

The Caco-2 design is commonly used in the drug discovery industry for predicting drug permeability [32]. However, there have been inconsistencies in using Caco-2 measurements to predict human permeability, leading to inaccuracies in the screening process. There have been suggested guidelines and recommendations to increase the precision of these models [33]. For the purpose of forecasting Caco-2 cell permeability, a trustworthy "Quantitative Structure-Property Relationship" (QSPR) model based on molecular size, polarity, and lipophilicity of molecules has been created [20]. Molecule permeability has been reported to be influenced by molecular weight and flexibility, with smaller molecules and those with fewer rotatable bonds exhibiting higher permeability [34]. A novel method based on molecular dynamics simulations has also been developed to find a value that closely matches chemical permeability as discovered by the Caco-2 cell line experiment [22]; this value is based on the Caco-2 cell line experiment. These techniques can be useful for *in silico* screening, "hit-to-lead and lead-to-drug optimisation," and "DMPK/ADMET" property optimisation during the drug discovery and development process.

3. SOLUBILITY PREDICTION MODEL

Many scientists are still interested in the prediction of solubility. In the review index, several prediction models have been published. The intrinsic solubility prediction models are categorized into a wide range of subcategories, including statistically determined models based on "2-D or 3-D chemical descriptors, fragment or group contribution-based models", and models based on statistical data [35]. Yet it is still difficult to forecast solubility accurately. The majority of organizations continue to prioritize and make decisions on compounds based on experimental solubility data [36]. Screen for solubility and measurement techniques in the early stages of drug development, high-throughput (HT) solubility screening techniques are frequently employed to characterize solubility. Standardized practices in chemical handling and delivery are also necessary due to the enormous number of compounds that must be analyzed in different tests [37]. They are typically carried out in 96- or 384-well microtiter plates. The chemicals are often administered using two techniques for early HT solubility experiments. Dimethyl sulfoxide (DMSO) stock solutions are one way to introduce the chemicals into an aqueous medium.

Alternatively, the substances can be dispensed as dimethyl sulfoxide (DMSO) solutions, which are subsequently extracted from the plate, and aqueous media are introduced [38]. In no instance is the ultimate pH or solid form assessed or described. Following the elimination of particles through filtration or centrifugation, the concentration of the saturated solutions can be determined using UV-*Vis* spectroscopy, liquid chromatography (UV or LC), or precipitation can be identified using UV or a nephelometric turbidity detector. Due to the presence of residual organic solvent and/or the generation of amorphous material following precipitation from DMSO stock solution, these high-throughput experiments often lead to an overestimation of solubility [39]. Nevertheless, requiring only a minimal quantity of compounds, this method proves to be an immensely valuable tool for ranking compounds and for the early detection of solubility issues.

These prediction models leverage quantitative structure-property relationships (QSPR) and machine learning techniques to correlate the molecular structure and physicochemical properties of a compound with its solubility behavior. Here are some key aspects of solubility prediction models:

3.1. Data Sources

Solubility prediction models require a comprehensive and diverse dataset containing information about the solubility of different compounds under varying conditions. These datasets are often curated from experimental measurements available in the literature or generated through high-throughput screening techniques [40].

3.2. Descriptors

Molecular descriptors are numerical representations of a compound's structure and properties. These descriptors can include parameters such as molecular weight, partition coefficient (logP), hydrogen bond donors and acceptors, polar surface area, and more. These descriptors capture the key features of a molecule that influence its solubility.

3.3. Model Development

Various modeling techniques are employed to establish relationships between the molecular descriptors and the solubility data. These techniques range from traditional linear regression and multiple linear regression to more advanced methods like support vector machines, random forests, neural networks, and deep learning [41].

3.4. Feature Selection

In the process of model development, selecting the most relevant descriptors is crucial. Feature selection methods help identify descriptors that contribute significantly to the predictive accuracy of the model while avoiding overfitting.

3.5. Validation

After training the model on a training dataset, it is essential to validate its predictive performance on a separate validation dataset. Common validation metrics include mean absolute error (MAE), root mean squared error (RMSE), and coefficient of determination (R^2).

3.6. Applicability Domain

Solubility prediction models need to define an applicability domain, which specifies the range of compounds for which the model's predictions are reliable. This prevents the extrapolation of predictions to compounds that fall outside the defined range [42].

3.7. Data Quality

The accuracy and reliability of solubility predictions depend on the quality of the underlying data. Errors or inconsistencies in the experimental data can propagate into the predictions, potentially leading to inaccurate results.

3.8. External Validation

To further validate the predictive power of the model, it should be tested on external datasets that were not used during the model's training and validation stages. A robust model should demonstrate consistent performance across multiple datasets. Fig. (1) demarcates various novel and traditional treatments for solubility enhancement.

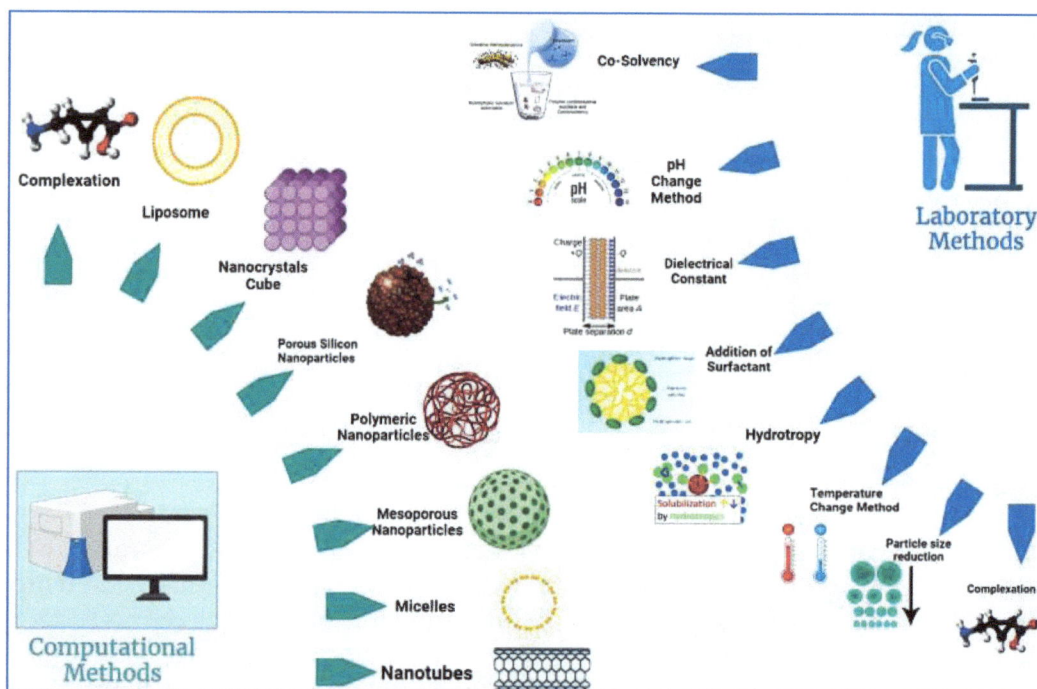

Fig. (1). Novel and traditional approaches for solubility enhancement.

3.9. Computer-aided Drug Discovery using Ligands (LB-CADD)

By examining ligands to identify molecules that interact with a target of interest, the ligand-based computer-aided drug discovery technique may be used to predict solubility [43]. By using reference structures, LB-CADD may depict compounds while keeping crucial physicochemical characteristics for desirable interactions. Compound selection based on chemical similarity and QSAR model generation are two LB-CADD methods. *In silico* screening, "hit-to-lead and lead-to-drug optimization, and DMPK/ADMET" property optimization may all be done using LB-CADD techniques [44]. The ignoring of the ligands' dynamic character is one drawback of these techniques [45].

3.10. Quantum Mechanics

While methods based on quantum mechanics can be used to foretell changes in a molecule's structure brought on by chemical reactivity [46], *in silico* methods such as QSPR and QSAR are commonly used to predict drug solubility [47]. These methods take into account various factors such as lipophilicity, solute-solvent interactions, and molecular size to predict a drug's solubility and permeability properties. The ability to accurately predict these properties is crucial in drug development as they play a critical character in determining a drug's bioavailability and efficacy, particularly when administered orally [47]. However, it is important to note that experimental validation is essential to confirm the accuracy of these predictions, as the results obtained from *in silico* methods may not always be reliable [48].

3.11. Quantum Mechanical Methods

3.11.1. ADF COSMO-RS Program

Calculations are used in the solubility prediction of drugs using the ADF COSMO-RS program [49]. The COSMO-RS approach uses intermediate outcomes from quantum mechanical computations on individual molecules to forecast the solubility of mixtures and other thermodynamic features [50]. The COSMO-RS approach, which was first created for the prediction of liquid-liquid and liquid-vapor equilibrium constants, has been expanded to include the prediction of solubility.

3.11.2. MFPCP Method

In this strategy, the descriptors comprised physicochemical attributes and molecular fingerprinting. The most prevalent method for representing molecules is through the SMILES format. In this context, molecular fingerprinting was utilized to convert the chemical structures of the molecules into formats understandable by computers [51]. There are various ways to create molecular fingerprints; however, the approach employed in this study involved using the 166-dimensional Molecular ACCESS System (MACCS) keys. RDKit simplified the generation of MACCS fingerprints, which were directly derived from SMILES [52]. The acquired solvent and solute fingerprints were then utilized as descriptors in the prediction model. This methodology facilitated the conversion of intricate molecular structures into formats amenable to computational analysis.

3.11.3. QSAR Method for Solubility Prediction of Drug

The QSAR method involves developing a mathematical model that correlates chemical structures with their biological activities. This approach can also be used for the solubility prediction of drugs by analyzing ligands to find compounds that interact with a target of interest. Chemometric techniques can be used to generate 2D QSAR models, as demonstrated by a study that developed a model for the role of Rho kinases in neurological treatment using urea-based scaffolds from aniline and benzylamine analogs [53]. "Quantitative structure-property relationship (QSPR) models" for water solubility have also been studied in several papers [54]. These methods can be helpful in the drug discovery and development process for-hit-to-lead and lead-to-drug optimization, and DMPK/ADMET" property optimization.

Table. (1). Comparison of methods for predicting medicinal compounds' solubility.

Tools	Duration of the Computation (in min)	References
Quantitative structure-activity relationship	60 – 120	[57]
Quantitative structure-property relationship	60 – 120	[58]
Molecular Dynamics	720 – 1440	[59]
Conductor-like screening model	60 – 120	[60]
Solubility Parameters	120 – 180	[61]
Free energy calculations	60 – 120	[58]
Flory–Huggins parameter	60 – 120	[62]

3.11.4. QSPR Technique for Solubility Prediction of Drug

The QSPR "quantitative structure-property relationship" model is an integrated technique used to calculate a drug's solubility. It is a mathematical model that connects a molecule's structural characteristics to its solubility. To anticipate a drug's solubility and improve its DMPK/ADMET characteristics, QSPR models have been frequently employed in the drug research and development process. However, it is significant to note that QSPR models can still overfit data, especially when prediction errors are lower than experimental errors [55]. Furthermore, QSAR "quantitative structure-activity relationship" models, which also use mathematical equations to predict molecular properties, have shown useful applications in drug discovery and molecular modeling, including predicting solubility and optimizing drug properties [56]. Table **1** enlists the compilation of different methods for the prediction of drug solubility.

3.12. Critical Factors Affecting Solubility and Permeability

3.12.1. Factors Affecting the Solubility of the Drug

The solubility of medications can be impacted by a number of circumstances. These include the amount of force exerted by the solute on the solvent, electronic variables, solute-solvent interactions, and steric variables [63]. Additionally, the presence of food can impact drug binding and blood supply, which can affect drug absorption It's important to note that most substances are endothermic, implying they absorb heat during the dissolution process [64]. As a result, the solubility of these chemicals can increase with a rise in temperature. However, it's essential to remember that predicting drug solubility can be intricate and might necessitate the application of diverse methods, such as Quantitative Structure-Property Relationship (QSPR), Quantitative Structure-Activity Relationship (QSAR), and quantum mechanical calculations. These methods can prove valuable for *in silico* screening, as well as for processes like "hit-to-lead" and "lead-to-drug" optimization., and DMPK/ADMET" property optimization in drug discovery and development. However, they can still have limitations and rely on reliable experimental data.

3.12.2. Factors Impacting Drug Permeability

Lipophilicity, molecular weight, and polarity are among the characteristics that affect membrane permeability, along with factors like membrane thickness and composition. Lipophilicity, in particular, holds significance as it determines how much of a drug will traverse the lipid membrane, which serves as a fundamental determinant of membrane permeability [65]. Moreover, insufficient lipophilicity can detrimentally influence the potency and effectiveness of a drug, resulting in reduced bioavailability [66].

3.12.3. Relationship Between the Drug's Solubility and Permeability

A drug's permeability and solubility are intricately linked and hold a pivotal role in determining both its bioavailability and effectiveness [67]. The term "solubility" refers to a substance's ability to dissolve in a solution, while permeability signifies the extent to which a substance can traverse a membrane [68]. In the realm of drug development, the prediction of solubility and permeability stands as a vital process for the optimization of drug characteristics and the assurance of efficacy [69]. While QSAR and QSPR methods are commonly employed for predicting drug solubility and permeability, they heavily rely on accurate experimental data [70]. Various factors that influence drug solubility and permeability encompass interactions between solute and solvent, steric considerations, and lipophilicity, among others [68]. These techniques are useful for "silico screening, hit-to-lead, lead-to-drug optimization, and DMPK/ADMET" property optimization in drug discovery and development [71]. However, they can have limitations, and experimental validation is necessary to confirm their predictions [72].

3.12.4. Solubility and Permeability Impacts on Drug Bioavailability

A drug's bioavailability is significantly influenced by its solubility and permeability, especially when administered orally [73]. The permeability of a drug refers to its ability to pass through a biological membrane, while solubility refers to its ability to dissolve in water. The membrane/aqueous partition coefficient divided by the diffusion coefficient against the membrane can be used to express the connection between the two. Predicting a medication's solubility and permeability during drug development is crucial for enhancing its formulation and efficacy [74, 75]. *In silico* methods such as for predicting these features, "quantitative structure-activity relationship (QSAR) and quantitative structure-property relationship (QSPR) models" are frequently utilized. However, experimental validation is necessary to confirm the accuracy of these predictions, since these methods have limitations and may not account for all factors that affect drug solubility and permeability.

CONCLUSION

With these computational techniques, the permeability of molecules and solubility predictions have become a vital module during drug development. A highly predictive approach called ligand-based computer-aided drug discovery employs techniques including pharmacophores, molecular descriptors, and quantitative structure-activity connections. Efficient and reliable computational methods are necessary for predicting biopharmaceutical properties such as permeability in addition to solubility. PAMPA, Caco-2, and rat intestinal canals are examples of *in vitro* techniques that are useful for researching solute-membrane interactions, while molecular surface areas have been used as descriptors of permeability and solubility. Cell culture techniques, such as those for nasal drug permeability, are also useful. While multiscale biomolecular simulations are necessary to disclose the action mechanism of a drug from the molecular structure level to cellular tissue, computational tools have substantially sped up and decreased the cost and time of drug discovery.

ACKNOWLEDGEMENT

The authors want to acknowledge research facilities and e-resources provided by the institute during the entire study process.

REFERENCES

[1]   Vo CLN, Park C, Lee BJ. Current trends and future perspectives of solid dispersions containing poorly water-soluble drugs. Eur J Pharm Biopharm 2013; 85(3): 799-813.
[http://dx.doi.org/10.1016/j.ejpb.2013.09.007] [PMID: 24056053]

[2]   Boyd BJ, Bergström CAS, Vinarov Z, *et al.* Successful oral delivery of poorly water-soluble drugs both depends on the intraluminal behavior of drugs and of appropriate advanced drug delivery systems. Eur J Pharm Sci 2019; 137: 104967.
[http://dx.doi.org/10.1016/j.ejps.2019.104967] [PMID: 31252052]

[3]   Sliwoski G, Kothiwale S, Meiler J, Lowe EW Jr. Computational methods in drug discovery. Pharmacol Rev 2014; 66(1): 334-95.
[http://dx.doi.org/10.1124/pr.112.007336] [PMID: 24381236]

[4]   Liu X, Testa B, Fahr A. Lipophilicity and its relationship with passive drug permeation. Pharm Res 2011; 28(5): 962-77.

[http://dx.doi.org/10.1007/s11095-010-0303-7] [PMID: 21052797]

[5]   Xie L, Ge X, Tan H, *et al.* Towards structural systems pharmacology to study complex diseases and personalized medicine. PLOS Comput Biol 2014; 10(5): e1003554.
[http://dx.doi.org/10.1371/journal.pcbi.1003554] [PMID: 24830652]

[6]   Shreya Shweta. Virtual screening of phytochemicals for drug discovery. Phytochem comput tools databases. Drug Discov 2023; 1: 149-79.

[7]   Yee S. *In vitro* permeability across Caco-2 cells (colonic) can predict *in vivo* (small intestinal) absorption in man-fact or myth. Pharm Res 1997; 14(6): 763-6.
[http://dx.doi.org/10.1023/A:1012102522787] [PMID: 9210194]

[8]   Bennion BJ, Be NA, McNerney MW, *et al.* Predicting a Drug's membrane permeability: A computational model validated with *in vitro* permeability assay data. J Phys Chem B 2017; 121(20): 5228-37.
[http://dx.doi.org/10.1021/acs.jpcb.7b02914] [PMID: 28453293]

[9]   Sliwoski GR. 3D enantioselective descriptors for ligand-based computer-aided drug design. 2012.

[10]   Kothiwale SK, Meiler J, Hess A, Pozzi A, Lybrand T. A novel knowledge based conformation sampling algorithm and applications in drug discovery 2016.

[11]   Makhouri FR, Ghasemi JB. Combating diseases with computational strategies used for drug design and discovery. Curr Top Med Chem 2019; 18(32): 2743-73.
[http://dx.doi.org/10.2174/1568026619666190121125106] [PMID: 30663568]

[12]   Stillwell W. Membrane transport. An Introd to Biol Membr 2013; 1: 305-37.
[http://dx.doi.org/10.1016/B978-0-444-52153-8.00014-3]

[13]   Fong CW. Permeability of the blood–brain barrier: Molecular mechanism of transport of drugs and physiologically important compounds. J Membr Biol 2015; 248(4): 651-69.
[http://dx.doi.org/10.1007/s00232-015-9778-9] [PMID: 25675910]

[14]   Drummond DC, Noble CO, Hayes ME, Park JW, Kirpotin DB. Pharmacokinetics and *in vivo* drug release rates in liposomal nanocarrier development. J Pharm Sci 2008; 97(11): 4696-740.
[http://dx.doi.org/10.1002/jps.21358] [PMID: 18351638]

[15]   Carpenter TS, Kirshner DA, Lau EY, Wong SE, Nilmeier JP, Lightstone FC. A method to predict blood-brain barrier permeability of drug-like compounds using molecular dynamics simulations. Biophys J 2014; 107(3): 630-41.
[http://dx.doi.org/10.1016/j.bpj.2014.06.024] [PMID: 25099802]

[16]   Moss GP, Dearden JC, Patel H, Cronin MTD. Quantitative structure–permeability relationships (QSPRs) for percutaneous absorption. Toxicol. *In Vitro* 2002; 16(3): 299-317.
[http://dx.doi.org/10.1016/S0887-2333(02)00003-6] [PMID: 12020604]

[17]   Wadhwa R, Yadav NS, Katiyar SP, Yaguchi T, Lee C, Ahn H, *et al.* Molecular dynamics simulations and experimental studies reveal differential permeability of withaferin-A and withanone across the model cell membrane. Sci Reports 2021; 11(1): 1-15.
[http://dx.doi.org/10.1038/s41598-021-81729-z]

[18]   Dimitrov S, Dimitrova G, Pavlov T, *et al.* A stepwise approach for defining the applicability domain of SAR and QSAR models. J Chem Inf Model 2005; 45(4): 839-49.
[http://dx.doi.org/10.1021/ci0500381] [PMID: 16045276]

[19]   Taillardat-Bertschinger A, Carrupt PA, Barbato F, Testa B. Immobilized artificial membrane HPLC in drug research. J Med Chem 2003; 46(5): 655-65.
[http://dx.doi.org/10.1021/jm020265j] [PMID: 12593643]

[20]   Wang NN, Dong J, Deng YH, *et al.* ADME properties evaluation in drug discovery: prediction of Caco-2 cell permeability using a combination of NSGA-II and boosting. J Chem Inf Model 2016; 56(4): 763-73.

[http://dx.doi.org/10.1021/acs.jcim.5b00642] [PMID: 27018227]

[21]   Briganti A, Passoni N, Ferrari M, *et al.* When to perform bone scan in patients with newly diagnosed prostate cancer: external validation of the currently available guidelines and proposal of a novel risk stratification tool. Eur Urol 2010; 57(4): 551-8.
[http://dx.doi.org/10.1016/j.eururo.2009.12.023] [PMID: 20034730]

[22]   Hou TJ, Zhang W, Xia K, Qiao XB, Xu XJ. ADME evaluation in drug discovery. 5. Correlation of Caco-2 permeation with simple molecular properties. J Chem Inf Comput Sci 2004; 44(5): 1585-600.
[http://dx.doi.org/10.1021/ci049884m] [PMID: 15446816]

[23]   Pidgeon C, Ong S, Liu H, *et al.* IAM chromatography: An *in vitro* screen for predicting drug membrane permeability. J Med Chem 1995; 38(4): 590-4.
[http://dx.doi.org/10.1021/jm00004a004] [PMID: 7861406]

[24]   Tsopelas F, Vallianatou T, Tsantili-Kakoulidou A. Advances in immobilized artificial membrane (IAM) chromatography for novel drug discovery. 2016; 11(5): 473-88.

[25]   Kerns EH. High throughput physicochemical profiling for drug discovery. J Pharm Sci 2001; 90(11): 1838-58.
[http://dx.doi.org/10.1002/jps.1134] [PMID: 11745742]

[26]   Carrasco-Correa EJ, Ruiz-Allica J, Rodríguez-Fernández JF, Miró M. Human artificial membranes in (bio)analytical science: Potential for *in vitro* prediction of intestinal absorption-A review. Trends Analyt Chem 2021; 145: 116446.
[http://dx.doi.org/10.1016/j.trac.2021.116446]

[27]   Bertucci C, Bartolini M, Gotti R, Andrisano V. Drug affinity to immobilized target bio-polymers by high-performance liquid chromatography and capillary electrophoresis. J Chromatogr B Analyt Technol Biomed Life Sci 2003; 797(1-2): 111-29.
[http://dx.doi.org/10.1016/j.jchromb.2003.08.033] [PMID: 14630146]

[28]   Flaten GE, Dhanikula AB, Luthman K, Brandl M. Drug permeability across a phospholipid vesicle based barrier: A novel approach for studying passive diffusion. Eur J Pharm Sci 2006; 27(1): 80-90.
[http://dx.doi.org/10.1016/j.ejps.2005.08.007] [PMID: 16246536]

[29]   Zhang C, Li J, Xu L, Shi ZG. Fast immobilized liposome chromatography based on penetrable silica microspheres for screening and analysis of permeable compounds. J Chromatogr A 2012; 1233: 78-84.
[http://dx.doi.org/10.1016/j.chroma.2012.02.013] [PMID: 22381893]

[30]   Lundahl P, Beigi F. Immobilized liposome chromatography of drugs for model analysis of drug-membrane interactions. Adv Drug Deliv Rev 1997; 23(1-3): 221-7.
[http://dx.doi.org/10.1016/S0169-409X(96)00437-1]

[31]   Liu XY, Nakamura C, Yang Q, Kamo N, Miyake J. Immobilized liposome chromatography to study drug–membrane interactions. J Chromatogr A 2002; 961(1): 113-8.
[http://dx.doi.org/10.1016/S0021-9673(02)00505-8] [PMID: 12186381]

[32]   Larregieu CA, Benet LZ. Distinguishing between the permeability relationships with absorption and metabolism to improve BCS and BDDCS predictions in early drug discovery. Mol Pharm 2014; 11(4): 1335-44.
[http://dx.doi.org/10.1021/mp4007858] [PMID: 24628254]

[33]   Larregieu CA, Benet LZ. Drug discovery and regulatory considerations for improving *in silico* and *in vitro* predictions that use Caco-2 as a surrogate for human intestinal permeability measurements. AAPS J 2013; 15(2): 483-97.

[34]   Veber DF, Johnson SR, Cheng HY, Smith BR, Ward KW, Kopple KD. Molecular properties that influence the oral bioavailability of drug candidates. J Med Chem 2002; 45(12): 2615-23.
[http://dx.doi.org/10.1021/jm020017n] [PMID: 12036371]

[35]   Elder D, Holm R. Aqueous solubility: Simple predictive methods (*in silico*, *in vitro* and bio-relevant approaches). Int J Pharm 2013; 453(1): 3-11.

[http://dx.doi.org/10.1016/j.ijpharm.2012.10.041] [PMID: 23124107]

[36] Alelyunas YW, Liu R, Pelosi-Kilby L, Shen C. Application of a Dried-DMSO rapid throughput 24-h equilibrium solubility in advancing discovery candidates. Eur J Pharm Sci 2009; 37(2): 172-82.
[http://dx.doi.org/10.1016/j.ejps.2009.02.007] [PMID: 19429424]

[37] Alsenz J, Kansy M. High throughput solubility measurement in drug discovery and development. Adv Drug Deliv Rev 2007; 59(7): 546-67.
[http://dx.doi.org/10.1016/j.addr.2007.05.007] [PMID: 17604872]

[38] Avdeef A, Fuguet E, Llinàs A, *et al.* Equilibrium solubility measurement of ionizable drugs – consensus recommendations for improving data quality. ADMET DMPK 2016; 4(2): 117-78.
[http://dx.doi.org/10.5599/admet.4.2.292]

[39] Hassellöv M, Readman JW, Ranville JF, Tiede K. Nanoparticle analysis and characterization methodologies in environmental risk assessment of engineered nanoparticles. Ecotoxicology 2008; 17(5): 344-61.
[http://dx.doi.org/10.1007/s10646-008-0225-x] [PMID: 18483764]

[40] Huang L, Tong WQ. Impact of solid state properties on developability assessment of drug candidates. Adv Drug Deliv Rev 2004; 56(3): 321-34.
[http://dx.doi.org/10.1016/j.addr.2003.10.007] [PMID: 14962584]

[41] Byrn SR, Pfeiffer RR, Stephenson G, Grant DJW, Gleason WB. Solid-state pharmaceutical chemistry. Chem Mater 1994; 6(8): 1148-58.
[http://dx.doi.org/10.1021/cm00044a013]

[42] Parikh T, Gupta SS, Meena AK, Vitez I, Mahajan N, Serajuddin ATM. Application of film-casting technique to investigate drug-polymer miscibility in solid dispersion and hot-melt extrudate. J Pharm Sci 2015; 104(7): 2142-52.
[http://dx.doi.org/10.1002/jps.24446] [PMID: 25917333]

[43] Zhou L, Yang L, Tilton S, Wang J. Development of a high throughput equilibrium solubility assay using miniaturized shake-flask method in early drug discovery. J Pharm Sci 2007; 96(11): 3052-71.
[http://dx.doi.org/10.1002/jps.20913] [PMID: 17722003]

[44] Kumar KK, Swathi M, Srinivas L, Basha SN. Formulation and evaluation of floating *in situ* gelling system of losartan potassium. Pharm Lett 2015; 7: 98-112.

[45] Gupta S, Kershaw SV, Rogach AL. 25th anniversary article: Ion exchange in colloidal nanocrystals. Adv Mater 2013; 25(48): 6923-44.
[http://dx.doi.org/10.1002/adma.201302400] [PMID: 24108549]

[46] Acharya C, Coop A, Polli JE, Mackerell AD Jr. Recent advances in ligand-based drug design: Relevance and utility of the conformationally sampled pharmacophore approach. Curr Computeraided Drug Des 2011; 7(1): 10-22.
[http://dx.doi.org/10.2174/157340911793743547] [PMID: 20807187]

[47] Salmaso V, Moro S. Bridging molecular docking to molecular dynamics in exploring ligand-protein recognition process: An overview. Front Pharmacol 2018; 9(Aug): 923.
[http://dx.doi.org/10.3389/fphar.2018.00923] [PMID: 30186166]

[48] Eros D, Kövesdi I, Orfi L, Takács-Novák K, Acsády G, Kéri G. Reliability of logP predictions based on calculated molecular descriptors: A critical review. Curr Med Chem 2002; 9(20): 1819-29.
[http://dx.doi.org/10.2174/0929867023369042] [PMID: 12369880]

[49] Bououden W, Benguerba Y, Darwish AS, *et al.* Surface adsorption of Crizotinib on carbon and boron nitride nanotubes as Anti-Cancer drug Carriers: COSMO-RS and DFT molecular insights. J Mol Liq 2021; 338: 116666.
[http://dx.doi.org/10.1016/j.molliq.2021.116666]

[50] Scheffczyk J, Schäfer P, Fleitmann L, *et al.* COSMO-CAMPD: A framework for integrated design of molecules and processes based on COSMO-RS. Mol Syst Des Eng 2018; 3(4): 645-57.

[http://dx.doi.org/10.1039/C7ME00125H]

[51]   Livingstone DJ. The characterization of chemical structures using molecular properties. A survey. J Chem Inf Comput Sci 1999; 40(2): 195-209.

[52]   Lee S, Lee M, Gyak KW, Kim SD, Kim MJ, Min K. Novel solubility prediction models: Molecular fingerprints and physicochemical features *vs* graph convolutional neural networks. ACS Omega 2022; 7(14): 12268-77.
[http://dx.doi.org/10.1021/acsomega.2c00697] [PMID: 35449985]

[53]   Lo YC, Rensi SE, Torng W, Altman RB. Machine learning in chemoinformatics and drug discovery. Drug Discov Today 2018; 23(8): 1538-46.
[http://dx.doi.org/10.1016/j.drudis.2018.05.010] [PMID: 29750902]

[54]   Grover M, Singh B, Bakshi M, Singh S. Quantitative structure–property relationships in pharmaceutical research – Part 2. Pharm Sci Technol Today 2000; 3(2): 50-7.
[http://dx.doi.org/10.1016/S1461-5347(99)00215-1] [PMID: 10664573]

[55]   Dai Y, Yang D, Zhu F, Wu L, Yang X, Li J. The QSPR (quantitative structure–property relationship) study about the anaerobic biodegradation of chlorophenols. Chemosphere 2006; 65(11): 2427-33.
[http://dx.doi.org/10.1016/j.chemosphere.2006.04.052] [PMID: 16750555]

[56]   Verma J, Khedkar V, Coutinho E. 3D-QSAR in drug design: A review. Curr Top Med Chem 2010; 10(1): 95-115.
[http://dx.doi.org/10.2174/156802610790232260] [PMID: 19929826]

[57]   Neves BJ, Braga RC, Melo-Filho CC, Moreira-Filho JT, Muratov EN, Andrade CH. QSAR-based virtual screening: Advances and applications in drug discovery. Front Pharmacol 2018; 9(Nov): 1275.
[http://dx.doi.org/10.3389/fphar.2018.01275] [PMID: 30524275]

[58]   Westergren J, Lindfors L, Höglund T, Lüder K, Nordholm S, Kjellander R. *In silico* prediction of drug solubility: 1. Free energy of hydration. J Phys Chem B 2007; 111(7): 1872-82.
[http://dx.doi.org/10.1021/jp064220w] [PMID: 17266351]

[59]   Filipe HAL, Loura LMS. Molecular dynamics simulations: Advances and applications. Molecules 2022; 27(7): 2105.
[http://dx.doi.org/10.3390/molecules27072105] [PMID: 35408504]

[60]   Klamt A, Eckert F, Arlt W. COSMO-RS: An alternative to simulation for calculating thermodynamic properties of liquid mixtures. Annu Rev Chem Biomol Eng 2010; 1(1): 101-22.
[http://dx.doi.org/10.1146/annurev-chembioeng-073009-100903] [PMID: 22432575]

[61]   Hancock B, York P, Rowe RC. The use of solubility parameters in pharmaceutical dosage form design. Int J Pharm 1997; 148(1): 1-21.
[http://dx.doi.org/10.1016/S0378-5173(96)04828-4]

[62]   Schotsch K, Wolf BA, Jeberien HE, Klein J. Concentration dependence of the Flory-Huggins parameter at different thermodynamic conditions. Die Makromolekulare Chemie: Macromolecular Chemistry and Physics. 1984 Oct; 185(10): 2169-81.

[63]   Shelby RA, Smith DR, Schultz S. Experimental Verification of a Negative Index of Refraction. Science (80-). 2001 Apr 6; 292(5514): 77–9.

[64]   Chen G, Liang J, Han J, Zhao H. Solubility modeling, solute–solvent interactions, and thermodynamic dissolution properties of p-nitrophenylacetonitrile in sixteen monosolvents at temperatures ranging from 278.15 to 333.15 K. Journal of Chemical & Engineering Data. 2018 Dec 19; 64(1): 315-23.

[65]   Xie L, Ge X, Tan H, *et al.* Towards structural systems pharmacology to study complex diseases and personalized medicine. PLOS Comput Biol 2014; 10(5): e1003554.
[http://dx.doi.org/10.1371/journal.pcbi.1003554] [PMID: 24830652]

[66]   Casares D, Escribá PV, Rosselló CA. Membrane lipid composition: Effect on membrane and organelle structure, function and compartmentalization and therapeutic avenues. International journal of

molecular sciences. 2019 May 1; 20(9): 2167.

[67] Shreya Shweta. Virtual screening of phytochemicals for drug discovery. Phytochem Comput Tools Databases Drug Discov 2023; 1: 149-79.

[68] Salib RJ, Howarth PH. Safety and tolerability profiles of intranasal antihistamines and intranasal corticosteroids in the treatment of allergic rhinitis. Drug Safety. 2003 Oct; 26: 863-93.

[69] Avdeef A. Physicochemical profiling (solubility, permeability and charge state). Current topics in medicinal chemistry. 2001 Sep 1; 1(4): 277-351.

[70] Kothiwale SK, Meiler J, Hess A, Pozzi A, Lybrand T. A novel knowledge based conformation sampling algorithm and applications in drug discovery 2016.

[71] Kothiwale SK, Meiler J, Hess A, Pozzi A, Lybrand T. A novel knowledge based conformation sampling algorithm and applications in drug discovery 2016.

[72] Stillwell W. Membrane transport. An Introd to Biol Membr 2013; 1: 305-37.
     [http://dx.doi.org/10.1016/B978-0-444-52153-8.00014-3]

[73] Henninot A, Collins JC, Nuss JM. The current state of peptide drug discovery: back to the future?. Journal of medicinal chemistry. 2018 Feb 22; 61(4): 1382-414.

[74] Drummond DC, Noble CO, Hayes ME, Park JW, Kirpotin DB. Pharmacokinetics and *in vivo* drug release rates in liposomal nanocarrier development. J Pharm Sci 2008; 97(11): 4696-740.

[75] Joshi T, Sharma P, Joshi T, Mathpal S, Pandey SC, Pandey A, *et al.* Recent advances on computational approach towards potential drug discovery against leishmaniasis. Pathog Treat Prev Leishmaniasis. 2021 Jan 1;63–84.

CHAPTER 7

Pharmacokinetic and Pharmacodynamic Modeling (PK/PD) in Pharmaceutical Research: Current Research and Advances

Richa Sood[1,*] and **Anita A.**[1]

[1] *College of Pharmaceutical Sciences, Dayananda Sagar University, Bengaluru, Karnataka-560078, India*

Abstract: The development of more intricately constructed molecules and drug delivery systems as a result of technological breakthroughs has increased our understanding of the complexities of disease and allowed us to identify a wide range of therapeutic targets. New drug combinations can be designed by correctly using dynamical systems-based PK/PD models. The unswerving approach that offers a better knowledge and understanding of therapeutic efficacy and safety is the use of pharmacokinetic-pharmacodynamic (PK-PD) modeling in drug research. *In vivo,* animal testing or *in vitro* bioassay is used to forecast efficacy and safety in people. Model-based simulation using primary pharmacodynamic models for direct and indirect responses is used to elucidate the assumption of a fictitious minimal effective concentration or threshold in the exposure-response relationship of many medicines. In this current review, we have abridged the basic PK-PD modeling concepts of drug delivery and documented how they can be used in current research and development.

Keywords: Clinical development, Modeling, Pharmacokinetics, Pharmacodynamics, Preclinical.

1. INTRODUCTION

Recent challenges for novel medication development have included mounting drug development costs and a decline in mass productivity. This outcome has been due to different possible causes. The use of alternative technologies is encouraged to obtain answers regarding efficacy and safety and cost-effectiveness. Alternative approaches to drug development include pharmacokinetic/pharmacodynamic (PK/PD) modeling and simulation, adaptive trial models, higher dependence on biomarkers, and the production of individualized drugs [1]. PK/PD modeling and simulation are helpful at all phases of drug development.

* **Corresponding author Richa Sood:** College of Pharmaceutical Sciences, Dayananda Sagar University, Bengaluru, Karnataka-560078, India; E-mail: richa88sood@gmail.com

Dilpreet Singh and Prashant Tiwari (Eds.)

Only experts in the respective field (*i.e.*, pharmacokinetics) have a thorough knowledge of the value and application of PK/PD modeling. However, modeling concepts need to be understood and embraced by both drug development team members and regulators to assist in drug development [2]. To simplify the issues and enhance comprehension of these drug delivery systems' *in vivo* behavior, mechanism-based pharmacokinetic-pharmacodynamic (PK-PD) modeling should be applied [3]. The standard drug development process includes the following stages: drug discovery process, preclinical development phase, exploratory clinical development phase, full clinical development, and regulatory filings [4]. PK/PD modeling is widely recognized as an important technique for choosing promising substances and figuring out appropriate doses and dosing regimens in people [5]. PK/PD modeling at this point enables the sponsor to eliminate unfavorable candidates early, when drug development costs are low, to more accurately predict the clinical profile of the drug, and to arrange candidates with a higher likelihood of success during the confirmatory late-phase development [6]. Pharmacokinetic-pharmacodynamic (PK-PD) modeling is a scientific approach to analyzing pharmacokinetics (PK), pharmacodynamics (PD), and their relationship, and is a significant part of the drug discovery phase and drug development process. The mechanism-based PK-PD model is applicable throughout the drug development process, as shown in Fig. (**1**). The process of drug ADME in the body is explicitly and quantitatively described by PK modeling. Indeed, PD modeling is used to assess the time course of the pharmacological effects of medications with time [7]. In addition, PK and PD modelling can be used to quantify the link between drug exposure and response and to characterize the influence of drug-relevant properties, delivery mechanisms, physiological systems, and pathological systems on this relationship [8]. Drug-specific metrics (receptor binding affinity and drug clearance rate) illustrate how a given medication affects the body's biological machinery. Clearance, release rate, and carrier internalization rate are examples of carrier features that are peculiar to a given drug delivery method. The physiological values represented by the physiological system-specific parameters include blood flow, cell longevity, expression of enzymes, and transporter expression [3]. The separation of drug-specific and system-specific characteristics in PK-PD modelling would allow for the evaluation and facilitation of the effects of various delivery system properties on the *in vivo* drug effect. The mechanism-based PK-PD models, which were created based on the PK-PD data from preclinical studies, can be used to forecast the human dosage schedule and optimize the drug delivery system, as seen in the bottom panel of Fig. (**1**). Once they are available, the clinical PK-PD data can be included in the PK-PD models to improve their design. To assist the ultimate approval, the PK-PD modeling can also develop concurrently with clinical development. The modeling method is currently used in medication delivery systems and modified big molecules [9].

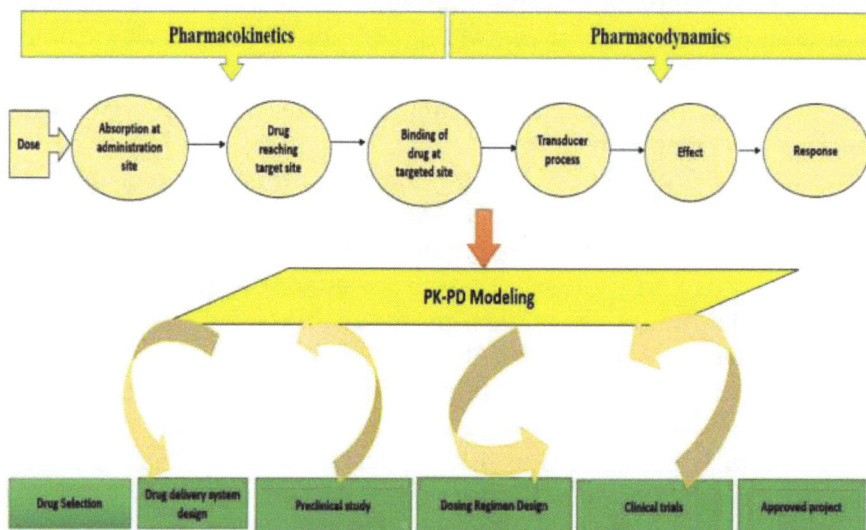

Fig. (1). Pictorial representation of PK-PD modeling in the drug delivery system development.

PK-PD modeling could direct formulation design and dosage regimen selection during the development of the drug delivery system based on preclinical and clinical data [10]. The pharmacological dose and the physiological reaction are linked by this system. A series of actions demonstrate the progression from administration to drug binding at the target location, receptor activation and binding, signal transmission, and impact on physiological response. In this review, we have outlined M&S approaches and discussed regulatory viewpoints on the use of PK/PD M&S in the drug development process.

2. PHARMACOKINETIC MODELS

Pharmacokinetic models compute useful pharmacokinetic parameters and shed light on how long it takes a medicine (or drugs) to travel throughout the body. In certain situations, pharmacokinetic models are helpful, such as in defining drug-induced patient behaviors with any dose regimen, predicting the drug's concentration in different bodily fluids, determining the best optimum dosage regimen for specific individuals, assessing the toxicity, determining the bioequivalence/bio inequivalence among various preparations of the same drug, and describing drug interactions [11, 12].

2.1. Methods of Pharmacokinetic Study on Experimental Data

2.1.1. Non-compartmental Analysis (NCA)

Non-compartmental analysis is a straightforward method used to calculate pharmacokinetic parameters without assuming a specific compartmental model. It is particularly useful when the underlying biological processes are not well understood or when the data does not conform to compartmental models. NCA involves the following steps:

2.1.1.1. Calculation of Area Under the Curve (AUC)

AUC represents the total exposure of the drug in the bloodstream over time. It is calculated using numerical integration methods like the trapezoidal rule [13].

2.1.1.2. Calculation of Clearance (CL)

Clearance is the rate at which the drug is eliminated from the body. It is calculated as the dose divided by AUC.

2.1.1.3. Calculation of Volume of Distribution (V_d)

V_d represents the apparent volume into which a drug is distributed in the body. It is calculated as the dose divided by the product of AUC and the elimination rate constant.

2.1.1.4. Calculation of Half-Life ($t\frac{1}{2}$)

Half-life is the time taken for the drug concentration in the body to reduce by half. It is calculated using the equation:

$t\frac{1}{2}$ = ln(2) / elimination rate constant.

NCA provides valuable information about drug exposure and basic pharmacokinetic parameters, making it a suitable method when detailed compartmental modeling is not feasible [14].

2.1.2. Compartmental Modeling

Compartmental modeling assumes that the body can be represented by distinct compartments, each with its physiological characteristics. The movement of the drug between these compartments is described using mathematical equations based on mass balance principles. The most commonly used compartmental models are the one-compartment and multi-compartment models [15]:

2.1.2.1. One-compartment Model

In this model, the body is considered as a single compartment where the drug is distributed uniformly [16].

2.1.2.2. Multi-compartment Model

More complex than the one-compartment model, this approach considers multiple interconnected compartments representing different tissues or organs. Differential equations are used that require specialized software for solutions. Fitting compartmental models to experimental data involves estimating model parameters, such as absorption rate constant, distribution rate constants, and elimination rate constants. Software tools like NONMEM, WinNonlin, and Phoenix are commonly used for this purpose.

2.1.3. Population Pharmacokinetic Modeling

Population pharmacokinetic modeling takes into account variability among individuals within a population. This approach is suitable for large datasets and allows for understanding how factors such as age, gender, genetics, and disease state impact drug pharmacokinetics. The process involves:

2.1.3.1. Data Collection

Gathering pharmacokinetic data from a diverse group of individuals undergoing the same treatment.

2.1.3.2. Model Development

Creating a model that describes both the average drug behavior and the variability among individuals. This involves incorporating fixed effects (average values) and random effects (variability) into the model equations.

2.1.3.3. Parameter Estimation

Using specialized software, the model is fitted to the data to estimate both fixed and random effects parameters.

2.1.3.4. Model Validation

Validating the model's predictive performance using separate datasets or internal cross-validation techniques. Population pharmacokinetic modeling provides insights into the sources of variability in drug response and aids in tailoring dosing strategies for different patient populations. In summary, pharmacokinetic studies use methods like Non-compartmental Analysis, Compartmental Modeling, and Population Pharmacokinetic Modeling to understand how drugs are absorbed, distributed, metabolized, and excreted in the body. The choice of method depends on the complexity of the drug's pharmacokinetics, available data, and the specific research objectives [17].

3. METHODOLOGY OF PK/PD MODELING SIMULATION

3.1. Non-compartmental and Compartmental Pharmacokinetic Analysis

In order to estimate important PK parameters like peak concentration (C_{max}), time to reach peak concentration (T_{max}), area under the concentration-time curve in plasma or serum, clearance (CL), volume of distribution (V_d), half-life ($t_{1/2}$), and others, non-compartmental analysis (NCA) is frequently used. Compartmental analysis has the disadvantage of requiring more assumptions than NCA, which has the virtue of being model-independent [18]. However, a fundamental drawback of NCA is that it cannot model or predict PK profiles when the dosage schedule is altered or nonlinearity takes place. By assuming that the body is divided into one or more compartments through which the delivered medication is distributed or removed, compartmental PK approaches use kinetic models to define and forecast the concentration-time curve. One advantage of the compartmental approach over NCA is the ability to predict the drug concentration at various times and across dose ranges [19].

3.1.1. Population Approach

Population pharmacokinetics (popPK), according to the U.S. FDA, is "the study of the sources and correlates of variability in drug concentrations among individuals who are the target patient population receiving clinically relevant doses of a drug of interest" [20]. The population model specifies a hierarchy with at least two tiers. At the most basic level, pharmacokinetic observations in an individual are viewed as deriving from an individual probability model, whose mean is given by a pharmacokinetic model (for example, a bi-exponential model) quantified by individual-specific parameters, which may vary depending on the value of individual-specific time-varying covariates. Indeed, additional individual-specific pharmacokinetic parameters are used to model the variance of specific pharmacokinetic findings (intrasubject variance). The population model makes use of specific inferential techniques that concentrate on offering estimates of some or all of the variables that make up variability in addition to mean parameter estimations. The probability distribution of the individual parameters (typically the mean and variance, *i.e.*, intersubject variance) is modeled at the second level as a function of the covariates that are distinct to each individual [21]. Population pharmacokinetics involves the utilization of models, their parameter values, and study designs and data analysis methods designed to elucidate population pharmacokinetic models and their associated parameter values. The two-stage approach and the nonlinear mixed-effects modeling methodology are two frequently used techniques for estimating the fixed effect (mean) and variability.

3.2. Approahes for popPK Data

3.2.1. Two-Stage Approach

Two stages make up the standard process for assessing pharmacokinetic data. In the initial phase of this strategy, dense concentration-time data from an individual are used to estimate pharmacokinetic parameters using nonlinear regression (data-rich situation). In the second stage, descriptive summary statistics on the sample are calculated, often using the mean parameter estimates, variance, and covariance of the individual parameter estimates that were acquired in the first stage as input data. An investigation of parameter and covariate dependencies using traditional statistical methods (linear stepwise regression, covariance analysis, cluster analysis) can be included in the second stage. The two-stage method, when applied properly, can yield precise estimates of demographic characteristics. Random effects (variance and covariance) are typically exaggerated in all actual scenarios, despite the fact that mean parameter estimations are typically unbiased. Bias correction for random effects covariance and differential weighting of individual data based on data quality and quantity have been presented as improvements to the two-stage technique.

3.2.2. Nonlinear Mixed-Effects Method (NLME)

A single-stage approach, such as nonlinear mixed-effects modeling, should be utilized in sparse data where the conventional two-stage approach is irrelevant because estimates of individual parameters are, a priori, out of reach. In the context of drug evaluation, the nonlinear mixed-effects modeling approach emerged from the understanding that data should be collected under less strict and restrictive design conditions if pharmacokinetics and pharmacodynamics are to be investigated in patients. For the estimate of the distribution of parameters and their connections with covariates within the population, this method uses the population study sample rather than the person as the unit of analysis. In addition to, or instead of, conventional pharmacokinetic data from traditional pharmacokinetic studies distinguished by rigid and extensive sampling design (dense data situation), the approach uses individual pharmacokinetic data of the observational (experimental) type, which may be sparse, unbalanced, and fragmentary. Estimates of the population features that characterize the population distribution of the pharmacokinetic (and/or pharmacodynamic) parameter are provided by the nonlinear mixed-effects model analysis. In the context of mixed-effects modeling, the set of population characteristics consists of the population mean values (derived from fixed-effects parameters) and the population's variability (often the variance-covariance values generated from random-effects parameters). A technique for population analysis of pharmacokinetic data using nonlinear mixed-effects modeling comprises explicitly computing population parameters from the complete set of individual concentration values. Each subject's distinctiveness is kept and taken into account even in the absence of much data. A more thorough explanation of the mixed-effects modeling approach is provided by the population PK approach [22].

3.3. Bayesian Method

In pharmacokinetics, the Bayesian method is a potent statistical technique that combines previous information and observed data to estimate model parameters, make predictions, and measure uncertainty. Given that it enables the use of prior knowledge to improve parameter estimate, it is especially helpful when working with sparse or noisy data.

3.3.1. Prior Distribution

A prior distribution is used in the Bayesian framework to describe previous knowledge about the parameters of the model. Before any data were observed, this distribution represents the earliest opinions or assumptions about the parameters. Priors may be influenced by earlier research, professional judgment, or other pertinent data. They may be non-informative (broad) or instructive (narrow).

3.3.2. Likelihood Function

The likelihood function explains the likelihood of detecting the data given particular parameter values. The relationship between the model's parameters and the data seen is represented mathematically. In pharmacokinetics, this entails simulating how the drug concentration alters over time based on many factors such as the rate of absorption, the amount of clearance, and the volume of distribution.

3.3.3. Bayesian Inference

To derive the posterior distribution, or the updated views about the parameters after taking the data into account, Bayesian inference entails updating the prior distribution with the observed data. By connecting the prior distribution, likelihood function, and observed data, Bayes' theorem is used to do this.

3.3.4. Posterior Distribution

Given the observed data, the posterior distribution combines the prior distribution and likelihood function to produce a probability distribution of the model parameters. Taking into account both the data and the prior knowledge, this distribution depicts the uncertainty in parameter estimates.

3.3.5. Parameter Estimation

In Bayesian pharmacokinetics, parameter estimation entails summing the posterior distribution to provide point estimates (like the mean, median, or mode) and credible intervals, which are comparable to confidence intervals in that they express the degree of uncertainty surrounding these estimates.

3.3.6. Model Validation

In Bayesian pharmacokinetics, model validation entails evaluating the goodness-of-fit between model predictions and observed data. By doing this, one can make sure that the estimated model accurately captures the pharmacokinetic behaviour that has been observed.

3.3.7. Predictions and Decision Making

The posterior distribution can be used to forecast future drug concentrations, dose modifications, or ideal dosing schedules once it has been acquired and validated. By using uncertainty metrics from the posterior distribution, decision-making can be directed.

4. APPLICATIONS OF MODELING AND SIMULATION IN DRUG DEVELOPMENT

Mathematical and computational methodologies are heavily reliant on the identification of drug targets and the associated active moiety [23]. These methods enable chemical and structural elucidation of the compounds, which lowers the possibility of problems and failures with drug prospects. The cost and duration of conventional experimental studies for drug design and development can be reduced by employing a variety of quantitative models and computational tools. As a result, the use of computational tools and statistical modeling for molecular modeling and drug finding is growing daily [24]. The body's response to drugs can be assessed by using the tools listed in Table **1**.

Table 1. ADME analysis tools used in computational drug design.

Sr. No.	Program	Description	References
1.	DSSTox	It is a public database of searchable distributed structure toxicity.	[25]
2.	Metabase	It is an Excel-based low-cost radio analytical LIM in ADME/PK research.	[26]
3.	PreADMET	It estimates the chances of carcinogenicity as well as poisonous potency.	[27]
4.	ChemTree	It is used to forecast ADMETox characteristics.	
5.	ADMETlab	ADMET in a systematic manner utilizing the ADMET database.	[28]
6.	TOPKAT	Used in toxicology prediction.	[29]
7.	Swiss ADME	It estimates physicochemical parameter and ADME prediction.	[30]

4.1. Preclinical Development

In preclinical research, drug candidates are assessed for their physicochemical parameters, Pharmacokinetic properties such as absorption, metabolism, distribution, and excretion (ADME), as well as their effectiveness and safety. A variety of modeling techniques, including PK, PK/PD, and allometric scaling, are frequently employed for drug molecule selection, optimization, and human dose estimations [31]. The goal of modeling at this point is essentially to "learn" about the properties of the molecule, which are later "confirmed" by information gathered during the early stages of clinical development [3]. Candidate optimization and selection, intrinsic activity, *in vivo* potency, drug interactions, the development of surrogate indicators, and animal models for efficacy/toxicity are currently the main objectives of modeling methodologies [32].

For example, candidate evaluations and human dose selection could be done using a model of PK/PD binding based on a mechanism. The modeling technique was used to predict the minimum acceptable biological effect level (MABEL) for an FIH experiment in cynomolgus monkeys using PK, receptor occupancy, and cell dynamics for a novel monoclonal antibody. Modeling and simulation enable the comparison of various new chemical entities (NCEs) and the selection of the most promising clinical candidate when it comes to NCEs that have a wealth of past knowledge from other medications in their therapeutic class. The choice of dose for phase I investigations could then be aided by PK/PD modeling [33]. The unified preclinical evaluation of a drug was discussed by Wong *et al.* [34] and included modeling and simulation tools. Additionally, it has been shown that the preclinical phase of an antihypertensive medication can predict anticipated therapeutic doses using comparator data [35]. A PBPK model was also created to characterize the absorption of a lipophilic BCS Class II drugs when it was delivered as a nanosuspension formulation. After optimization, bile micelle solubilization and colonic absorption were included in the preclinical rat model. The sensitivity analysis also called "what if analysis" [36] revealed that the absorption parameter fugut was a crucial factor in human PK and DDI. Additionally, in xenografted mice, tumour growth inhibition following the administration of anticancer drugs is predicted using the PK/PD model. It is a cutting-edge and straightforward method for determining the additive effects of anticancer drugs administered in *in-vivo* combination trials. The PK/PD parameters of each medication are estimated using the control and single-agent arms. The expected result is then predicted using the additive combination model based on their estimates [37].

4.2. Clinical Development

Clinical development can make use of modeling and simulation to enhance dosage regimens, plan upcoming clinical studies, and effectively analyze information to support label requirements [38]. Furthermore, it is considered crucial in the design and dosing guidance for certain populations like children, elderly, and obese patients. In addition, dosage forms and various routes of administration, dietary results, drug-disease interactions, and tolerance development can all be assessed using modeling and simulation [39]. For instance, model-based, optimal design and adaptive execution were utilized in the first-in-human study of olokizumab, a monoclonal antibody for the treating rheumatoid arthritis [40]. Population PK/PD models for the P-gp inhibitor zosuquidar recommended examining a 6-hour intravenous infusion schedule rather than a 24-hour schedule in upcoming trials with daunorubicin. The P-gp inhibition was increased while the PK interaction and toxicity level are reduced by the shorter infusion period. For optimum P-gp inhibition for the longest possible time with the least amount of PK interaction and toxicity, a variable infusion regimen can be developed. In addition, specific population studies could be performed using modeling and simulation techniques. The PK of the quetiapine ER formulation in children and adolescents was quantitatively predicted using PBPK modeling based on the PK profiles of immediate-release (IR) formulations in children and adults as well as the PK profile of the ER formulation in adults [41, 42]. These findings were approved by the FDA in place of additional clinical research, and they were useful in determining pediatric dosing schedules. In children aged 2 to 10 with newly diagnosed epilepsy, the PK/PD model of topiramate revealed an efficient monotherapy dose regimen without the requirement for additional studies [43]. Studies have shown that the PBPK model can be used to describe and explain the underlying mechanisms impacting drug absorption and disposition in various organs and tissues in cases of renal failure [44]. Using mathematical modeling, the PKPD relationships between systemic exposure to crizotinib, ALK or MET inhibition, and tumour growth inhibition (TGI) in human cancer xenograft models were comprehensively described. According to the results, >50% TGI requires >50% MET inhibition, while >50% TGI requires >50% ALK inhibition. Additionally, PKPD modeling predicted that patients' tumors would show >75% ALK inhibition and >95% MET inhibition when treated with the clinically advised regimen, which included twice-daily doses of 250 mg of crizotinib (500 mg/day) [45].

4.3. Lifecycle Assessment

The use of modeling and simulation may be beneficial after the preclinical and clinical phases of drug development, such as during the NDA filing and review, post-marketing surveillance, and during the development of new formulations. Modeling techniques incorporate data from both preclinical and clinical trials, encompassing different subpopulations. Thus, during the review process, modeling enables comparisons of the dose-concentration-effect relationship across species and subpopulations [46]. A richer comprehension of the chemical and justification for dosage selection for the reviewer would result from simulations for various scenarios enabled by well-defined PK/PD models. Thus, modeling strategies can aid in the creation of drugs based on the New Drug Model. Approaches for PK/PD analysis and modeling may also be helpful for post-marketing surveillance. To identify DDI, drug-disease interactions, or other factors like demography or genetics that affect a drug's effect or toxicity, population PK/PD modeling under structured monitoring may be helpful [47, 48]. For instance, a population PK method demonstrated a correlation between average plasma concentrations and the occurrence of adverse events following moclobemide medication [49]. The process of creating novel formulations, like an ER formulation, is typically thought to be time-consuming and expensive. An IVIVC is regarded as the best option in formulation development since it allows for the prediction of a formulation or manufacturing adjustment in the clinical performance. For carbamazepine in solid dose forms, the value of IVIVC and gastrointestinal simulation has been demonstrated for the justification of biowaivers [50].

5. REGULATORY ASPECTS

The development of safe and effective therapies for those who need them is a shared objective of drug sponsors and regulatory bodies. The Food and Drug Administration (FDA) outlines Guidance for Industry in providing clinical evidence of effectiveness for human drugs and biological products: 1. when efficacy can be extrapolated entirely from existing efficacy studies, and 2. when one adequate and well-controlled study is sufficient. In the latter scenario, the advice may additionally refers to further supporting facts and information, like knowledge and comprehension of the dose-systemic exposure-response link [1, 50]. Even though the guidance is imprecise in this area, PK/PD data and/or knowledge of PK/PD correlations can help clarify any doubts or concerns about the safety of drug doses and dosing regimens in specific patient populations while also supporting the case for a medication's efficacy [51]. Other regulatory decisions made during the approval process, such as providing a basis for logical label language related to doses and dosage adjustments when there is a significant increase or decrease in systemic exposure due to patient-intrinsic or extrinsic factors, may benefit from the use of PK/PD information in addition to the agency's determination that effectiveness has been sufficiently demonstrated in clinical outcome trials. PD may deliver more information on the risk/benefit analysis ratio, which may also be used to report exact issues. By demonstrating their receptivity to these ideas, federal regulatory bodies can have an impact on the adoption of PK/PD modeling and simulation [51].

FUTURE PROSPECTIVES AND CONCLUSION

It is undeniable that pharmaceutical companies and regulatory organizations have accepted the usage of PBPK modeling to maximize medication clinical potential, and its potential may grow even more in the future. It is assumed that PBPK modeling is an exhaustive and data-intensive process. More complex models will be created with an understanding of physiology and biochemical mechanisms, particularly in the context of various disease states, increases. Therefore, regardless of the user's degree of knowledge, PBPK modeling demands ongoing education as the models change. It is essential to have enough resources, proper training in applying the models, and in-depth knowledge of the ADME facts needed to power the models in order to successfully use PBPK modeling in the drug development process. The creation of reliable PBPK models, which can then be utilized prospectively to address various drug development-related concerns, is made possible by effective communication between personnel involved in preclinical drug discovery and clinical drug development. New insights into disease mechanisms demand rationally designed and fixed drug combinations, which can provide exciting new opportunities for new drug discovery and development, including new chemical entities and existing molecules that have long been used as medicines. This will result in emerging "systems therapeutics," with PK/PD modeling serving as the most important systematic *in vivo* tool to evaluate its rationality. The extensive acceptance and growing application of modeling PK/PD relationships in the industry, academia, and regulatory agencies will, in the future, not only provide opportunities for optimizing applied pharmacotherapy but will also speed up the process of drug development. Additionally, it will serve as a potent catalyst for clinical pharmacology's mission and aims as it enters a new era.

REFERENCES

[1]   Li C, Wang J, Wang Y, *et al.* Recent progress in drug delivery. Acta Pharm Sin B 2019; 9(6): 1145-62.
[http://dx.doi.org/10.1016/j.apsb.2019.08.003] [PMID: 31867161]

[2]   Ågerfalk PJ. Artificial intelligence as digital agency. Eur J Inf Syst 2020; 29(1): 1-8.
[http://dx.doi.org/10.1080/0960085X.2020.1721947]

[3]   Wang W, Ouyang D. Opportunities and challenges of physiologically based pharmacokinetic modeling in drug delivery. Drug Discov Today 2022; 27(8): 2100-20.
[http://dx.doi.org/10.1016/j.drudis.2022.04.015] [PMID: 35452792]

[4]   Hinder M, Hartl D. Translational Medicine–The Bridging Discipline. Role and tools in the drug development process. Principles Biomed Sci Ind 2022; pp. 119-38.

[5]   Chen B, Dong JQ, Pan WJ, Ruiz A. Pharmacokinetics/pharmacodynamics model-supported early drug development. Curr Pharm Biotechnol 2012; 13(7): 1360-75.
[http://dx.doi.org/10.2174/138920112800624436] [PMID: 22201585]

[6]   Lesko LJ, Rowland M, Peck CC, Blaschke TF. Optimizing the science of drug development: Opportunities for better candidate selection and accelerated evaluation in humans. J Clin Pharmacol 2000; 40(8): 803-14.
[http://dx.doi.org/10.1177/00912700022009530] [PMID: 10934664]

[7]   Kuepfer L, Fuellen G, Stahnke T. Quantitative systems pharmacology of the eye: Tools and data for ocular QSP. CPT Pharmacometrics Syst Pharmacol 2023; 12(3): 288-99.
[http://dx.doi.org/10.1002/psp4.12918] [PMID: 36708082]

[8]   Azad I, Khan T, Ahmad N, Khan AR, Akhter Y. Updates on drug designing approach through computational strategies: A review. Future Sci OA 2023; 9(5): FSO862.
[http://dx.doi.org/10.2144/fsoa-2022-0085] [PMID: 37180609]

[9]   Zhang L, Xie H, Wang Y, Wang H, Hu J, Zhang G. Pharmacodynamic parameters of pharmacokinetic/pharmacodynamic (PK/PD) integration models. Front Vet Sci 2022; 9: 860472.
[http://dx.doi.org/10.3389/fvets.2022.860472] [PMID: 35400105]

[10]   Chien JY, Friedrich S, Heathman MA, de Alwis DP, Sinha V. Pharmacokinetics/pharmacodynamics and the stages of drug development: Role of modeling and simulation. AAPS J 2005; 7(3): E544-59.
[http://dx.doi.org/10.1208/aapsj070355] [PMID: 16353932]

[11]   Sherwin CMT, Kiang TKL, Spigarelli MG, Ensom MHH. Fundamentals of population pharmacokinetic modelling: Validation methods. Clin Pharmacokinet 2012; 51(9): 573-90.
[http://dx.doi.org/10.1007/BF03261932] [PMID: 22799590]

[12]   Wang Y, Xing J, Xu Y, *et al. In silico* ADME/T modelling for rational drug design. Q Rev Biophys 2015; 48(4): 488-515.
[http://dx.doi.org/10.1017/S0033583515000190] [PMID: 26328949]

[13]   Holz M, Fahr A. Compartment modeling. Adv Drug Deliv Rev 2001; 48(2-3): 249-64.
[http://dx.doi.org/10.1016/S0169-409X(01)00118-1] [PMID: 11369085]

[14]   Leung HW. Use of physiologically based pharmacokinetic models to establish biological exposure indexes. Am Ind Hyg Assoc J 1992; 53(6): 369-74.
[http://dx.doi.org/10.1080/15298669291359799] [PMID: 1605109]

[15]   Khanday MA, Rafiq A, Nazir K. Mathematical models for drug diffusion through the compartments of blood and tissue medium. Alex J Med 2017; 53(3): 245-9.
[http://dx.doi.org/10.1016/j.ajme.2016.03.005]

[16]   Gabrielsson J, Weiner D. Non-compartmental Analysis. Methods Mol Biol 2012; 929: 377-89.
[http://dx.doi.org/10.1007/978-1-62703-050-2_16] [PMID: 23007438]

[17]   Lavi O, Gottesman MM, Levy D. The dynamics of drug resistance: A mathematical perspective. Drug Resist Updat 2012; 15(1-2): 90-7.
[http://dx.doi.org/10.1016/j.drup.2012.01.003] [PMID: 22387162]

[18]   Acharya C, Hooker AC, Türkyılmaz GY, Jönsson S, Karlsson MO. A diagnostic tool for population models using non-compartmental analysis: The ncappc package for R. Comput Methods Programs Biomed 2016; 127: 83-93.
[http://dx.doi.org/10.1016/j.cmpb.2016.01.013] [PMID: 27000291]

[19]   Park MY, Bae S, Heo JA, *et al.* Safety, tolerability, pharmacokinetic/pharmacodynamic characteristics of bersiporocin, a novel prolyl-tRNA synthetase inhibitor, in healthy subjects. Clin Transl Sci 2023; 16(7): 1163-76.
[http://dx.doi.org/10.1111/cts.13518] [PMID: 37095713]

[20]   Ette EI, Williams PJ. Population pharmacokinetics I: Background, concepts, and models. Ann Pharmacother 2004; 38(10): 1702-6.
[http://dx.doi.org/10.1345/aph.1D374] [PMID: 15328391]

[21]   Hsieh NH, Bois FY, Tsakalozou E, *et al.* A Bayesian population physiologically based pharmacokinetic absorption modeling approach to support generic drug development: Application to bupropion hydrochloride oral dosage forms. J Pharmacokinet Pharmacodyn 2021; 48(6): 893-908.
[http://dx.doi.org/10.1007/s10928-021-09778-5] [PMID: 34553275]

[22]   Hardiansyah D, Riana A, Eiber M, Beer AJ, Glatting G. Population-based model selection for an accurate estimation of time-integrated activity using non-linear mixed-effects modeling. Z Med Phys 2023; 1-9.

[23]   Hardiansyah D, Riana A, Beer AJ, Glatting G. Single-time-point estimation of absorbed doses in PRRT using a non-linear mixed-effects model. Z Med Phys 2023; 33(1): 70-81.
[http://dx.doi.org/10.1016/j.zemedi.2022.06.004] [PMID: 35961809]

[24]   Van de Schoot R, Depaoli S, King R, *et al.* Bayesian statistics and modelling. Nature Reviews Methods Primers 2021; 1(1): 1-60.
[http://dx.doi.org/10.1038/s43586-020-00001-2]

[25]   Sucharitha P, Reddy KR, Satyanarayana SV, Garg T. Absorption, distribution, metabolism, excretion, and toxicity assessment of drugs using computational tools. In Computational Approaches for Novel Therapeutic and Diagnostic Designing to Mitigate SARS-CoV2 Infection 2022; 1: 335-5.
[http://dx.doi.org/10.1016/B978-0-323-91172-6.00012-1]

[26]   Agamah FE, Mazandu GK, Hassan R, *et al.* Computational/*in silico* methods in drug target and lead prediction. Brief Bioinform 2020; 21(5): 1663-75.
[http://dx.doi.org/10.1093/bib/bbz103] [PMID: 31711157]

[27]   Hasan MR, Alsaiari AA, Fakhurji BZ, *et al.* Application of mathematical modeling and computational tools in the modern drug design and development process. Molecules 2022; 27(13): 4169.
[http://dx.doi.org/10.3390/molecules27134169] [PMID: 35807415]

[28]   Richard AM, Williams CR. Distributed structure-searchable toxicity (DSSTox) public database network: A proposal. Mutat Res 2002; 499(1): 27-52.
[http://dx.doi.org/10.1016/S0027-5107(01)00289-5] [PMID: 11804603]

[29]   Kesharwani CK, Vishwakarma VK, Keservani RK, Singh P, Katiyar N, Tripathi S. Role of ADMET tools in current scenario: Application and limitations. Comput-Aided Drug Des 2020; pp. 71-87.

[30]   Agour MA, Hamed AA, Ghareeb MA, Abdel-Hamid EAA, Ibrahim MK. Bioactive secondary metabolites from marine Actinomyces sp. AW6 with an evaluation of ADME-related physicochemical properties. Arch Microbiol 2022; 204(8): 537.
[http://dx.doi.org/10.1007/s00203-022-03092-5] [PMID: 35913539]

[31]   Bolleddula J, Brady K, Bruin G, *et al.* Absorption, distribution, metabolism, and excretion of therapeutic proteins: Current industry practices and future perspectives. Drug Metab Dispos 2022; 50(6): 837-45.
[http://dx.doi.org/10.1124/dmd.121.000461] [PMID: 35149541]

[32]   Mamadalieva NZ, Youssef FS, Hussain H, *et al.* Validation of the antioxidant and enzyme inhibitory potential of selected triterpenes using *in vitro* and *in silico* studies, and the evaluation of their ADMET properties. Molecules 2021; 26(21): 6331.
[http://dx.doi.org/10.3390/molecules26216331] [PMID: 34770739]

[33]   Daina A, Michielin O, Zoete V. SwissADME: A free web tool to evaluate pharmacokinetics, drug-likeness and medicinal chemistry friendliness of small molecules. Sci Rep 2017; 7(1): 42717.
[http://dx.doi.org/10.1038/srep42717] [PMID: 28256516]

[34]   Wong H, Gould SE, Budha N, *et al.* Learning and confirming with preclinical studies: Modeling and simulation in the discovery of GDC-0917, an inhibitor of apoptosis proteins antagonist. Drug Metab Dispos 2013; 41(12): 2104-13.
[http://dx.doi.org/10.1124/dmd.113.053926] [PMID: 24041744]

[35]   Bolser DM, Chibon PY, Palopoli N, *et al.* MetaBase: The wiki-database of biological databases. Nucleic Acids Res 2012; 40(D1): D1250-4.
[http://dx.doi.org/10.1093/nar/gkr1099] [PMID: 22139927]

[36]   Franco YL, Da Silva L, Charbe N, Kinvig H, Kim S, Cristofoletti R. Integrating forward and reverse translation in pbpk modeling to predict food effect on oral absorption of weakly basic drugs. Pharm Res 2023; 40(2): 405-18.
[http://dx.doi.org/10.1007/s11095-023-03478-0] [PMID: 36788156]

[37]   Yu J, Karcher H, Feire AL, Lowe PJ. From target selection to the minimum acceptable biological effect level for human study: Use of mechanism-based PK/PD modeling to design safe and efficacious biologics. AAPS J 2011; 13(2): 169-78.
[http://dx.doi.org/10.1208/s12248-011-9256-y] [PMID: 21336535]

[38]   Balani S, Miwa G, Gan LS, Wu JT, Lee F. Strategy of utilizing *in vitro* and *in vivo* ADME tools for lead optimization and drug candidate selection. Curr Top Med Chem 2005; 5(11): 1033-8.
[http://dx.doi.org/10.2174/1568026605774297038] [PMID: 16181128]

[39]   Sinha VK, Snoeys J, Osselaer NV, Peer AV, Mackie C, Heald D. From preclinical to human – prediction of oral absorption and drug–drug interaction potential using physiologically based pharmacokinetic (PBPK) modeling approach in an industrial setting: A workflow by using case example. Biopharm Drug Dispos 2012; 33(2): 111-21.
[http://dx.doi.org/10.1002/bdd.1782] [PMID: 22383166]

[40]   Kretsos K, Jullion A, Zamacona M, *et al.* Model-based optimal design and execution of the first-inpatient trial of the anti-IL-6, olokizumab. CPT Pharmacometrics Syst Pharmacol 2014; 3(6): 1-8.
[http://dx.doi.org/10.1038/psp.2014.17] [PMID: 24941311]

[41]   Adjei A, Teuscher NS, Kupper RJ, *et al.* Single-dose pharmacokinetics of methylphenidate extended-release multiple layer beads administered as intact capsule or sprinkles *versus* methylphenidate immediate-release tablets (Ritalin(®)) in healthy adult volunteers. J Child Adolesc Psychopharmacol 2014; 24(10): 570-8.
[http://dx.doi.org/10.1089/cap.2013.0135] [PMID: 25514542]

[42]   Wang Y, Zhu H, Madabushi R, Liu Q, Huang SM, Zineh I. Development of physiologically based pharmacokinetic model to evaluate the relative systemic exposure to quetiapine after administration of IR and XR formulations to adults L.Wang Y, Zhu H, Madabushi R, Liu Q, Huang SM, Zineh I. Model-informed drug development: Current US regulatory practice and future considerations. Clin Pharmacol Ther 2019; 105(4): 899-911.
[http://dx.doi.org/10.1002/cpt.1363] [PMID: 30653670]

[43]   Girgis IG, Nandy P, Nye JS, *et al.* Pharmacokinetic–pharmacodynamic assessment of topiramate

dosing regimens for children with epilepsy 2 to <10 years of age. Epilepsia 2010; 51(10): 1954-62.
[http://dx.doi.org/10.1111/j.1528-1167.2010.02598.x] [PMID: 20880232]

[44]   Min JS, Bae SK. Prediction of drug–drug interaction potential using physiologically based pharmacokinetic modeling. Arch Pharm Res 2017; 40(12): 1356-79.
[http://dx.doi.org/10.1007/s12272-017-0976-0] [PMID: 29079968]

[45]   Heal DJ, Gosden J. What pharmacological interventions are effective in binge-eating disorder? Insights from a critical evaluation of the evidence from clinical trials. Int J Obes 2022; 46(4): 677-95.
[http://dx.doi.org/10.1038/s41366-021-01032-9] [PMID: 34992243]

[46]   Varma MV, Pang KS, Isoherranen N, Zhao P. Dealing with the complex drug–drug interactions: Towards mechanistic models. Biopharm Drug Dispos 2015; 36(2): 71-92.
[http://dx.doi.org/10.1002/bdd.1934] [PMID: 25545151]

[47]   Fujita K, Masnoon N, Mach J, O'Donnell LK, Hilmer SN. Polypharmacy and precision medicine. Cambridge Prisms. Precision Medicine 2023; 1: 1-22.

[48]   Toutain PL, Lees P. Integration and modelling of pharmacokinetic and pharmacodynamic data to optimize dosage regimens in veterinary medicine. J Vet Pharmacol Ther 2004; 27(6): 467-77.
[http://dx.doi.org/10.1111/j.1365-2885.2004.00613.x] [PMID: 15601441]

[49]   Bandeira LC, Pinto L, Carneiro CM. Pharmacometrics: The already-present future of precision pharmacology. Ther Innov Regul Sci 2023; 57(1): 57-69.
[http://dx.doi.org/10.1007/s43441-022-00439-4] [PMID: 35984633]

[50]   Kovačević I, Parojčić J, Homšek I, Tubić-Grozdanis M, Langguth P. Justification of biowaiver for carbamazepine, a low soluble high permeable compound, in solid dosage forms based on IVIVC and gastrointestinal simulation. Mol Pharm 2009; 6(1): 40-7.
[http://dx.doi.org/10.1021/mp800128y] [PMID: 19248231]

[51]   Mitra A, Wang Y. Applications of model informed drug development (MIDD) in drug development lifecycle and regulatory review. Pharm Res 2022; 39(8): 1663-7.
[http://dx.doi.org/10.1007/s11095-022-03327-6] [PMID: 35790617]

Experimental Tools as an "Alternative to Animal Research" in Pharmacology

Kunjbihari Sulakhiya[1,*], Rishi Paliwal[2], Anglina Kisku[1], Madhavi Sahu[1], Shivam Aditya[1], Pranay Soni[3] and Saurabh Maru[4]

[1] *Neuro Pharmacology Research Laboratory (NPRL), Department of Pharmacy, Indira Gandhi National Tribal University, Amarkantak, Madhya Pradesh, India*

[2] *Nanomedicine and Bioengineering Research Laboratory (NBRL), Department of Pharmacy, Indira Gandhi National Tribal University, Amarkantak, Madhya Pradesh, India*

[3] *Department of Pharmacy, Indira Gandhi National Tribal University, Amarkantak, Madhya Pradesh, India*

[4] *Department of Pharmacology, School of Pharmacy and Technology Management, SVKM's Narsee Monjee Institute of Management Studies, Maharashtra, India*

Abstract: Experimental tools have emerged as a promising alternative to animal research in pharmacology. With growing ethical concerns and regulatory restrictions surrounding animal experimentation, researchers are increasingly turning towards *in vitro* and *in silico* methods to develop new drugs and evaluate their safety and efficacy. *In vitro* tools include cell culture systems, 3D organoid models, and microfluidic devices replicating complex physiological conditions, such as the blood-brain barrier or the liver microenvironment. These systems can provide more accurate and predictive results than animal models, reducing ethical concerns and experimental costs. *In silico* methods, such as computer modelling, simulation, and artificial intelligence, enable researchers to predict the drug-target interactions, toxicity, and pharmacokinetic and pharmacodynamic properties of new drugs without animal testing. Experimental tools have several advantages over animal research, including more accurate and predictive results, lower costs, higher throughput, and reduced ethical concerns. However, the limitations of these tools must also be acknowledged, such as the inability to fully replicate the complexity of a living organism, which requires further validation. These tools offer a promising avenue for advancing pharmacological research while reducing the reliance on animal experimentation. In conclusion, experimental tools provide an excellent alternative to animal research in pharmacology to identify and avoid potential toxicities early in the drug discovery process and have the potential to revolutionize drug discovery and development. This chapter mainly focuses on the numerous *in vitro*, *in silico*, non-animal *in vivo*, and emerging experimental tools and their regulatory perspectives on validation, acceptance, and implementation of the alternative methods used in pharmacological research.

* **Corresponding author Kunjbihari Sulakhiya:** Neuro Pharmacology Research Laboratory (NPRL), Department of Pharmacy, Indira Gandhi National Tribal University, Amarkantak, Madhya Pradesh, India; E-mails: niperkunj@gmail.com, kunj@igntu.ac.in

Keywords: Animal research, Cell culture, Computer modelling, Emerging tools, Microfluidic devices, Organ-on-a-Chip technology, Pharmacology, Regulatory guidelines.

1. INTRODUCTION

The use of animals for various purposes is as old as human beings. However, the ethical implications of animal testing have led to the development of experimental tools that can serve as an alternative to animal testing [1]. These tools are based on the 3Rs: reduction, refinement, and replacement of animal testing. Reduction aims to minimize the number of animals used in testing, while refinement aims to improve the welfare of animals. Replacement aims to replace animal testing with alternative methods, such as human cell-based assays, 3D cell cultures and organoids, and human volunteer studies [2]. These tools can provide valuable information on drug efficacy and toxicity, while human volunteer studies can provide information on the pharmacokinetics and pharmacodynamics of drugs. There is also a focus on the fourth R, which stands for the Rehabilitation of animals after usage [3]. Animals were initially used in research and instruction when people started exploring cures and preventative measures for the disease. Using animals in research makes it possible to find the bulk of current medications. The laws governing the use of animals in research and education still need to be clarified as they are contradictory. Each laboratory and educational facility, however, has a unique policy regarding the use and treatment of animals. The use of animals in irrelevant experiments has paid attention to and raised concerns due to the widespread use of animals in toxicity research and the testing of dermatological treatments. Conversely, other scientists believe using animals in research is a responsible and suitable way to progress discoveries.

For a very long time, medical institutes in India have included animal experiments as a necessary component of their pharmacology curriculum. Despite attempts by activists and concerned instructors to lower this number, thousands of animals are still utilized annually in educational facilities. In India and other nations, several medical institutions have either offered alternatives to these trials or are debating this divisive subject. Some people believe that medicine can only be taught or learned with exposure to wards and clinics, and so can pharmacology without animal experiments. However, with evolving educational ideas and practices, there is a growing consensus that animals shouldn't be killed to learn experimental procedures and abilities. These tests are costly, time-consuming, and tedious. In this evolving situation, the development of alternatives to animal testing is the need of the day [3].

2. BRIEF HISTORY OF ANIMAL RESEARCH IN PHARMACOLOGY

The use of animals in biomedical research dates back to ancient Greek times, with Aristotle, Erasistratus, and Galen, all were performing experiments on living animals. In the Renaissance, philosopher Francis Bacon valued anatomist Vesalius and Animal physiological experiments as a teaching and learning tool [4]. The Age of Enlightenment saw continued physiological experiments on animals. René Descartes described animals as "machine-like," Immanuel Kant rejected Cartesian mechanistic views and acknowledged sentience to other animals [5]. The eighteenth century saw notable contributions to experimental physiology from polymaths Stephen Hales and Albrecht von Haller. Opposition to vivisection became more prominent in the second half of the century, particularly in northern Europe [6]. In the twentieth century, animal research played a role in several significant medical discoveries, but it also led to disasters such as the mass poisoning caused by the elixir sulphanilamide. The Federal Food, Drug, and Cosmetic Act of 1938, which mandated that pharmaceuticals undergo animal safety testing before being approved for human consumption, was passed in response to public outcry [4]. The development of alternative methods to animal research has been illustrated in Fig. (1).

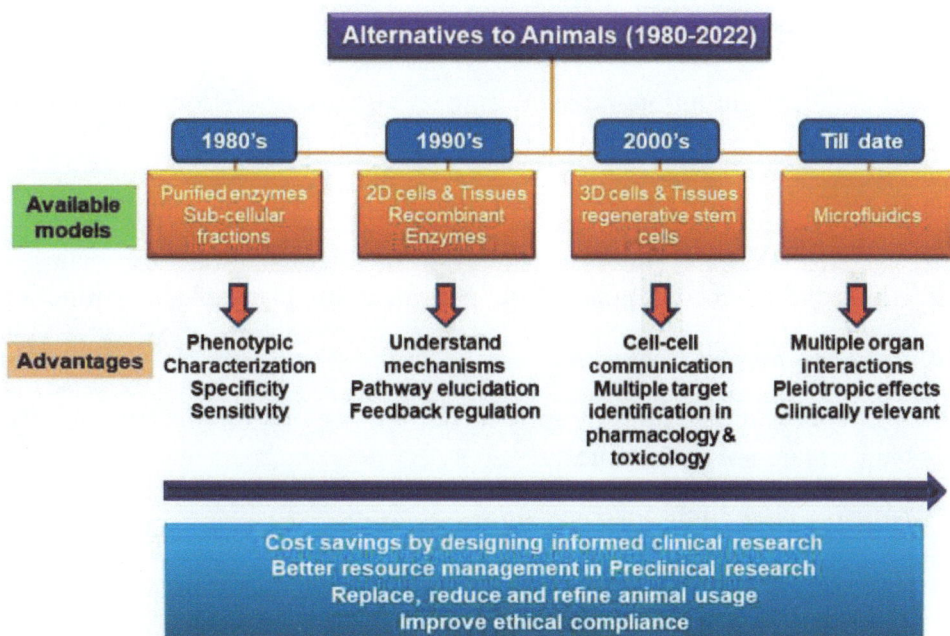

Fig. (1). Development of alternative research methods during the past 50 years.

2.1. Overview of Alternative Experimental Tools and Techniques

Various experimental tools and techniques can be used as alternatives to animal testing in biomedical research. Some of these include:

2.1.1. Cell Cultures

Researchers can grow human or animal cells in a laboratory setting to study how they react to different treatments or stimuli [7].

2.1.2. Computer Modeling and Simulation

With the help of advanced computer models, researchers can simulate the drug's effects and other treatments on the human body, reducing the need for animal testing.

2.1.3. Microfluidic Sevices

These small gadgets that imitate the composition and operation of human organs and tissues enable researchers to conduct more accurate and realistic medication and therapy trials.

2.1.4. Human Tissue Samples

Researchers can get human tissue samples from operations or autopsies to examine the effects of medications and other therapies.

2.1.5. In vitro Assays

These lab tests utilize biochemical or molecular methods to examine how medications and other treatments affect certain cellular or molecular functions.

2.1.6. Epidemiological Research

These studies examine illness trends in human populations to find possible risk factors and create novel therapies.

2.1.7. Non-invasive Imaging Techniques

These techniques, such as MRI and PET scans, allow researchers to study the drugs' effects and other treatments for humans in a non-invasive way.

2.1.8. High-throughput Screening

This technique uses automated systems to rapidly test many compounds' potential effects on specific cellular or molecular processes [8].

These and other alternative methods can help reduce the number of animals used in biomedical research while allowing researchers to make important discoveries and develop new treatments for human diseases (Fig. **2**).

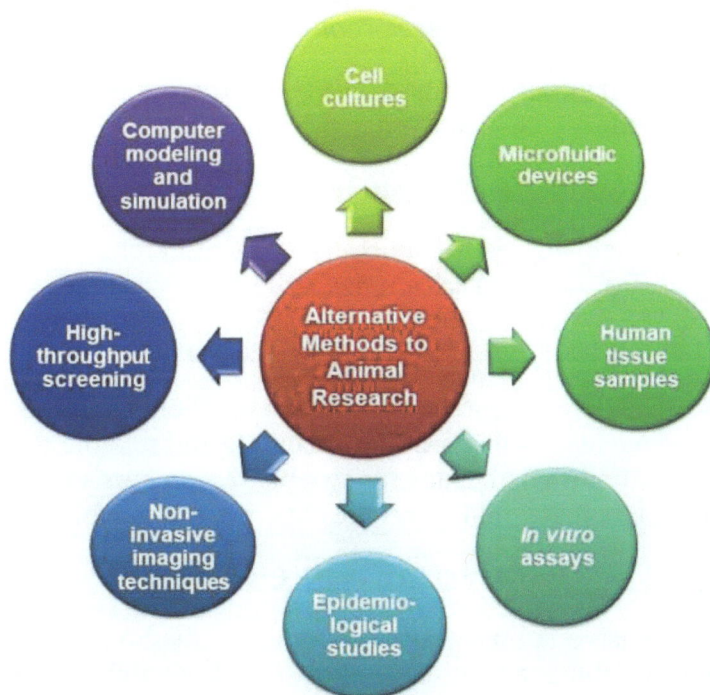

Fig. (2). Diagrammatic representation of alternative methods to animal research.

2.2. Need for Alternative Methods to Animal Research in Pharmacology

Animals' suffering, mortality, and misery during scientific studies have long been controversial. Alternatives are required for initial, intermediate, and final *in vivo* testing at all phases [1]. The recent creation of 3D human-derived models for translation into non-animal technologies was a notable advancement [9]. Additionally, simultaneous technological breakthroughs such as those in systems biology tools are assisting in discovering new biomarkers. *In silico*, computer-based models for predicting human pharmacology, toxicology, and other biological characteristics are also developing quickly. Future human correlations should be created using human-derived, non-animal technologies and dynamic flow models replicating human physiology, biochemistry, pharmacogenetics, and other pertinent factors. We should obtain more vitally human-relevant data that can be used for clinical trial design and enhancing effective clinical outcomes by optimizing these human-derived models with -omics and *in silico* techniques. In conclusion, the demand for animal research in pharmacology is now unabated.

While we aspire to a day when our knowledge will be so complete that animal testing will no longer be necessary, we must acknowledge that we are still far from that state of affairs [10].

2.2.1. Benefits of Using Alternative Methods

There are many potential benefits of using alternative methods in research and experimentation, some of which include the following:

2.2.1.1. Reduced Reliance on Animal Testing

Alternatives can help to reduce animal use in research, testing, and teaching, minimizing pain and suffering for animals used.

2.2.1.2. Greater Accuracy and Dependability of Results

Modern techniques like computer simulations, mathematical models, and *in vitro* testing can produce results that are more accurate and reliable than those obtained through the use of animals because they give researchers more control over the variables and factors that could affect the experiment's outcome.

2.2.1.3. Research that is Quicker and more Cost-effective

By avoiding extensive animal testing, alternative methods enable researchers to examine many possible medication candidates and other chemicals swiftly.

2.2.1.4. More Ethical Research

Using cutting-edge techniques like human tissue engineering, organ-on-a-chip systems, and stem cell research, which maximizes accuracy and reliability while minimizing harm to human subjects, alternatives can assist researchers in overcoming ethical and legal challenges related to human experimentation.

2.2.1.5. The Safety and Effectiveness of Medications and other Chemicals

Alternatives can contribute to the safety and efficacy of drugs and other substances by giving researchers access to more precise and dependable information on the possible impacts of these compounds on human and animal health.

2.2.1.6. More Sustainable Research Practices

Alternatives can help promote more sustainable research practices by reducing the use of resources such as animals, reducing waste and pollution, and promoting the use of renewable energy sources.

Overall, alternative methods used in research and experimentation can provide many benefits, including improved accuracy and reliability of results, reduced reliance on animal testing, faster and more cost-effective research, more ethical research practices, improved safety and efficacy of drugs and other substances, and more sustainable research practices [11].

2.3. Ethical Concerns and Criticisms of Animal Research

More animals are used in research as medical technology advances. Millions of experimental animals are used annually all around the world. For example, 3.71 million animals were used in a study conducted in 2011 (www.rspca.org.uk). As estimated by the research study, 1,131,076 animals were used in the United States in 2009 compared to 2.13 million in Germany in 2001. Animals are taken away from their social groupings and used as instruments in clinical testing facilities despite their natural tendencies. The experimental techniques can use the complete animal or its organs and tissues [1, 12]. For this reason, defined protocols are used to euthanize (kill) animals. At the end of an experiment, the animals who survive the clinical testing are typically put to death to protect them from further suffering. In certain situations, the investigation may result in animal deaths (for instance, in LD 50 analyses). Many statutes and legislation have been put into place to lessen animal suffering during research and restrict the unethical use of animals. For instance, the group for animal rights was established in 1824 by the Royal Society for the Prevention of Cruelty to Animals. Animal cruelty was outlawed in the UK in 1876 according to law. It emerged in 1960, 1963, and 1966 in India, France, and the USA, respectively. Numerous rules and regulations are followed worldwide to safeguard animals from mistreatment and cruelty. Guidelines for animal housing, breeding, feeding, transportation, and primarily their use in scientific experiments are provided by organizations like ICH (International Conference on Harmonization of Technical Requirements for Registration of Pharmaceuticals for Human Use), CCSEA (Committee for Control and Supervision of Experiments on Animals) NIH (National Institutes of Health), and OECD [1, 12].

Aside from the primary ethical concern, other limitations of animal research include the requirement for knowledgeable/trained staff and time-consuming procedures. Another drawback is the outrageous cost of living, breeding, and the elaborated protocols used in animal experiments. Alternatives in experimental design and techniques are used to provide researchers with different ways of conducting experiments and improve the accuracy and reliability of results while minimizing the use of animals, humans, or other resources. One of the primary purposes of using alternative methods is to reduce the number of animals used in research, testing, and teaching, and to minimize pain and distress for animals used. This is often done using alternative methods involving *in vitro* (outside the living organism), computer simulation, or mathematical models that can provide information on the efficacy, toxicity, and safety of drugs, chemicals, and other substances without the need for animal testing. Alternative methods can also help improve the efficiency and cost-effectiveness of the drug discovery and development processes by enabling researchers to quickly screen a large number of potential drug candidates and other substances in a shorter amount of time without going through a time-consuming animal testing process. Further, cutting-edge techniques like stem cell research, human tissue engineering, and human organ-on-a-chip systems give researchers more accurate and reliable results while minimizing harm to human participants.

Various methods can also be used to overcome ethical and legal challenges associated with human experimentation. They can reduce the number of animals used in research, the pain and suffering of those animals, and increase the safety and effectiveness of medications and other chemicals. Ultimately, this advances scientific understanding and safeguards human and animal health.

2.4. *In vitro* Methods for Alternative to Animal Research

The *in vitro* methods are performed outside of biological cells, that are often regarded as test-tube experiments or research, and are generally performed in labware, *e.g.*, flasks, test tubes, microtiter plates, and Petri dishes. The most famous example is *in vitro* fertilization (or test tube babies) in hospitals to overcome fertilization issues among couples and bless them with a baby. These methods are often used to investigate the absorption, distribution, metabolism, excretion, and toxicity (ADMET, the efficacy of drugs, and targeting the biomolecules of the living system to decrease or eliminate the animal modelling based-research and minimize the cost of clinical trials and biological phenomenon.

The *in vitro* studies enable convenient and in-depth analysis of various biological mechanisms and help to predict cellular mechanisms, including ADMET analysis of pharmaceutical drugs and cellular chemicals [13]. The European Centre for the Validation of Alternative Methods (ECVAM) is actively engaged in developing and assessing alternative reliable protocols related to *in vitro*, Quantitative structure-activity relationship (QSAR), and *in vivo* studies. The clinical trials have five stages; drug discovery and development, pre-clinical examination, clinical trials, FDA approval, and post-market drug safety monitoring by the regulatory agencies. Clinical trials deal with the evaluation of the safety and effectiveness of the drug against the standard treatment available. Drug approval often fails due to insufficient quantitative data related to safety measures [14]. This clinical trial research is an expensive affair that can be overcome through *in vitro* and *in vivo* extrapolations (IVIVE) models that have been suggested to predict drug metabolism and pharmacokinetics (DMPK) [15]. Presently, consistent improvements have been made to develop more dynamic three-dimensional (3D) models of the organs, *i.e.*, the liver and kidney.

2.4.1. Types of In vitro Models

Various *in vitro* models are being developed to appraise the pharmacodynamics (PDs) and pharmacokinetics (PKs) of drugs. Such models are also examined towards substituting animal model-based traditional clinical trials to cell lines and 3D cell models-based modern clinical trial approaches [16]. Additionally, organoids, 3D tissue models, and organ-on-a-chip are evaluated to study PDs and PKs of drug candidates and disease modelling. Some *in vitro* models for pharmacological research are:

2.4.1.1. 3D Models of Cells and Tissues

Are scaffold-based 3D materials that are the *in vitro* co-culture of diversified and dynamic cells. These cells can migrate and grow over a 3D scaffold. The 3D cell model maintains the cellular microenvironment and can establish a research model for studying target cellular phenomena under a controlled microenvironment. It is often used to screen, validate and evaluate drug toxicity (*i.e.*, genotoxin and neurotoxin). Examples include the humanized recombinant 3D-skin-cell model [17] as an alternative to mammalian cells, which is recognized by the International Workgroup of Genetic Toxicology (IWGT).

2.4.1.2. Organoids

Are stem cells derived from self-organizing *in vitro* organ models that mimic entire physiological processes and characteristics of the native organs of living organisms. The organoids have very close configuration and behaviour to native organs and stable genetic makeup, makings them suitable for high-throughput genetic screening and disease modelling [18]. The organoids are relatively modest compared to sophisticated animal models that offer convenient disease modelling, screening, and evolution drugs and demonstrate drug toxicity, and other therapeutic applications [19].

2.4.1.3. Artificial Membranes

Make drug screening more simple, efficient, rapid, and customized as per research demand. The parallel artificial membrane permeability (PAMP) assay technique is presently used for drug screening. Artificial phospholipids-based biofilms are developed that mimic the natural drug's transmembrane barrier and membrane permeation, which makes them reliable for *in vitro* drug screening. Some artificial membrane-based models are blood-brain barrier and skin absorption models for the screening of drugs. These models provide flexible, cost-effective, and reproducible drug screening and characterization.

2.4.1.4. 3D Tissue Models

Offer a precise and accurate simulation of drug interaction in the cellular microenvironment of the human body. Such models can effectively be used to explore the effects of various drug candidates on human cells, characterize the micro-anatomy of cells, and establish cell-to-cell interaction models to define specific cellular phenomena. Recently, 3D-tissue models have been used to discover effective antiviral vaccines [20].

2.4.1.5. Organ-On-a-Chip

Is the sophisticated 3D-microfluidic miniaturized organ model embedded in a chip and used to simulate physiological responses of target molecules towards organs. New frontier technologies are used to develop numerous stem cell-based, self-organizing miniature organoids to mimic structural and functional features *in vivo* counterparts of target organs. The emergence of interdisciplinary research opens up the field of innovative engineering for biological research. The use of engineering principals for the production of organoids and in-depth insides of their microenvironment have been consistently investigated to reconstruct the physiologically and mechanically functional *in vitro,* precisely controlled organ models with smooth fluids movement (microfluidics) and dynamic interfacing ability at cells or tissue level. The dynamic *in vitro* organ models are prepared by advanced biological engineering technologies that are very close to *in vivo* organs and are broadly applicable in life sciences, healthcare, pharmaceuticals, drug development, clinical trials, development of personalized medicine, toxicity assessment, and biological defence research studies [21]. A recent example includes the fabrication of an oxygen-sensitive intestine (O2SI)-on-a-chip model to assess the oxygen-sensitive bacterial interaction with the intestine [22]. Likewise, the *in vitro* model, namely, the environmental-enteric-dysfunction-chip, has been used to screen intestinal drugs and to predict possible therapies for intestinal dysfunction caused by environmental stress [23].

2.4.1.6. Cell Cultures/Tissue Cultures

The embryonic stem cells (ESCs), induced pluripotent stem cells (iPSC), Epstein-Barr virus-transformed lymphoblastoid cell lines (E-BvTLCs), human lymphoblastoid cell lines (HLCs), and modified tumour cell lines (TCs) are frequently used towards pharmaceutical research models. The ESCs and iPSCs are generally used to synthesize normal human cell lines for drug screening and toxicity prediction [24]. The immortalized E-BvTLCs cell lines have also been widely accepted to examine the consequence related to germline genetic variation while evaluating drug potency and toxicity prediction [25]. The genetic and epigenetic biomarkers related to metformin for anticancer response [26] and genetic variants associated with Homoharringtonine response [27] have been evaluated using HLCs. The modified TCs have been used to reveal cancer biology and screen anticancer drugs [28].

2.4.1.7. Tissue Engineering

Tissue engineering (TE) emerged in the 1980s to develop biological substitutes to restore, replace, or regenerate damaged and defective tissues using *in vitro* methods [29, 30]. The TE deals with developing functional cell or tissue constructs using biological principles and advanced engineering that can restore injured tissues or organs. The example includes artificial skin and connective (cartilage) tissues, which are approved by the Food and Drug Administration (FDA). The engineering triad, viz., cells, scaffolds, and growth-factor signals, are the primary components of constructing engineered tissue [31]. Scaffolds are designed of biopolymeric materials which provide structural support to the cell's attachment, growth, and proliferation. The technical aspects of TE comprise the selection of target cells, isolation, strategies of progenitor cell synthesis, stimulating cell differentiation using suitable growth factors, and scaffold design [32].

2.4.1.7.1. Applications of Tissue Engineering

The central applied area of TE includes tissue-engineered bone and skin, nano-biomaterials for drug delivery, drug-molecular signalling, cell reprogramming, synthetic soft tissue, artificial blood vessels, engineered heart models for the study of cardiovascular diseases, *i.e.*, arrhythmogenic cardiomyopathy using iPSC, development of organ-on-chip, 3-D matrix, and *in vitro* organ models. Tissue engineering (TE) principal and techniques help to construct artificial tissues to repair damaged tissues, investigate the behaviour of stem cells, and develop models for the analysis of severe diseases. Recent developments exhibit that there is the potential to integrate with the synthesis of regenerative medicines for cancer patients along with chemotherapy, gene therapy, and immunotherapy to encourage *in vivo* tissue healing and regeneration [33], and controlled-drug release to ensure minimal side effects [34 - 36].

2.4.1.7.2. Advantages and Disadvantages of Tissue Engineering

The advantageous context of TE includes effective scaffolds almost similar to *in vivo* extracellular matrix that could facilitate attachment and proliferation of cells. It provides numerous optimization and surface modification opportunities to predict several *in vivo* therapeutic measures [37]. TE works to improve healthy life, solving numerous complex medical problems using *in vitro* tissue- or organ models [38], and developing regenerative medicines [39]. The artificial organs have been developed using TE to screen, predict, and simulate several medically essential biomolecules and drugs, along with disease modelling [40].

The major disadvantages associated with TE are early degradation, insufficient mechanical strength of the scaffold, sudden uncertain cellular interaction, incompatible with *in vivo* cellular microenvironments, growth factors associated issues, *e.g.*, ectopic bone, osteolysis, and swelling [38, 41]. Tissue engineering depends on a mechanically potent scaffold for target cell or tissue attachment and proliferation. The problem encountered like, the ϵ-Caprolactone-based scaffold has excellent mechanical potency and *in vivo* biocompatibility and does not need any reinforcement. Still, it does not meet the ideal scaffold properties for tissue engineering, *e.g.*, due to a low absorption rate [42]. The scaffold has enough properties to facilitate its fixation to target *in vivo* cell microenvironments. Another issue arises is that animal models are often used to assess scaffold efficacy. As discussed earlier, the physiological microenvironment and mechanical properties have differed from species to species or species to genus—moreover, tumorigenesis, graft rejection, histological and cell migration [43].

2.4.2. Advantages and Disadvantages of In vitro Methods

Cell-based *in vitro* assays are valued tools in drug development studies due to their cost-effectiveness, reliability, and rapid results [44]. These assays are used to investigate cellular and functional characteristics of the disease and assist in better therapeutic interventions [45]. The advantageous part includes *in vitro* methods controlled cellular microenvironment, high throughput, and minimal use of animals [46]. Sophisticated laboratories and skilled personnel are required as compared to *in vivo* methods. Substantially, *in vitro* methods are the first and foremost choice of the pharmaceutical sector to produce large-scale cell-based products (*e.g.*, red blood carpels, monoclonal antibodies) for medical applications due to most minor ethical concerns, *i.e.*, ethical clearance from the American Association for Laboratory Animal Science (AALAS) and Institutional Animal Care and Use Committee (IACUC). However, small-scale production of cell-based products using *in vitro* culture methods is usually expensive [47].

However, *in vitro* methods cannot accurately predict the whole organism, but a satisfactory and significant conclusion can be drawn regarding the same [48]. The composition of the extracellular matrix, matrix stiffness, concentration gradients, and stromal cells are the major microenvironmental factors that define cellular phenotypic characteristics and affect drug response inside the living organism (*in vivo* system). As mentioned earlier, developing *in vitro* models with active biological interaction between factors seems tough to incorporate.

2.4.3. In vitro Models Applied in Pharmacological Research

As mentioned earlier, the *in vitro* synthesized cell lines and models generally fail to replicate and reproduce in the native cellular microenvironment (*in vivo*). To eradicate limitations as mentioned earlier, the United States-based company BMSEED has developed Micro-Electrode Array Stretching Stimulating und Recording Equipment (M-EASSRE) research tool [46] such as, Electrophysiology Module (EM) to control electrophysiology, Mechanics Module (MM) to facilitate cell stretch for neurological analysis, tissue engineering, and drug screening assays, Mechanics & Imaging Module (MIM) to predict strain profiles and various imaging tools to enhance resolution for cell culture studies. Some tissue engineering-based organoids, organ- or tissue-, and cell-on-chip models for pharmaceutical drug screening and clinical trials are mentioned in Table **1**.

Table 1. Tissue engineering-based models for pharmaceuticals.

Model	Pharmaceutical Application Area	References
Cardiac-organoids	Cardiovascular drug screening, trials, and disease modelling.	[49, 50]
Disease Models	Predict the possible therapy for disease.	[51]
Pancreatic cancer organoids	Organoids used to evaluate potential drug and therapeutic opportunities.	[52]
Liver cancer organoids	Organoids used to evaluate potential drug and therapeutic opportunities.	[40]
3-D neurosphere model	*In vitro* neurotoxicity of bioactive agents, *e.g.*, harmane.	[53, 54]
Functional vasculature	Disease modelling and drug testing.	[18]
Numerical Simulation	Predicts the behaviors of *in vitro* organoids using mathematical and statistical models.	[55]
Miniaturized gut–liver–brain axis bio-platform	*In vitro* prediction of a cellular phenomenon and dysfunction associated with the brain, gut, and liver.	[56]
Female reproductive bioengineered models	Cervix, fallopian tubes, ovaries, uterus, and vagina dysfunction modelling and prediction of possible therapy.	[57, 58]
Multitissue-on-a-chip platform	Able to predict multicellular biological phenomena and drug interaction.	[59]

(Table 1) cont.....

Model	Pharmaceutical Application Area	References
Cell and tissue-based physiological micro-models	High throughput *in vitro* pre-clinical trials of a drug candidate.	[60]
Mini-gut *in vitro* model	*In vitro* model to screen intestinal drugs and to predict possible therapies for intestinal dysfunction.	[61]
Immune Organs-on-Chip	*In vitro* drug development research.	[62]
HuMiX model	Human–microbial crosstalk model is used to explore cellular integration and signalling between humans and microbes, and based on the research findings, possible drugs and therapy can be prescribed.	[63]
Intestine-on-a-chip	The miniaturized chip-based model that is used to screen intestinal drugs and to predict possible therapies for intestinal dysfunction.	[64]
3-D tissue models	*In vitro* drug ADMET study.	[65]
Intestine organoids	3-D models used to screen intestinal drugs and to predict possible therapies for intestinal dysfunction.	[66]

2.4.4. Comparison with Animal Studies

The *in vitro* cultivated primary cells are mostly different from the *in vivo* corresponding cell type in an organism and limit the outcome of *in vitro* data towards the prediction of *in vivo* physiological behaviour. The animal models often fail to define species-specific differences, which can be overcome by using stem cells derived self-organization-enabled organoids [19]. However, more precise fabrication and simulation must still be introduced *in vitro* modelling and tissue engineering to make it close to the *in vivo* microenvironment.

3. *IN SILICO* METHODS (COMPUTER MODELING AND SIMULATION)

In silico models, which simulate complex systems using equations or rules, are becoming more and more common in clinical research and drug development. Comparing computational models and simulations to human-based clinical trials offers significant benefits regarding both operational aspects and therapeutic outcomes [67]. Applications for *in silico* models range from optimizing medica-tion candidates to forecasting the toxicity of contaminants or agrochemicals to environmental species. Now, there is more awareness of the potential of *in silico* tools to give information for regulatory submissions to fulfill legislative expectations since the personal care products, (agro)chemicals, and food sectors all employ a variety of *in silico* tools in product development [68].

The strength of *in silico* models is that all predictions are based only on the structures of the relevant molecules [68]. With the help of *in silico* computational models, researchers can compare treatments for specific diseases on a qualitative and quantitative level, along with testing a wider variety of different hypotheses (*e.g.*, dosing) (Fig. **3**). These models depict human illnesses in an abstract way, an idea that is frequently constrained by *in-vitro/in-vivo* techniques [67].

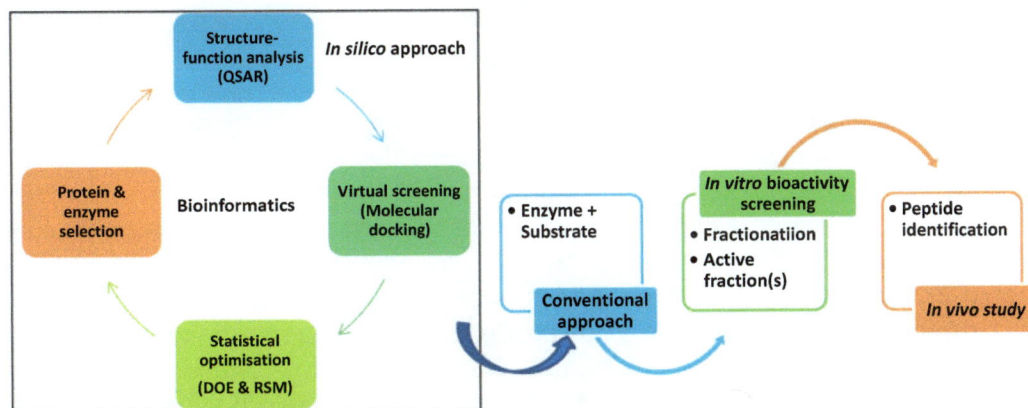

Fig. (3). Diagrammatic illustration of *in silico* and conventional approach for computer modeling and simulation techniques. QSAR: quantitative structure-activity relationship; DOE: design of experiment; RSM: response surface methodology [69].

The use of animals in research could eventually be reduced or even replaced entirely by computer-based models and methods, such as *in silico*. The pharmaceutical industry has focused on the transition from animal models to computational versions for many crucial reasons, including:

3.1. Animal Eelfare

Concern for animals' mental and physical well-being has led several companies to diversify in alternative ways of simulating disease to decrease the use of animals in drug development, even though animal models continue to show significant value in preclinical studies [70].

3.2. Cost and Time

Caring for animals used in research can be time-consuming and expensive, involving maintaining them in large, pricey buildings and regularly providing them with supplies like food. Additionally, some animal models can take a long time to develop, which delays and raises the cost of drug development. For example, it may take up to a year for the brains of epilepsy rodent models to undergo pathological changes before experiments can start [70].

In silico medicine has made significant strides in simulating how drugs interact therapeutically with virtual organs and bodily systems. Patients are still required to best model tolerability and efficacy in late-stage studies. However, *in silico* trials will be able to speed up and make risk assessments more cost-effective, lowering the total number of human subjects.

3.3. Molecular Modeling and Simulations

In silico modelling, simulation, and visualization, also called computational medicine, can assist in predicting the likelihood that medicinal compounds and drugs will be successful while highlighting potential negative side effects during the drug discovery process [71]. Clinical trials are carried out to ascertain whether a product is secure and efficient. However, no justifications and suggestions are offered to improve outcomes, if the results of the clinical trials are unsuccessful. Even products that have potential or promise are ultimately abandoned without doing suggested corrections. Such a strict, all-or-nothing mindset discourages inquiry and stifles innovation. So, fewer new biomedical products come to the market, and overall development costs rise, which exposes investors to more risk. The expenses associated with a failed clinical trial are sobering. A clinical trial may be well into its testing phase when harmful side effects or other problems are found or apparent. Around 90% of the candidates fall short of the Phase I trial and legislative clearance stages.

An average of 17% of pharmaceutical companies' annual revenue is set aside for R&D. Pharmaceutical manufacturers with a high research and development share their revenue allocation in 2019 built in Swedish company AstraZeneca (26%), US manufacturer Eli Lilly (22%), and French company Roche (21%). Then, costs associated with abandoned trials could add another $1 billion to $3 billion to the development of the new drug [72, 73]. This is also the reason why it is becoming more and more challenging for businesses to develop new products that focus on rare conditions, either because the expenses involved cannot be justified given the poor investment returns or because the new drug's ultimate selling price would be so expensive. New products that target uncommon diseases are not being developed and it would be difficult for universal healthcare systems to afford.

3.3.1. Quantitative Structure-Activity Relationship (QSAR)

Quantitative Structure-Activity Relationships (QSAR) is another facet of computer modelling." The basis for any structure-activity relationship is that the biological activity of a new or untested chemical can be inferred from the molecular structure, or properties, of similar compounds whose activities have already been assessed," according to The Journal of Molecular Structure: THEOCHEM. The journal further claims that this relationship provides "yes-or-no or, at best, ranking information" and is more of a general rule [74]. Quantitative structure-activity relationships predict whether something is harmful or hazardous based on the results of other, comparably-sized substances and our fundamental knowledge of how our systems work.

3.3.2. Virtual Screening

Virtual patients (VPs) are needed for *in silico* clinical trials as a computational model. Utilizing the Virtual Physiological Human (VPH) is the first step in developing a virtual patient population. The VPH is a framework that many organizations use to integrate computer models of the technical, physiological, and biochemical functions of a living human body. The VPH is a European business that allows teams to study every part of the human body, down to the genetic level. The target cohort's parameters are described in quantitative VPH models to create VPs. The relevant human physiology's qualitative information is then programmed into these models. Compared to human volunteers, VPs offer several advantages. Considering the COVID-19 pandemic, without needing expensive and lingering tests on live candidates, VPs could have predicted the likelihood that particular vaccines would be effective with no potential side effects [75].

3.4. Applications of Computer Modeling and Simulation in Pharmacological Research

3.4.1. Drug Development

Drug development still needs help in getting FDA approval and dealing with rising costs. To more accurately identify drug targets and predict efficacy, novel approaches are required. Molecular docking is an example of an *in silico* technique that can help with the current issues in drug development.

Computer-aided drug design (CADD) is a widely used technique to reduce the expense and time of the drug creation process. The advantage of CADD methods is that they increase the likelihood of identifying compounds with the desired properties, increasing a compound's chance of overcoming preclinical testing's challenges. Structure-based virtual screening (SBVS) is reliable and practical in this situation. In an attempt to predict the best method for two molecules to connect and form a stable complex, SBVS employs scoring functions to gauge the intensity of the non-covalent interactions between a ligand and a molecular target [76].

The knowledge contained in known active ligands is used in ligand-based virtual screening techniques rather than the target protein's structure for lead optimization and discovery. Only ligand-based techniques are used when there is no 3D structure of the target protein. An effective similarity metric and a trustworthy scoring technique are crucial components of a ligand-based algorithmic approach. The complexity of molecular dynamics is one of the main drawbacks of pharmacophore-LBVS, as it takes hundreds to thousands of nanoseconds to complete an analysis and is computationally demanding [77].

3.4.2. Drug Repurposing

In addition to optimizing drug formulation, *in silico* models are used in medication repurposing. Drug tasking, drug redirection, drug profiling, and drug recycling are other names for the exciting strategy known as drug repurposing. It provides several options for drug candidates with tested formulations, extensive pharmacokinetics, toxicology, clinical trials, and post-marketing surveillance safety data. Recent advancements in statistics and computational-based system biology have benefited the creation of novel medicines. Systems biology techniques for drug repositioning mainly use the knowledge from drug-associated data, which will aid in understanding the molecular basis of disease and the mode of action of medications.

In recent research, network-based analysis was used to simulate how human disease diagnosis progresses. The research developed a network using "claims data," including risk factors and past diagnoses for genetic and non-genetic illnesses. Additionally, research claims that the data present the patients' sequential medical records. The developed network allowed for a comprehensive examination of symptom correlations and a better understanding of the relationship between illnesses [78].

3.4.3. Molecular Docking

The conformation and orientation of molecules, collectively called their "pose," as they enter the binding site of a macromolecular target, are analyzed using molecular docking. Search algorithms generate various poses, which are then scored using rating functions. Docking tools have also been used in the hit-to-lead optimization process. Still, this application has been challenging due to the difficulty in accurately predicting relative binding affinities for related molecules, which has been a weakness since the early development of docking software. However, docking can still be used in hit-to-lead refinement to assess whether modified versions of a hit molecule have improved interactions with the target. When dealing with more extended and more flexible ligands, the efficacy of docking can significantly decrease, especially for shallow binding sites that lack distinctive chemical features, such as those found in polymer-binding proteins like peptidases and glycosidases. Scoring functions based on force fields face a fundamental problem of computing binding affinities from reduced interaction energies, which is necessary to ensure fast computation for managing large compound libraries during docking calculations [79].

3.4.4. In Silico Imaging in Clinical Trials

Typical clinical trials can tell whether a technology or product is risky or ineffective, but they frequently fall short of describing why or suggesting ways to improve it. Computer simulation is used *in silico* clinical trials to develop or assess a medical device, intervention, or product. Developing algorithms that recognize errors or model potential improvements get around the problems with traditional clinical trials. For example, industry and regulators can use simulation tools to forecast the performance of new technology and better comprehend changes to current devices. Digital breast tomosynthesis (DBT) as a substitute for digital mammography (DM) was the subject of the *in silico* clinical imaging study known as VICTRE (Virtual Imaging Clinical Trial for Regulatory Evaluation). The outcomes of the simulated trial were contrasted with those of a prior human clinical study that involved exposing more than 400 women to both methods twice and having radiologists analyze the pictures. Determining the scientific value of imaging technology compared to the norm of care is the primary goal of these clinical trials. Three major categories—detection, diagnosis, and directing/monitoring of disease treatment—are used to evaluate imaging techniques. R&D has improved over the past ten years, particularly in imaging studies like MRIs [80].

3.5. Advantages and Limitations of *In Silico* Methods

The current drug discovery and testing system allows for incremental technological advancements or the integration of brand-new ones. *In silico* clinical trials are anticipated to have significant advantages over existing *in vivo* clinical trials, where testing is conducted on living organisms, such as animals or humans, even though fully simulated clinical trials are not currently feasible given the state of technology.

The primary benefit of *in silico* trials is that the drugs are tested in a simulated or virtual environment with virtual patients rather than on live people or animals. The potential effects of a drug regimen—therapeutic advantages and adverse effects—can be seen and even predicted using computer modelling and simulation. *In silico* studies could thus safeguard the public's health by sparing users from harmful side effects or unfavorable drug interactions. By enabling physicians to experiment with different treatment options, *in silico* can also advance more individualized medicine. Unlike conventional testing that uses live subjects, virtual human models offer significant cost savings because they can be used indefinitely.

The US Food and Drug Administration (FDA) already promotes the use of *in silico* modelling and simulation because it can aid in creating new safe, and effective therapeutics [81].

The benefits of *in silico* clinical research are:

• Enabling the collection of additional evidence before the beginning of bench or animal studies.

• Including patient phenotypes with uncommon, extreme, or challenging recruitment in the trial cohort.

• Comparison of two different treatments in the same virtual population directly (reducing the observed effect variance).

• Device evaluation under physiologically demanding conditions that might represent extreme but realistic applications (off-label use).

• Minimizing suffering by lowering the number of test subjects—both humans and animals—and by improving long-term studies.

Clinical research is just starting with using *in silico* models to create virtual populations. Additionally, there are still very few databases for newly developed drugs available, which restrict the use of *in silico* modelling, especially in the context of pharmacokinetics, and is one of the significant challenges of integrating *silico* models into drug development. A growing number of businesses are using *in silico* to reduce the number of animals used in preclinical research for drug development. The results of recent *in silico* trials offer hope for addressing some of the problems that arise in conventional clinical trials.

Clinical research could be revolutionized, and computational-based methods could optimize drug development. *In silico* trials shorten study duration and cut costs by eliminating the need for human cohorts and animal models. The network-level modelling of diseases also makes it possible to personalize a treatment or device before it needs to be given to or implanted in a patient. *In silico* modelling may allow precision medicine development for diseases with complex treatment profiles across the patient population [82].

3.6. Comparison with Animal Studies

In silico modelling offers more practical, cost-effective experiments than *in-vivo* techniques in complete organisms. Also, computational methods limit the use of animals as research models, which is in line with the case for creating novel, safe therapeutic candidates. For example, computer modelling systems have largely supplanted preclinical research on the effect of measurement errors on glycemia, particularly in type 1 diabetes (T1D) [83]. In another example of the simulation model, the pharmacokinetics and disposition parameters from *in vivo* experiments were combined with other parameters (such as permeability and solubility) projected by the GastroPlus software. The ACAT model was then used to simulate absorption in the intestines. The model used the expected physicochemical and dispositional factors [84].

4. *IN VIVO* NON-ANIMAL METHODS (HUMAN-BASED METHODS)

Recently, animal welfare has opposed animal research for the proper treatment of animals. Government of India has developed the Committee for Control and Supervision of Experiments on Animals (CPCSEA) to oversee animal experiments through ethics committees established in the relevant institutions. Economic factors involving the cost of animals, the upkeep of animal facilities and equipment, and the availability of qualified employees are also required [8, 85].

Clinical trials mean "Clinical research" refers to studies and "trials" that end with a man. In human clinical trials, a drug should be carefully and ethically evaluated for safety and efficacy before use for therapeutic purposes, such as; safety in humans refers to clinical trials. After completing preclinical drug testing, the company filed an investigational new drug (IND) application with regulatory authority for permission to test the drug in humans. A clinical trial with five phases (0-IV) and information obtained from one step is analyzed before proceeding to the next. ***Phase 0 (Micro dosing)*** involves testing the investigational chemical entity at microdose level directly in healthy human volunteers to assess its safety and efficacy. ***Phase I consists of*** testing small doses of the medication in healthy volunteers (10-100) under controlled conditions. The main objective is to determine the medication's safety, the ADME, the mechanisms of action, and the maximum tolerated dose. ***Phase II*** (exploratory therapeutic phase) consists of the participation of 50–500 healthy volunteers to determine the drug's effectiveness and effective dose range. ***Phase III*** (confirmatory therapeutic phase) aims to confirm the drug's efficacy in many patients of either sex. Generally, randomization and double-blind comparative studies are carried out with many volunteers. After the successful conduction of this phase, if the investigational molecule is safe and efficacious, then permission for market authorization is granted. ***Phase IV*(*Post-marketing surveillance)*** involves closely monitoring long-term safety in a larger population. It also has data on thousands of patients with treated drugs collected and analysed to extend the range of indications for which the drug had been used. This phase helps estimate the adverse drug reactions, which detects previously unknown negative responses and identifies the risk factor—the details of the drug discovery and development process are given in Fig. (**4**) below.

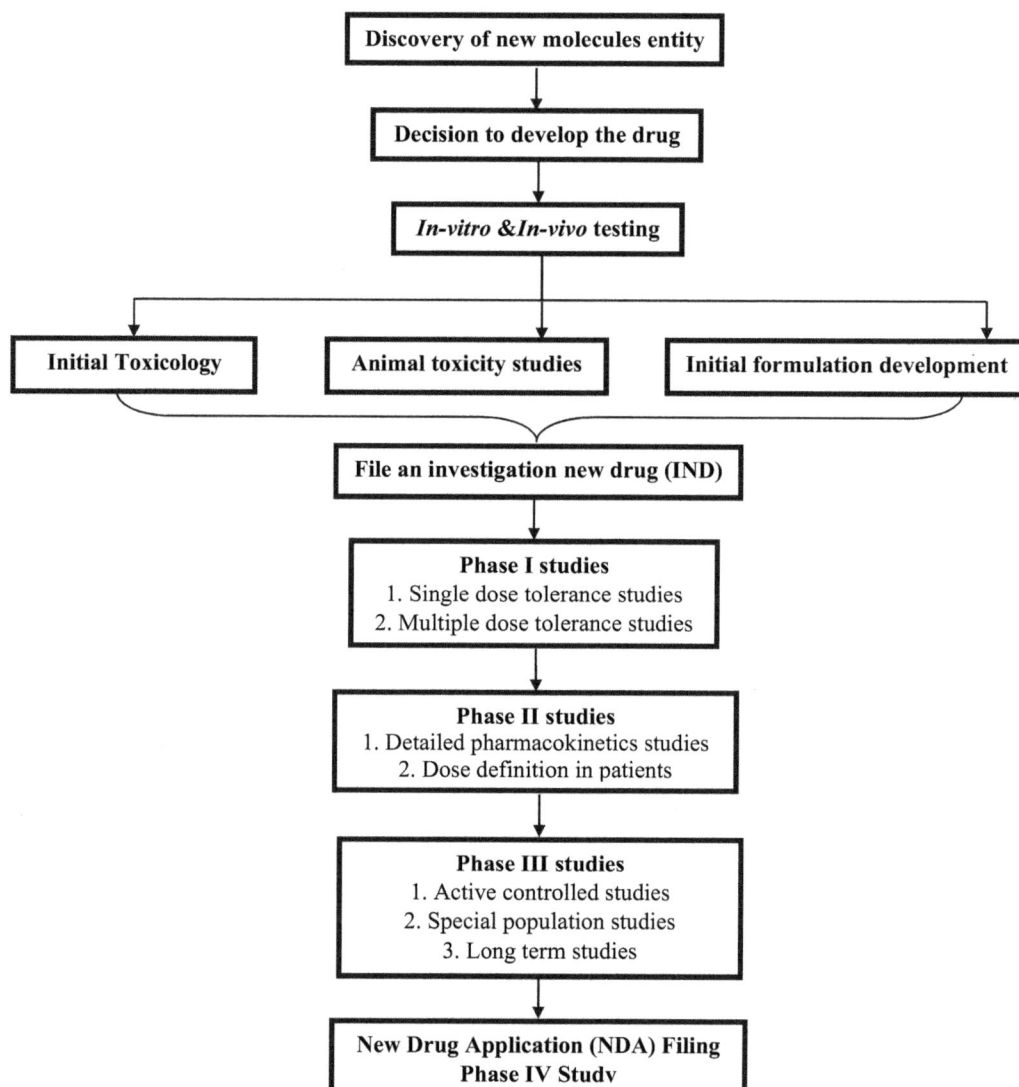

Fig. (4). Schematic representation of steps involved in human clinical trials.

Epidemiology studies have demonstrated the distribution, aetiology, and consequences of health and disease in patient groups and the connections between environmental factors or therapeutic interventions and clinical outcomes. Some research specialties are developing evidence-based practice and policy guidelines, applying practices and methodologies in health systems, systematic evaluation, and disease screening and prevention [8, 85]. There is an urgent need for new medications for many clinical disorders. Still, sadly, as development costs rise, fewer medications are getting marketing approval. Medicine development is also a protracted, challenging, and expensive process that usually takes 10–15 years and costs between US$500 million and US$1 billion for each marketed medicine. Drug concentrations that are too low at the target organ for too short of a time can result in effectiveness failures, while incorrect concentrations reaching the false organs for too long time exposure can result in toxicity—consequently, a novel experimental method [86, 87]. Further, the various advantages and limitations of of non-animal (human) methods have been highlighted in Table **2**.

Table 2. Advantages and limitations of *in vivo* non-animal methods.

Sr. No.	Advantages	Limitations
1.	Reduction in the number of animals.	Primary research requires an animal's metabolic response to metabolized drugs inside the organisms.
2.	Ability to obtain results more quickly.	Non-animal alternatives cannot provide complete information about the subject.
3.	Cost reduction for the studies.	Before using novel surgical techniques on people and non-animals, animal models must be used for testing.
4.	Flexibility in modifying experiment conditions.	Idiosyncratic responses to a substance produce an allergic reaction.
5.	Develop new treatments and research protocols.	Transplant studies involving organ and tissue substitutions are limited to *in vivo*, non-animal methods.
6.	*In vivo* methods and live volunteers need to be confirmed.	In a non-animal model, we cannot test the unpredicted response.
7.	Cells derived from animals or cell lines have an infinite lifespan in this model.	It is relatively high in cost.
8.	Relatively cheap, simple to procure, and efficient with the rapid result.	-

4.1. Significance of Micro-Dosing

• This is a novel method with the help of sophisticated instrumentation to get data on human pharmacokinetics before phase I safety procedure performed through phase 0 (micro-dosing).

• Reduces or replaces the extensive animal testing for pharmacokinetics.

• Utilizes minimal doses of medications that are not meant to have any pharmacologic effects on humans and that do not necessarily have to have negative effects to provide important pharmacokinetic data for a compound's development.

• Ultrasensitive and specific analytical methods are required to measure drug and metabolite concentrations in the low range (LC-MS-MS) [86, 87].

4.2. Advantages of Micro-dosing

• It requires small and safe quantities of the drug for testing.

• For a small dose, when given to human subjects, it does not have any pharmacologic effect; as a result, there is a lower side effect.

• Human screening of compounds is done earlier in drug trials, and fewer animal studies before Phase I clinical studies are necessary. Thus, it is possible to prevent further animal research with compounds having unsuitable pharmacokinetic profiles.

• Comparative human microdose studies can determine the identical pharmaco-kinetic substances that are analyzed during drug development. These pharmacokinetic data can be used in other ways to select the ideal candidate drug and first in the drug's human dose.

• Potent analytical tool for human micro-dosing.

• From a future perspective, phase 0 trials are becoming more complicated, and human micro-dosing is applied to various drugs that could potentially be administered.

4.3. Human-Based Methods Employed in Pharmacological Research [8]

4.3.1. Microbiological Systems

As a screening system, it must be present in animal studies, such as using fungi in studies for the metabolism of drugs. The use of fungi could generally decrease the use of laboratory animals.

4.3.2. Tissue/Organ Culture Preparation

For ADME, tissues and bodily fluids from healthy humans and animals are used to measure various drugs'. Human dopaminergic neurons replace animal models of brain disorders and are transgenic with altered gene expression.

4.3.3. Human Dopaminergic Neurons

These can be used to research the mechanisms behind degeneration and ageing and the efficacy of drugs. The LUHMES will enhance tyrosine hydroxylase expression and improve in a human neural model cell line.

4.3.4. Plant Analysis

The substitution of plants had inadequate achievement in animal research, and human-related effects of various drugs have been observed.

4.3.5. Stem Cells in Toxicological Research

Stem cells might be helpful as an alternate approach to utilizing animals in both *in vitro* disease models and toxicity testing. Human disease tissues that can test possible treatments are created by inserting disease genes into embryonic stem cells.

5. EMERGING EXPERIMENTAL TOOLS

The growing scientific knowledge and contemporary high-tech developments in fabrication methods for micro and nanoscales have had a considerable impact on several scientific sectors. The recent advancement of various medications enhances and lengthens human life. Also, creating novel, efficient remedies encourages significantly lower healthcare expenses. There is a need to have more clarity regarding the requirements for ongoing drug development caused by rising drug resistance. Nevertheless, creating new, effective medications requires a time-consuming procedure that costs millions over the years (Table 3).

Table 3. Applications and advantages of emerging experimental tools in pharmacological research.

Sr. No.	Emerging Experimental Tools	Applications	Advantages
1.	Microfluidic devices	• Human lead compounds are more reliable in predicting a drug's efficacy and toxicity. • For screening of drug analysis, determination, metabolism, and toxicity. • Multiple types of functional cell co-culture. • Bioartificial organs cultivated culture provide a favourable environment. • In drug metabolism, it is combined with other aspects of research. • In novel pharmaceuticals, the evaluation of drug toxicity for safe development. • *In vitro* models with human skin can be used for drug toxicity testing and disease study. • In drug metabolism, optical detection is rapid.	• Microstructures directly mimic the extracellular environment. • It provides the flexibility of drug concentration. • It allows serial dilutions that monitor a drug's dose response. • These devices can isolate single cells for further culturing. • Microfluidic provides a platform to study the 3D structures of a cell. • In an *in-vivo* conditions, biomolecules should regulate cell growth and differentiation. • Drug screening is the development of combination therapies in microfluidic devices. • Microfluidic allows the monitoring of cell cultures over a long period.
2.	Organ-on-a-chip technology	• It has the potential as a new drug discovery and development tool. • It significantly improves functionality, integration, automation, and manufacturing. • Allow real-time, *in situ*, and dynamic cell maintenance. • It can monitor various parameters, such as shear stress, pH, cytokines, and chemokines. • For the manufacturing process, improvement, and standardization of the product. • It is based on patient-derived materials, such as patient tissue, biological systems, and other materials.	• The device design has flexibility. • It controls experimental flexibility. • In adequate quantity, having a low number of cells. • It has single-cell handling. • Real-time analysis with complete automation. • Controlled co-culture. • The reagent consumption.

5.1. Microfluidic Devices

A micro-level approach, "microfluidic technology," has quickly evolved as a potent tool for various applications to biomedical research. The first microfluidic device appeared about 30 years ago, primarily for reducing analytical procedures and enhancing analyte separation. It improves the design, manufacture, and use of microfluidic systems, which rapidly has the potential for widespread applications across biological systems (Table **3**). In recent years, microfluidic technology has advanced and found countless benefits in life science, and microfluidic is called a mature technology. It offers a toolbox for handling and modifying fluid samples, suspended cells, and particles. Many cutting-edge microfluidic devices are utilized annually to observe and overprotect cells, simulate organs, and find biomarkers. Microfluidics are automated, multiplexed in that cells are separately cultured on a similar chip to enable biochemical tests without using many reagents. Microfluidic devices on "Lab-on-a-chip" (LOC) are inherent in handling shallow volumes and are mainly used for biochemical analysis. Recently, however, it has been focused on diagnostics. LOC, also called "Lab-on-a-chip," is used to manage microfluidics (10-9 to 10-18 L) fluids utilizing channels that range in size from tens to hundreds of microns and are necessary for integrated pumps, electrodes, valves, electrical fields and microelectronics (Fig. **5**).

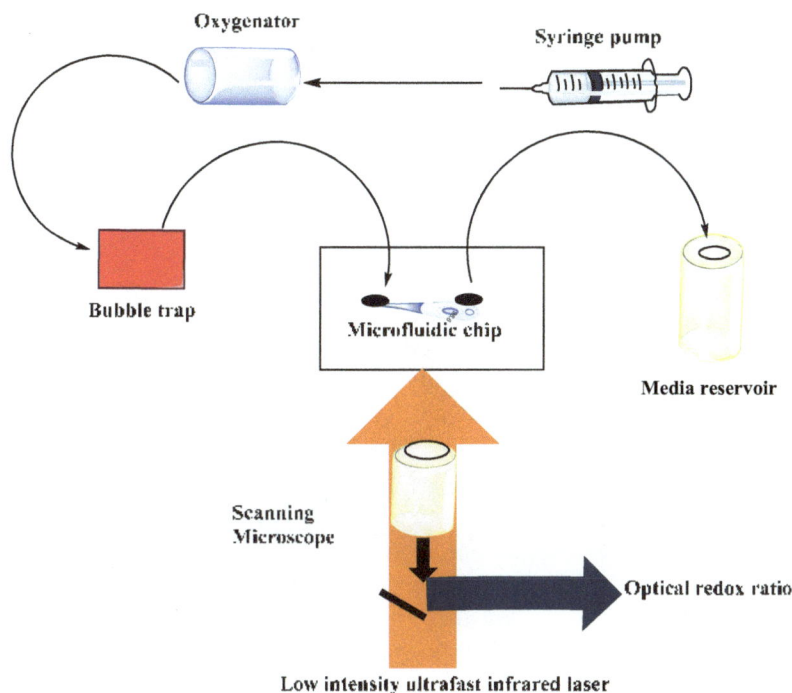

Fig. (5). Schematics of the microfluidic system.

5.2. Organ-on-a-chip Technology

Organ tissue structures may be appropriately modeled and constructed at the microscale due to the specific design and flow control. Micro-engineered biomimetic devices called "organs-on-chips" are crucial functional units of living human organs. Organs-on-a-chip is used to create accurate simulations that mimic the operation of biological organs or tissues. Cells are developed inside the chambers and channels to generate tissues or organs to emulate biology and modify physiology. *In vitro* models, including the liver, kidney, lung, and gut, have their functions mimicked related to organs and tissues. These systems may be utilized as *in vitro* models, enabling pharmacological regulation of intricate biological processes (Table 3). This device integrates many organ functions on a microfluidic device and, in a variety of cells, provides a biomimetic chip with similar physiological processes on the particular structure of the chip, which is closer to the actual external environment of the disease than the traditional single-cell culture model [88, 89].

They use micromachining cell biology to regulate the culture in internal and exterior cell habitats. Dynamic mechanical stress, fluid shear, and concentration gradients on the chip are just a few examples of the external factors it incorporates and how well they imitate physiological conditions.

5.2.1. Dynamic Mechanical Stress

This type of mechanical stress comprises blood, lung, and bone pressure and plays a big part in sustaining mechanically stressed tissues, including blood vessels, skeletal muscle, bone, and cartilage.

5.2.2. Fluid Shear

OOAC applies the required physical pressure to endothelial cells' typical biological processes by activating cell surface molecules and related signaling cascade. Similarly, the OOAC device's attachment to fluid enables physical examinations of individual organs.

5.2.3. Concentration Dradients

At the microscale level, fluid generally behaves as a laminar flow, causing a constant rise of biological molecules that can control space and time. Several biochemical signals in biological processes, such as angiogenesis, invasion, and migration, are held by concentration gradients [90].

6. REGULATORY PERSPECTIVES

Concerns over animal suffering and ethical issues have brought much interest to animal alternatives, commonly known as *in vitro* techniques or non-animal testing. As a result, there is an increasing demand for a robust regulatory framework and validation procedure to guarantee these alternative procedures' effectiveness, safety, and dependability. Regulatory frameworks and validation procedures for animal substitutes are anticipated to move in several important ways.

6.1. Current Regulatory Guidelines

In the past, the idea of "equal or superior" in the context of non-animal testing methods has relied on a direct comparison with existing animal test results. However, there are instances in which such a comparison is not feasible or appropriate, such as when there is no animal test method available for a specific endpoint, when the conventional animal test method is not reproducible, and when it does not provide information pertinent to human biology or toxicity mechanisms, or when it is not possible to conduct the comparison. A modern, flexible framework should consider these factors for establishing scientific confidence in new approach methodologies (NAMs), which should also be supported by qualities like independent scientific peer review, detailed protocols, sufficient test data, usefulness for risk assessment purposes, clearly identified strengths and limitations, robustness and transferability, time and cost-effectiveness, and harmonization with other testing requirements [91]. *In vivo* tests must only be used as a last resort under the European Regulation on Registration, Evaluation, Authorization, and Restriction of Chemicals (REACH). The European Chemicals Agency (ECHA) actively contributes to the development of substitutes in partnership with the OECD and the European Union Reference Laboratory for Alternatives to Animal Testing (EURL-ECVAM) [92]. Validated approaches must fit into the regulatory framework and be adaptable enough to consider pertinent scientific data during regulatory risk assessment for regulatory authorities to approve them [93].

6.2. Challenges to Regulatory Acceptance and Implementation

Uncertainty in the validation process can be attributed to several factors, including individual *v/s* combination approaches, scientific and legal considerations, and gold-standard animal data.

• The use of a single absolute objective number does not accurately represent statistical reality, according to science.

• It is unclear how many (proprietary) methods are desirable to provide the same information [92].

6.3. Recommendations for Overcoming these Challenges

The recommendations for regulatory acceptance are meant to help create a standardized procedure for assessing novel approaches to regulatory adoption.

Owing to the rapid development of science and technology, qualified scientific experience is crucial for evaluating a new direction. This knowledge is necessary for test techniques to be correctly approved or refused, which would postpone the adoption of new, scientifically sound methodologies. The processes leading to the regulatory adoption of novel test methods should consider the following factors to improve the effectiveness of evaluations of proposed new and amended methods and raise the possibility that new methods will get proper scientific assessment [93].

6.4. Future Directions for Regulatory Framework and Validation

Animal alternative techniques will probably be subject to more regulatory scrutiny and standardization. There is a need to have more uniformity in the validation and regulatory acceptability of these methodologies in many locations and nations. Regulatory organizations are anticipated to adopt more uniform standards and specifications to validate alternative animal techniques. Standardized testing, data gathering, reporting procedures, and requirements for using these techniques in regulatory decision-making may be part of this. Additionally, the regulatory and validation framework for animal substitutes may use computer models and artificial intelligence (AI) to a greater extent. There is less need for animal testing since computational models can simulate biological processes and forecast results. Predictive models may be developed for massive datasets and finding trends using AI algorithms. These technologies can make animal alternative approaches more accurate and reproducible while streamlining the validation process and lowering expenses. Regulatory framework and validation in animal alternatives are likely to move toward mechanical, human-relevant methods in the future, with the growing incorporation of computational models and AI. These developments might change the area of animal substitutes and open the door to future safety evaluation techniques that are both more efficient and compassionate. Nonetheless, it is imperative to guarantee that these new methods are thoroughly proven, open, and ethical [94].

CONCLUSION

In conclusion, the chapter on alternative approaches to pharmacology research made many vital comments. First, alternative techniques for animal testing in pharmacology are becoming more widely accepted and used. Examples include computer modelling, *in vitro* cell-based assays, and human-based clinical trials. These techniques decrease the usage of animals, are more economical, and have better human relevance. Going ahead, there is hope for future advancements and the employment of non-traditional approaches in pharmacology. Technology advancements like organs-on-a-chip, 3D printing of tissues, and customized medicine can significantly reduce the need for animal research and medication development. In summation, it is advised that the scientific community and decision-makers continue to encourage the creation and application of alternative pharmacological techniques in research. It involves advocating financing and regulatory support, establishing academic-industry-regulatory agency partnerships, and promoting education and training on alternative methods. These methods might result in more moral, economic, and human-relevant pharmacological research and medication development procedures.

ACKNOWLEDGEMENT

This work is supported by the Ministry of Tribal Affairs, Government of India, New Delhi, India, to Ms. Anglina Kisku for her PhD work (202122-NFST-J-A-01578). There is no direct funding for this work.

REFERENCES

[1]   Doke SK, Dhawale SC. Alternatives to animal testing: A review. Saudi Pharm J 2015; 23(3): 223-9.
        [http://dx.doi.org/10.1016/j.jsps.2013.11.002] [PMID: 26106269]

[2]   Richmond J. Refinement, reduction, and replacement of animal use for regulatory testing: Future improvements and implementation within the regulatory framework. ILAR J 2002; 43 (Suppl. 1): S63-8.
        [http://dx.doi.org/10.1093/ilar.43.Suppl_1.S63] [PMID: 12388854]

[3]   Badyal D, Desai C. Animal use in pharmacology education and research: The changing scenario. Indian J Pharmacol 2014; 46(3): 257-65.
        [http://dx.doi.org/10.4103/0253-7613.132153] [PMID: 24987170]

[4]   Hajar R. Animal testing and medicine. Heart Views 2011; 12(1): 42.
        [http://dx.doi.org/10.4103/1995-705X.81548] [PMID: 21731811]

[5]   Kant I. Lectures on ethics. Cambridge University Press 2001; Vol. 2.

[6]   Franco N. Animal experiments in biomedical research: A historical perspective. Animals 2013; 3(1): 238-73.
        [http://dx.doi.org/10.3390/ani3010238] [PMID: 26487317]

[7]   National Research Council (US) Committee on the Use of Laboratory Animals in Biomedical and Behavioral Research. Use of laboratory animals in biomedical and behavioral research 1988.

[8]   Sharma KK, Arora T, Joshi V, *et al.* Substitute of animals in drug research: An approach towards fulfillment of 4R's. Indian J Pharm Sci 2011; 73(1): 1-6.
        [http://dx.doi.org/10.4103/0250-474X.89750] [PMID: 22131615]

[9]   Alépée N, Bahinski A, Daneshian M, *et al.* State-of-the-art of 3D cultures (organs-on-a-chip) in safety testing and pathophysiology. Altern Anim Exp 2014; 31(4): 441-77.
        [PMID: 25027500]

[10]   Kaul R, Swaminathan S, Kumar V. Need for alternatives to animals in experimentation: An Indian perspective. Indian J Med Res 2019; 149(5): 584-92.
        [http://dx.doi.org/10.4103/ijmr.IJMR_2047_17] [PMID: 31417025]

[11]   Akkermans A, Chapsal JM, Coccia EM, *et al.* Animal testing for vaccines. Implementing replacement, reduction and refinement: Challenges and priorities. Biologicals 2020; 68: 92-107.
        [http://dx.doi.org/10.1016/j.biologicals.2020.07.010] [PMID: 33041187]

[12]   Homberg JR, Adan RAH, Alenina N, *et al.* The continued need for animals to advance brain research. Neuron 2021; 109(15): 2374-9.
[http://dx.doi.org/10.1016/j.neuron.2021.07.015] [PMID: 34352213]

[13]   Chung TDY, Terry DB, Smith LH. *In vitro and in vivo* assessment of ADME and PK properties during lead selection and lead optimization guidelines. benchmarks and rules of thumb 2015.

[14]   Langhans SA. Three-dimensional *in vitro* cell culture models in drug discovery and drug repositioning. Front Pharmacol 2018; 9: 6.
[http://dx.doi.org/10.3389/fphar.2018.00006] [PMID: 29410625]

[15]   Bell SM, Chang X, Wambaugh JF, *et al. In vitro* to *in vivo* extrapolation for high throughput prioritization and decision making. Toxicol. *In Vitro* 2018; 47: 213-27.
[http://dx.doi.org/10.1016/j.tiv.2017.11.016] [PMID: 29203341]

[16]   Lin A, Sved Skottvoll F, Rayner S, *et al.* 3D cell culture models and organ-on-a-chip: Meet separation science and mass spectrometry. Electrophoresis 2020; 41(1-2): 56-64.
[http://dx.doi.org/10.1002/elps.201900170] [PMID: 31544246]

[17]   Gledhill K, Guo Z, Umegaki-Arao N, Higgins CA, Itoh M, Christiano AM. Melanin transfer in human 3D skin equivalents generated exclusively from induced pluripotent stem cells. PLoS One 2015; 10(8): e0136713.
[http://dx.doi.org/10.1371/journal.pone.0136713] [PMID: 26308443]

[18]   Zhang S, Wan Z, Kamm RD. Vascularized organoids on a chip: Strategies for engineering organoids with functional vasculature. Lab Chip 2021; 21(3): 473-88.
[http://dx.doi.org/10.1039/D0LC01186J] [PMID: 33480945]

[19]   Wu L, Ai Y, Xie R, Xiong J, Wang Y, Liang Q. Organoids/organs-on-a-chip: New frontiers of intestinal pathophysiological models. Lab Chip 2023; 23(5): 1192-212.
[http://dx.doi.org/10.1039/D2LC00804A] [PMID: 36644984]

[20]   Lawko N, Plaskasovitis C, Stokes C, *et al.* 3D tissue models as an effective tool for studying viruses and vaccine development. Front Mater 2021; 8: 631373.
[http://dx.doi.org/10.3389/fmats.2021.631373]

[21]   Park SE, Georgescu A, Huh D. Organoids-on-a-chip. Science 2019; 364(6444): 960-5.
[http://dx.doi.org/10.1126/science.aaw7894] [PMID: 31171693]

[22]   Jalili-Firoozinezhad S, Gazzaniga FS, Calamari EL, *et al.* A complex human gut microbiome cultured in an anaerobic intestine-on-a-chip. Nat Biomed Eng 2019; 3(7): 520-31.
[http://dx.doi.org/10.1038/s41551-019-0397-0] [PMID: 31086325]

[23]   Bein A, Fadel CW, Swenor B, *et al.* Nutritional deficiency in an intestine-on-a-chip recapitulates injury hallmarks associated with environmental enteric dysfunction. Nat Biomed Eng 2022; 6(11): 1236-47.
[http://dx.doi.org/10.1038/s41551-022-00899-x] [PMID: 35739419]

[24]   Zhu MM, *et al.* Industrial production of therapeutic proteins: Cell lines, cell culture, and purification. Handbook of industrial chemistry and biotechnology 2017.

[25]   Niu N, Wang L. *In vitro* human cell line models to predict clinical response to anticancer drugs. Pharmacogenomics 2015; 16(3): 273-85.
[http://dx.doi.org/10.2217/pgs.14.170] [PMID: 25712190]

[26]   Niu N, Liu T, Cairns J, *et al.* Metformin pharmacogenomics: A genome-wide association study to identify genetic and epigenetic biomarkers involved in metformin anticancer response using human lymphoblastoid cell lines. Hum Mol Genet 2016; 25(21): ddw301.
[http://dx.doi.org/10.1093/hmg/ddw301] [PMID: 28173075]

[27]   Tong Y, Niu N, Jenkins G, *et al.* Identification of genetic variants or genes that are associated with Homoharringtonine (HHT) response through a genome-wide association study in human

lymphoblastoid cell lines (LCLs). Front Genet 2015; 5: 465.
[http://dx.doi.org/10.3389/fgene.2014.00465] [PMID: 25628645]

[28] Goodspeed A, Heiser LM, Gray JW, Costello JC. Tumor-derived cell lines as molecular models of cancer pharmacogenomics. Mol Cancer Res 2016; 14(1): 3-13.
[http://dx.doi.org/10.1158/1541-7786.MCR-15-0189] [PMID: 26248648]

[29] Karp JM, Langer R. Development and therapeutic applications of advanced biomaterials. Curr Opin Biotechnol 2007; 18(5): 454-9.
[http://dx.doi.org/10.1016/j.copbio.2007.09.008] [PMID: 17981454]

[30] Langer R, Tirrell DA. Designing materials for biology and medicine. Nature 2004; 428(6982): 487-92.
[http://dx.doi.org/10.1038/nature02388] [PMID: 15057821]

[31] Chan BP, Leong KW. Scaffolding in tissue engineering: General approaches and tissue-specific considerations. Eur Spine J 2008; 17(S4) (Suppl. 4): 467-79.
[http://dx.doi.org/10.1007/s00586-008-0745-3] [PMID: 19005702]

[32] Olson JK, Boldyrev AI. *Ab initio* characterization of the flexural anion found in the reversible dehydrogenation. Comput Theor Chem 2011; 967(1): 1-4.
[http://dx.doi.org/10.1016/j.comptc.2011.04.011]

[33] Mansouri V, Beheshtizadeh N, Gharibshahian M, Sabouri L, Varzandeh M, Rezaei N. Recent advances in regenerative medicine strategies for cancer treatment. Biomed Pharmacother 2021; 141: 111875.
[http://dx.doi.org/10.1016/j.biopha.2021.111875] [PMID: 34229250]

[34] Hu J, Ma PX. Nano-fibrous tissue engineering scaffolds capable of growth factor delivery. Pharm Res 2011; 28(6): 1273-81.
[http://dx.doi.org/10.1007/s11095-011-0367-z] [PMID: 21234657]

[35] Motamedian SR, Hosseinpour S, Ahsaie MG, Khojasteh A. Smart scaffolds in bone tissue engineering: A systematic review of literature. World J Stem Cells 2015; 7(3): 657-68.
[http://dx.doi.org/10.4252/wjsc.v7.i3.657] [PMID: 25914772]

[36] Nagai Y, Unsworth LD, Koutsopoulos S, Zhang S. Slow release of molecules in self-assembling peptide nanofiber scaffold. J Control Release 2006; 115(1): 18-25.
[http://dx.doi.org/10.1016/j.jconrel.2006.06.031] [PMID: 16962196]

[37] Hussey GS, Dziki JL, Badylak SF. Extracellular matrix-based materials for regenerative medicine. Nat Rev Mater 2018; 3(7): 159-73.
[http://dx.doi.org/10.1038/s41578-018-0023-x]

[38] Pogorielov M, Oleshko O, Hapchenko A. Tissue engineering: Challenges and selected application. Adv Tissue Eng Regen Med Open Access 2017; 3: 330-4.

[39] Han F, Wang J, Ding L, *et al.* Tissue engineering and regenerative medicine: Achievements, future, and sustainability in Asia. Front Bioeng Biotechnol 2020; 8: 83.
[http://dx.doi.org/10.3389/fbioe.2020.00083] [PMID: 32266221]

[40] De Siervi S, Turato C. Liver organoids as an *in vitro* model to study primary liver cancer. Int J Mol Sci 2023; 24(5): 4529.
[http://dx.doi.org/10.3390/ijms24054529] [PMID: 36901961]

[41] Guan N, Liu Z, Zhao Y, Li Q, Wang Y. Engineered biomaterial strategies for controlling growth factors in tissue engineering. Drug Deliv 2020; 27(1): 1438-51.
[http://dx.doi.org/10.1080/10717544.2020.1831104] [PMID: 33100031]

[42] Ikada Y. Challenges in tissue engineering. J R Soc Interface 2006; 3(10): 589-601.
[http://dx.doi.org/10.1098/rsif.2006.0124] [PMID: 16971328]

[43] Seppänen-Kaijansinkko R. Tissue engineering — pros and cons. Int J Oral Maxillofac Surg 2017; 46: 50.

[http://dx.doi.org/10.1016/j.ijom.2017.02.183]

[44]   Caballero D, *et al.* Forecast cancer: The importance of biomimetic 3D *in vitro* models in cancer drug testing/discovery and therapeutically. *In vitro* Models 2022; 1(2): 119-23.

[45]   Shingatgeri V, Dhawan A, Kwon S. Chapter 10 - Safety concerns using cell-based *in vitro* methods for toxicity assessment. *In vitro* Toxicology. Academic Press 2017; pp. 187-207.

[46]   Graudejus O, *et al.* Bridging the gap between *in vivo* and *in vitro* research: Reproducing *in vitro* the mechanical and electrical environment of cells *in vivo*. Front Cell Neurosci 2019; 12.

[47]   Jackson LR, Trudel LJ, Fox JG, Lipman NS. Evaluation of hollow fiber bioreactors as an alternative to murine ascites production for small scale monoclonal antibody production. J Immunol Methods 1996; 189(2): 217-31.
[http://dx.doi.org/10.1016/0022-1759(95)00251-0] [PMID: 8613673]

[48]   National Research Council (US) Committee on Methods of Producing Monoclonal Antibodies. Monoclonal antibody production 1999.

[49]   Li J, Yang J, Zhao D, Lei W, Hu S. Promises and challenges of cardiac organoids. Mamm Genome 2023; 34(2): 351-6.
[http://dx.doi.org/10.1007/s00335-023-09987-y] [PMID: 37016187]

[50]   Liu Y, Lin L, Qiao L. Recent developments in organ-on-a-chip technology for cardiovascular disease research. Anal Bioanal Chem 2023; 415(18): 3911-25.
[http://dx.doi.org/10.1007/s00216-023-04596-9] [PMID: 36867198]

[51]   Silva-Pedrosa R, Salgado AJ, Ferreira PE. Revolutionizing disease modeling: The emergence of organoids in cellular systems. Cells 2023; 12(6): 930.
[http://dx.doi.org/10.3390/cells12060930] [PMID: 36980271]

[52]   Sereti E, Papapostolou I, Dimas K. Pancreatic cancer organoids: An emerging platform for precision medicine? Biomedicines 2023; 11(3): 890.
[http://dx.doi.org/10.3390/biomedicines11030890] [PMID: 36979869]

[53]   Aro R, *et al.* Essential tremor: A three-dimensional neurosphere *in vitro* model to assess the neurotoxicity of harmane. J Tradit Chin Med Sci 2022; 10(1): 19-34.

[54]   Bedford R, Perkins E, Clements J, Hollings M. Recent advancements and application of *in vitro* models for predicting inhalation toxicity in humans. Toxicol. *In Vitro* 2022; 79: 105299.
[http://dx.doi.org/10.1016/j.tiv.2021.105299] [PMID: 34920082]

[55]   Zheng F, Xiao Y, Liu H, Fan Y, Dao M. Patient-specific organoid and organ-on-a-chip: 3D cell-culture meets 3D printing and numerical simulation. Adv Biol 2021; 5(6): 2000024.
[http://dx.doi.org/10.1002/adbi.202000024] [PMID: 33856745]

[56]   Trapecar M, *et al.* Human physiomimetic model integrating microphysiological systems of the gut, liver, and brain for studies of neurodegenerative diseases. Sci Adv 2021; 7(5): 1707.
[http://dx.doi.org/10.1126/sciadv.abd1707]

[57]   Heidari-Khoei H, Esfandiari F, Hajari MA, Ghorbaninejad Z, Piryaei A, Baharvand H. Organoid technology in female reproductive biomedicine. Reprod Biol Endocrinol 2020; 18(1): 64.
[http://dx.doi.org/10.1186/s12958-020-00621-z] [PMID: 32552764]

[58]   Zubizarreta ME, Xiao S. Bioengineering models of female reproduction. Biodes Manuf 2020; 3(3): 237-51.
[http://dx.doi.org/10.1007/s42242-020-00082-8] [PMID: 32774987]

[59]   Rajan SAP, Aleman J, Wan M, *et al.* Probing prodrug metabolism and reciprocal toxicity with an integrated and humanized multi-tissue organ-on-a-chip platform. Acta Biomater 2020; 106: 124-35.
[http://dx.doi.org/10.1016/j.actbio.2020.02.015] [PMID: 32068138]

[60]   Zhang H, Whalley RD, Ferreira AM, Dalgarno K. High throughput physiological micro-models for *in vitro* pre-clinical drug testing: A review of engineering systems approaches. Progress in Biomedical

Engineering 2020; 2(2): 022001.
[http://dx.doi.org/10.1088/2516-1091/ab7cc4]

[61] Nikolaev M, Mitrofanova O, Broguiere N, *et al.* Homeostatic mini-intestines through scaffold-guided organoid morphogenesis. Nature 2020; 585(7826): 574-8.
[http://dx.doi.org/10.1038/s41586-020-2724-8] [PMID: 32939089]

[62] Shanti A, Teo J, Stefanini C. *In vitro* immune organs-on-chip for drug development: A review. Pharmaceutics 2018; 10(4): 278.
[http://dx.doi.org/10.3390/pharmaceutics10040278] [PMID: 30558264]

[63] Shah P, Fritz JV, Glaab E, *et al.* A microfluidics-based *in vitro* model of the gastrointestinal human–microbe interface. Nat Commun 2016; 7(1): 11535.
[http://dx.doi.org/10.1038/ncomms11535] [PMID: 27168102]

[64] Kim HJ, Huh D, Hamilton G, Ingber DE. Human gut-on-a-chip inhabited by microbial flora that experiences intestinal peristalsis-like motions and flow. Lab Chip 2012; 12(12): 2165-74.
[http://dx.doi.org/10.1039/c2lc40074j] [PMID: 22434367]

[65] Elliott NT, Yuan F. A Review of Three-Dimensional In vitro. Tissue Models 2011.

[66] Sato T, Vries RG, Snippert HJ, *et al.* Single Lgr5 stem cells build crypt-villus structures *in vitro* without a mesenchymal niche. Nature 2009; 459(7244): 262-5.
[http://dx.doi.org/10.1038/nature07935] [PMID: 19329995]

[67] Amberg A. *In silico* methods. In: van de Waterbeemd H, Ed. Drug Discovery and Evaluation: Safety and Pharmacokinetic Assays. Springer 2013; pp. 1273-96.
[http://dx.doi.org/10.1007/978-3-642-25240-2_55]

[68] Madden JC, Enoch SJ, Paini A, Cronin MTD. A review of *in silico* tools as alternatives to animal testing: Principles, resources and applications. Altern Lab Anim 2020; 48(4): 146-72.
[http://dx.doi.org/10.1177/0261192920965977] [PMID: 33119417]

[69] FitzGerald RJ, Cermeño M, Khalesi M, Kleekayai T, Amigo-Benavent M. Application of *in silico* approaches for the generation of milk protein-derived bioactive peptides. J Funct Foods 2020; 64: 103636.
[http://dx.doi.org/10.1016/j.jff.2019.103636]

[70] Di Salvo C. How in Silico Modelling is Changing Our Approach to Drug Development and Clinical Research 2021.

[71] Gartner TE III, Jayaraman A. Modeling and simulations of polymers: A roadmap. Macromolecules 2019; 52(3): 755-86.
[http://dx.doi.org/10.1021/acs.macromol.8b01836]

[72] *In Silico* and AI: Computer simulation in drug discovery. 2023. Available from: https://vamstar.io/my-resources/in-silico-and-ai-computer-simulation-in-drug-discovery/

[73] Rennane S, Baker L, Mulcahy A. Estimating the cost of industry investment in drug research and development: A review of methods and results. Inquiry 2021; 58
[http://dx.doi.org/10.1177/00469580211059731] [PMID: 35170336]

[74] Perkins R, Fang H, Tong W, Welsh WJ. Quantitative structure-activity relationship methods: Perspectives on drug discovery and toxicology. Environ Toxicol Chem 2003; 22(8): 1666-79.
[http://dx.doi.org/10.1897/01-171] [PMID: 12924569]

[75] Kilinç H, Okur MR, Usta İ. Ä°lker U. The opinions of field experts on online test applications and test security during the COVID-19 pandemic. International Journal of Assessment Tools in Education 2021; 8(4): 975-90.
[http://dx.doi.org/10.21449/ijate.875293]

[76] Maia EHB, Assis LC, de Oliveira TA, da Silva AM, Taranto AG. Structure-based virtual screening: From classical to artificial intelligence. Front Chem 2020; 8: 343.

[http://dx.doi.org/10.3389/fchem.2020.00343] [PMID: 32411671]

[77]   Hamza A, Wei NN, Zhan CG. Ligand-based virtual screening approach using a new scoring function. J Chem Inf Model 2012; 52(4): 963-74.
[http://dx.doi.org/10.1021/ci200617d] [PMID: 22486340]

[78]   Pandita V, *et al.* System and network biology-based computational approaches for drug repositioning. Computational Approaches for Novel Therapeutic and Diagnostic Designing to Mitigate SARS-CoV2 Infection. Elsevier 2022; pp. 267-90.
[http://dx.doi.org/10.1016/B978-0-323-91172-6.00003-0]

[79]   Torres PHM, Sodero ACR, Jofily P, Silva-Jr FP. Key topics in molecular docking for drug design. Int J Mol Sci 2019; 20(18): 4574.
[http://dx.doi.org/10.3390/ijms20184574] [PMID: 31540192]

[80]   Badano A. *In silico* imaging clinical trials: Cheaper, faster, better, safer, and more scalable. Trials 2021; 22(1): 64.
[http://dx.doi.org/10.1186/s13063-020-05002-w] [PMID: 33468186]

[81]   Di Salvo C. A Critical Evaluation of the Advantages and Limitations of In Silico Methods in Clinical Research 2021. Available from: https://proventainternational.com/a-critical-evaluation-of-the-advantages-and-limitations-of-in-silico-methods-in-clinical-research/

[82]   Sasikumar AP, Ramaswamy S, Sudhir S. A scientific pharmacognosy on Gaucher's disease: An *in silico* analysis. Environ Sci Pollut Res Int 2022; 29(17): 25308-17.
[http://dx.doi.org/10.1007/s11356-021-17534-y] [PMID: 34839442]

[83]   Breton MD, Hinzmann R, Campos-Nañez E, Riddle S, Schoemaker M, Schmelzeisen-Redeker G. Analysis of the accuracy and performance of a continuous glucose monitoring sensor prototype: An *in-silico* study using the UVA/PADOVA type 1 diabetes simulator. J Diabetes Sci Technol 2017; 11(3): 545-52.
[http://dx.doi.org/10.1177/1932296816680633] [PMID: 28745098]

[84]   Ahmad A, Alqahtani S, Jan BL, Raish M, Rabba AK, Alkharfy KM. Gender effect on the pharmacokinetics of thymoquinone: Preclinical investigation and *in silico* modeling in male and female rats. Saudi Pharm J 2020; 28(4): 403-8.
[http://dx.doi.org/10.1016/j.jsps.2020.01.022] [PMID: 32273798]

[85]   Freires IA, Sardi JCO, de Castro RD, Rosalen PL. Alternative animal and non-animal models for drug discovery and development: Bonus or burden? Pharm Res 2017; 34(4): 681-6.
[http://dx.doi.org/10.1007/s11095-016-2069-z] [PMID: 27858217]

[86]   Burt T, Young G, Lee W, *et al.* Phase 0/microdosing approaches: Time for mainstream application in drug development? Nat Rev Drug Discov 2020; 19(11): 801-18.
[http://dx.doi.org/10.1038/s41573-020-0080-x] [PMID: 32901140]

[87]   Rani PU, Naidu MUR. Phase 0 - Microdosing strategy in clinical trials. Indian J Pharmacol 2008; 40(6): 240-2.
[http://dx.doi.org/10.4103/0253-7613.45147] [PMID: 21279177]

[88]   Ingber DE. Human organs-on-chips for disease modelling, drug development and personalized medicine. Nat Rev Genet 2022; 23(8): 467-91.
[http://dx.doi.org/10.1038/s41576-022-00466-9] [PMID: 35338360]

[89]   Sosa-Hernández JE, Villalba-Rodríguez AM, Romero-Castillo KD, *et al.* Organs-on-a-chip module: A review from the development and applications perspective. Micromachines 2018; 9(10): 536.
[http://dx.doi.org/10.3390/mi9100536] [PMID: 30424469]

[90]   Wu Q, Liu J, Wang X, *et al.* Organ-on-a-chip: Recent breakthroughs and future prospects. Biomed Eng Online 2020; 19(1): 9.
[http://dx.doi.org/10.1186/s12938-020-0752-0] [PMID: 32050989]

[91]   van der Zalm AJ, Barroso J, Browne P, *et al.* A framework for establishing scientific confidence in

new approach methodologies. Arch Toxicol 2022; 96(11): 2865-79.
[http://dx.doi.org/10.1007/s00204-022-03365-4] [PMID: 35987941]

[92]   Piersma AH, Burgdorf T, Louekari K, *et al.* Workshop on acceleration of the validation and regulatory acceptance of alternative methods and implementation of testing strategies. Toxicol. *In Vitro* 2018; 50: 62-74.
[http://dx.doi.org/10.1016/j.tiv.2018.02.018] [PMID: 29501630]

[93]   Stokes WS, Schechtman LM, Hill RN. The interagency coordinating committee on the validation of alternative methods (ICCVAM): A review of the ICCVAM test method evaluation process and current international collaborations with the european centre for the validation of alternative methods (ECVAM). Altern Lab Anim 2002; 30(2): 23-32.
[http://dx.doi.org/10.1177/026119290203002S04] [PMID: 12513648]

[94]   MacGregor JT. The future of regulatory toxicology: Impact of the biotechnology revolution. Toxicol Sci 2003; 75(2): 236-48.
[http://dx.doi.org/10.1093/toxsci/kfg197] [PMID: 12883082]

CHAPTER 9

Newer Screening Software for Computer Aided Herbal Drug Interactions and its Development

Sunil Kumar Kadiri[1,*] and **Prashant Tiwari[2]**

[1] *Department of Pharmacology, College of Pharmaceutical Sciences, Dayananda Sagar University, K.S Layout, Bengaluru-560111, Karnataka, India*

[2] *Department of Pharmacology, Dayananda Sagar University, Bengaluru, India*

Abstract: Self-diagnosis and treatment by consumers as a means of reducing medical costs contribute to the predicted continued growth in the usage of herbal products. Herbal products are notoriously difficult to evaluate for potential drug interactions because of the wide range of possible interactions, the lack of clarity surrounding the active components, and the often insufficient knowledge of the pharmacokinetics of the offending constituents. It is a standard practice for innovative drugs in development to identify particular components from herbal goods and describe their interaction potential as part of a systematic study of herbal product drug interaction risk. By cutting down on expenses and development times, computer-assisted drug design has helped speed up the drug discovery process. The natural origins and variety of traditional medicinal herbs make them an attractive area of study as a complement to modern pharmaceuticals. To better understand the pharmacological foundation of the actions of traditional medicinal plants, researchers have increasingly turned to *in silico* approaches, including virtual screening and network analysis. The combination of virtual screening and network pharmacology can reduce costs and improve efficiency in the identification of innovative drugs by increasing the proportion of active compounds among candidates and by providing an appropriate demonstration of the mechanism of action of medicinal plants. In this chapter, we propose a thorough technical route that utilizes several *in silico* approaches to discover the pharmacological foundation of the effects of medicinal plants. This involves discussing the software used in the prediction of herb-drug interaction with a suitable database.

Keywords: Candidates, Computer-assisted, Cost, Composition, Components, Drug discovery, Drug design, Expenditures, Efficiency, Herbal products, *In silico*, Interactions, Medical, Medicinal plant, Pharmaceutical, Pharmacokinetics, Self-diagnosis, Traditional, Treatment, Virtual screening.

---

[*] **Corresponding author Sunil Kumar Kadiri:** Department of Pharmacology, College of Pharmaceutical Sciences, DayanandaSagar University, K.S Layout, Bengaluru-560111, Karnataka, India; E-mail: sunil.cology@gmail.com

Dilpreet Singh and Prashant Tiwari (Eds.)

1. INTRODUCTION

A crucial component of our contemporary healthcare system is the use of pharmaceuticals and/or conventional treatments [1] which are referred to by a variety of names under different regulatory frameworks across the world. In contrast to traditional pharmaceuticals, the majority of herbs or their preparations have not undergone extensive testing for safety, effectiveness, and quality control before being marketed [2]. The co-administration of these herbs with Western medications, whether intentionally or accidentally, increases the risk of pharmacokinetic (PK) and/or pharmacodynamic (PD) herb-drug interactions (HDI) [3, 4]. This raises serious safety concerns, particularly for drugs like warfarin that have a narrow therapeutic index [5, 6], as well as for digoxin [7, 8]. Therefore, collecting and analyzing these reported HDIs would be extremely useful for primary care physicians and the public at large. Most studies on HDI, however, are case reports or poorly documented clinical notes. Reports of herb-drug interactions (HDI) are often disregarded as 'unable to be evaluated' in light of established criteria for determining the trustworthiness of studies for Western pharmaceuticals [7]. However, the development of HDIs does occur, and everyone is vulnerable to experiencing them at some point. Recognizing the therapeutic potential of HDI in the late 1990s, a small group of researchers and companies began building HDI databases using a range of information technologies(IT) [8-15]. These HDI databases are now divided into two groups, one including freely available resources and the other containing paid subscription information. Developing an HDI dataset follows the same three-step process as developing any other professional database. The initial two steps of creating a database are often the most time-consuming and complex since they necessitate the involvement of experts and professionals as well as technical assistance. However, the third phase contains the most time-consuming and labor-intensive jobs, which are also present during database creation and maintenance. Therefore, the building of databases is limited by the third stage of the process, which entails extracting structured data from texts written in natural languages. All kinds of natural language publications conceal HDI information. This covers summaries, journal articles, books, and reports on the assessment of medications. Researchers with backgrounds in medicine are the only ones equipped to do this sort of work. Without consistent funding, open-source databases seldom get updated after their original release. Successful commercial databases, on the other hand, might recuperate all costs involved with database creation and maintenance by charging users a subscription charge. Recent efforts to tackle labor and time-wasting problems in the construction of HDI databases have employed AI technologies to independently extract information on HDI from the literature. Subsequently, AI is utilized to present the collected data [16]. While the integration of AI brings clarity to the process of obtaining and evaluating HDI information from literature, there remains a considerable gap to bridge before AI can fully substitute for departing medical specialists. Fig. (**1**) provides a schematic representation of HDI with known chemical substances.

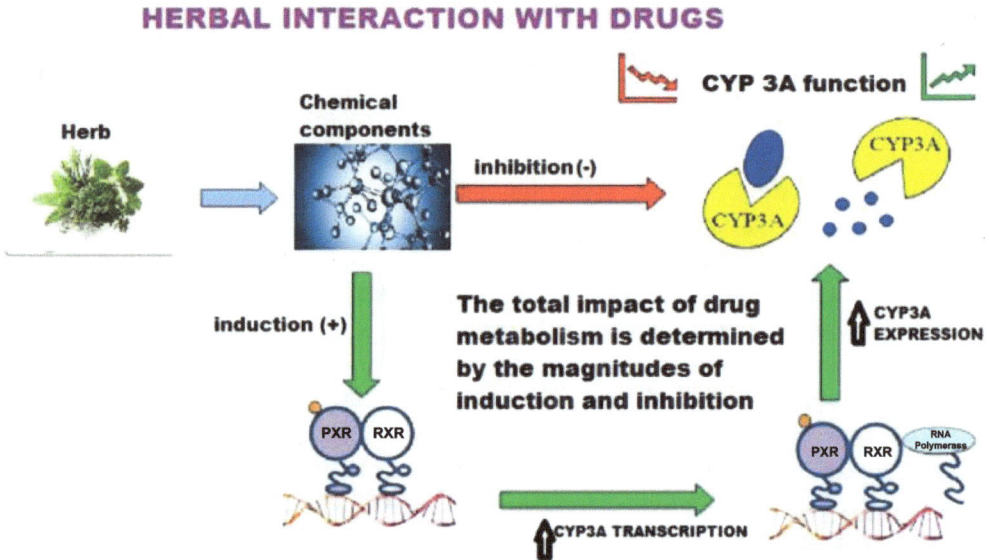

Fig. (1). Drug Interactions of herbs with chemical substances.

2. HDI ESTIMATIONS

2.1. Current Approaches

In contrast to qualitative descriptions of herb-drug interaction, future numerical forecasts of these interactions are still in their early stages. In the United States and Europe, at least, herbal products are not subject to the same level of regulation as conventional pharmaceuticals, therefore, there is often no requirement for a comprehensive assessment of HDI risk. When case reports documenting a potential interaction are received, or when results from *in vitro* experiments highlight a potential interaction, investigations into herb-drug interactions (HDI) are often initiated. A mechanistic understanding of HDIs could be enhanced through a prospective and systematic approach, which would enable the anticipation, reduction, and preferably avoidance of adverse HDIs [17].

2.2. Limitations of Current Strategies

Static equations are preferred over more advanced techniques because the latter are not adaptable enough to account for complicated interactions involving several components, such as PBPK modeling and simulation. The quality of provided data varies greatly, making it challenging to apply PBPK modeling methodologies [18]. This is due to the absence of standardization of herbal products (as already described) and the differences in experimental designs between laboratories. The diagrammatic representations of the procedures followed to assess and determine the toxicity of medicinal herbs are provided in Fig. (2).

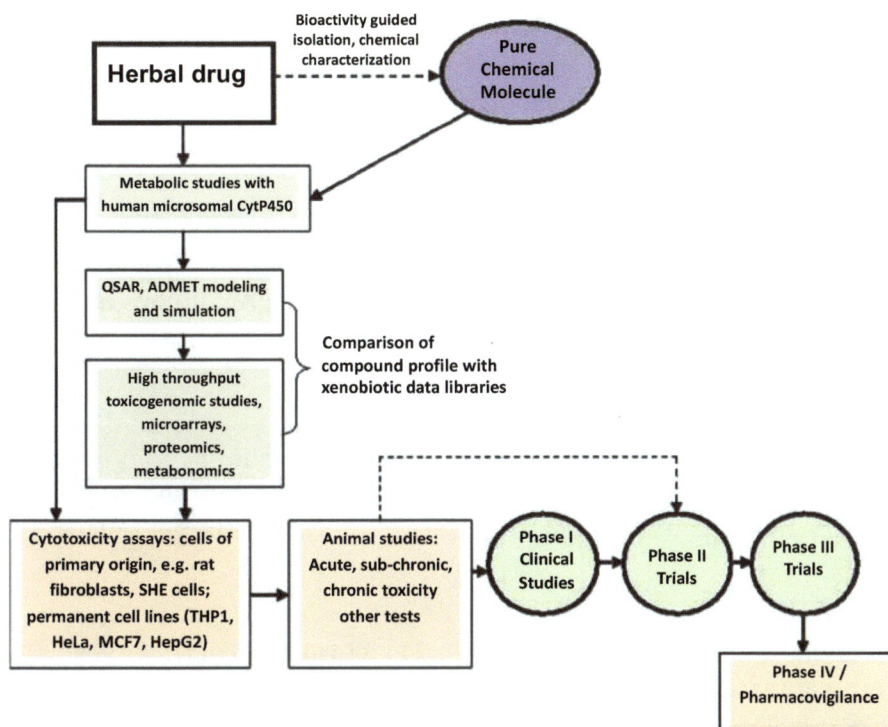

Fig. (2). Diagrammatic representations of the procedures followed to assess and determine the toxicity of medicinal herbs.

2.3. Databases and Web Services

2.3.1. Drugbank

It integrates rich target knowledge with substantial drug data, including pharmacological, chemical, and pharmaceutical information, such as genome, structure, and pathway data [19]. The database is updated frequently and currently contains information on 15,199 medication-target interactions and 7,759 drug entities [20]. Some datasets and papers omit these figures.

2.3.2. Supertarget

We have 332,828 drug-target interactions studied in SuperTarget, being our massive database. This database may be searched for information on drugs, targets, pathways, ontologies, and cytochromes P450s [21].

2.3.3. Database of Therapeutic Targets

With knowledge of these targets and the associated medications, especially those that are already in clinical trials and applications, drug development can be accelerated significantly. This database already has 714 medicines and leads, 210 drug structures, and 1755 biomarkers for 365 disease categories.

2.4. TDR Methods

The TDR Targets repository functions as a chemogenomics resource for neglected tropical diseases with the aim of finding and selecting medicines and therapeutic targets in neglected disease pathogens [22, 23].

2.5. MATADOR

The list of protein-substance interactions, including binding and non-binding, was compiled by a combination of computerized text mining and hand collecting. Using MATADOR, you may look for drugs or proteins to target [24].

2.6. PDTD

The Potential Pharmacological Target Database (PDTD) serves as both a structural database of known and potential pharmacological targets and an informatics database. This database primarily focuses on pharmacological targets for which 3D structures are already known, classified into 15 different types based on therapeutic use and 13 different types based on biochemical criteria [25]. Data from various studies, other databases, and literature are all included through STITCH, a database of chemical-protein interactions encompassing both known and anticipated interactions. Recent updates to STITCH have increased the number of human interactions between chemicals and proteins by 45% compared to the previous version [26].

2.7. Integrity

Many medicines are included in this database, and each one is annotated with information on the diseases they treat, the drugs they target, and the phases of treatment they are now in [27]. These numbers are regularly collected by hand from the scholarly literature. The database current availability of 5.4 million bioactivity measurements aids in drug discovery [28].

2.8. FAERS

Data on adverse event keywords (also known as side effect keywords) for drugs was compiled through reports of adverse events and medication errors.

2.9. ZINC

For ligand discovery and virtual screening, ZINC provides access to approximately 20 million commercially available molecules that may be searched for based on the targets they bind.

2.10. SIDER

A publicly available, machine-readable database includes information on currently available pharmaceuticals, recorded adverse drug reactions, and links to further resources such as drug-target relationships.

2.11. ChemBank

A database is a repository of data made public from small molecules and small-molecule screens, as well as a set of tools for analyzing those properties in order to gain biological and medical insights. There are three ways in which ChemBank is differentiated from other small-molecule databases: Its primary function is the preservation of raw screening information. It employs metadata and gives a formal description of screening experiments in terms of statistical hypothesis testing to hierarchically structure linked assays into screening projects.

2.12. CanSAR

CanSAR is one of the greatest freely available integrated resources, and it facilitates the rapid retrieval of information.

2.13. The IUPHAR/BPS Pharmacology Manual

The IUPHAR/BPS Pharmacology Guide is an easily available information library that details both approved targets and investigational drugs. The information in this database is particularly pertinent to the pharmaceutical, chemical, genetic, functional, and pathophysiological domains.

2.14. DCDB

DCDB stands for the Medication Combination Database, which was developed by researchers from Zhejiang University. It stands as the pioneering database of its kind, focusing on discovering novel approaches for combining medications. As it is today, there are almost 6000 references, 1363 approved or approved-pending drug combinations, 237 failed drug combinations, 904 individual drugs, and so on. The DCDB is used as a basis for drug interaction simulation and theoretical modeling by compiling patterns of beneficial drug interactions. Medication combinations in the database were curated manually using resources like PubMed and the US FDA Orange Book. Drug and target information is manually annotated using resources such as PubChem, Drugbank, UniProt, and pharmaceuticals. com, as well as the scholarly literature. The DCDB's web interface allows for searches to be conducted on drug names, drug combinations, diseases, and drug targets. It is also possible to search for drugs based on molecular similarities.

2.15. DINIES

Possible drug-target interaction networks are inferred using a web server called DINIES, which stands for network inference engine for drug-target interactions. This method relies on supervised investigation. DINIES can receive information on chemical structure, amino acid and protein domains, side effects, and more.

3. DIGITAL MODELING AND SIMULATION

Experimental procedures for predicting drug-target interactions are still exceedingly challenging and time-consuming despite the availability of high-throughput screening and other biological assays today. Large-scale implementations of molecular docking have been hampered by the fact that it is difficult to get the 3D structures of most drugs.

3.1. HDI Data Accessed Without Cost

3.1.1. Chi Mei's Indexing Service (CMSS)

Probot strives to consolidate all known facts concerning interactions between Chinese medicines and other medicines [29] in order to deliver actual information to clinicians and the general public. Hong Kong Innovation and Technology Council and Healthy Power Limited are actively contributing to the ongoing development of this database. There were 6,292 interactions between 193 botanicals and 726 Western medicines as of July 2021, based on the 4,342 references found. Probot facilitates bilingual Chinese and English play and exploration.

3.1.2. SUPP.AI

The database includes information from 22 million articles, including 60,000 interactions and 195,000 evidence phrases. Keyword searches, such as a herb's name, revealed 2,044 herb supplements and 2,842 drug interactions as of July 2021. On the results page for interactions, you'll see the pertinent evidence sentences along with likely HDIs. Medicinal plants and pharmaceuticals are highlighted in each supporting sentence. Along with the evidence sentence and citations from the semantic scholar database, you also get connections to extra information about the sources. One of its primary benefits is that it employs automation to gather evidence phrases, which not only reduces time and minimizes labor, but also allows users to manage data using purpose-built computer programs. This database contains information on several supplements and their interactions with prescription medications. The published database examines the gaps in current HDI data dissemination methods. SUPP.AI appears to be hampered in its ability to extract probable HDI information from literature due to the insufficiency of its natural language processing (NLP) approaches, the artificial difference between medicines and plants, as well as a lack of regulated herb nomenclature.

3.1.3. Information Stored in the PHYDGI Database

This list provides herbal entities, the strength of their pharmacokinetic interactions, and vice versa. This database, which will be available to doctors, phytodrug manufacturers, and dietary supplement companies, has the potential to lessen the likelihood of HDI and increase the prevalence of the secure use of botanical medicines and supplements [30]. Fig. (3) provides an illustration of the data repository from the PHYDGI database.

3.2. Commercially-available HDI Databases

3.2.1. Database of Drug Interactions at UW (DIDB)

Data that is relevant to the mediated mechanism(s), such as inhibition or induction of an enzyme or transporter, is collected and organized by hand in DIDB. Specific experimental conditions, study designs, and dosing schedules are among the selected factors for *in vitro* kinetics and clinical pharmacokinetics [31]. There are pharmacokinetic, pharmacodynamic, and safety considerations that go into the clinical outcome of each interaction. DIDB is always being checked and updated by experts [32]. In addition to general terms, the database may be searched using more particular ones, such as *in vitro* parameters, exposure changes, QT prolongation, and so on [33-35].

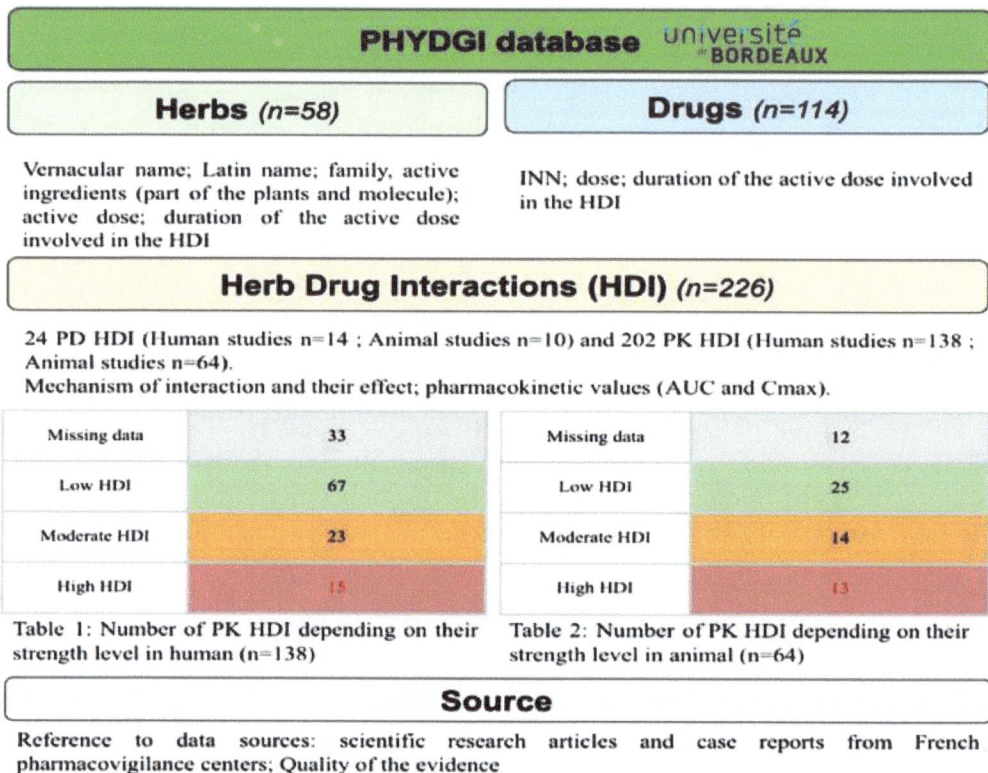

PHYDGI database université BORDEAUX

Herbs *(n=58)*

Vernacular name; Latin name; family, active ingredients (part of the plants and molecule); active dose; duration of the active dose involved in the HDI

Drugs *(n=114)*

INN; dose; duration of the active dose involved in the HDI

Herb Drug Interactions (HDI) *(n=226)*

24 PD HDI (Human studies n=14 ; Animal studies n=10) and 202 PK HDI (Human studies n=138 ; Animal studies n=64).
Mechanism of interaction and their effect; pharmacokinetic values (AUC and Cmax).

Missing data	33
Low HDI	67
Moderate HDI	23
High HDI	15

Missing data	12
Low HDI	25
Moderate HDI	14
High HDI	13

Table 1: Number of PK HDI depending on their strength level in human (n=138)

Table 2: Number of PK HDI depending on their strength level in animal (n=64)

Source

Reference to data sources: scientific research articles and case reports from French pharmacovigilance centers; Quality of the evidence

Fig. (3). An illustration of data from the PHYDGI database.

3.2.2. A One-Stop Source for Natural Health Products (NMCD)

Positioned as the primary resource for nutritional supplements, herbal remedies, and various forms of complementary and integrative medicine, this database presents itself as a comprehensive reference [36]. To access HDI data, users are required to navigate to the "Food, Herbs, and Supplements" database within NMCD. This particular database is projected to contain more than 1,200 entries of food, herbs, and supplements by July 2021.

3.2.3. Herbal Drug Interactions from Stockley's (SHMI)

With over 2000 references, the collection includes monographs for over 216 herbal drugs, nutritional supplements, and nutraceuticals. Each of the 216 HERBs has an associated summary page that covers synonyms, components, indications, pharmacokinetics, and interaction monographs [32]. Consumers can get some insight into the potential risks associated with herbal remedies thanks to the availability of educational information and relevant scientific ideas. At the point where databases for herbs and HDI collide, references with numbers greater than 0 were noted. The collection of the rugged database in herb-drug interactions from SHMI is depicted in Fig. (**4**).

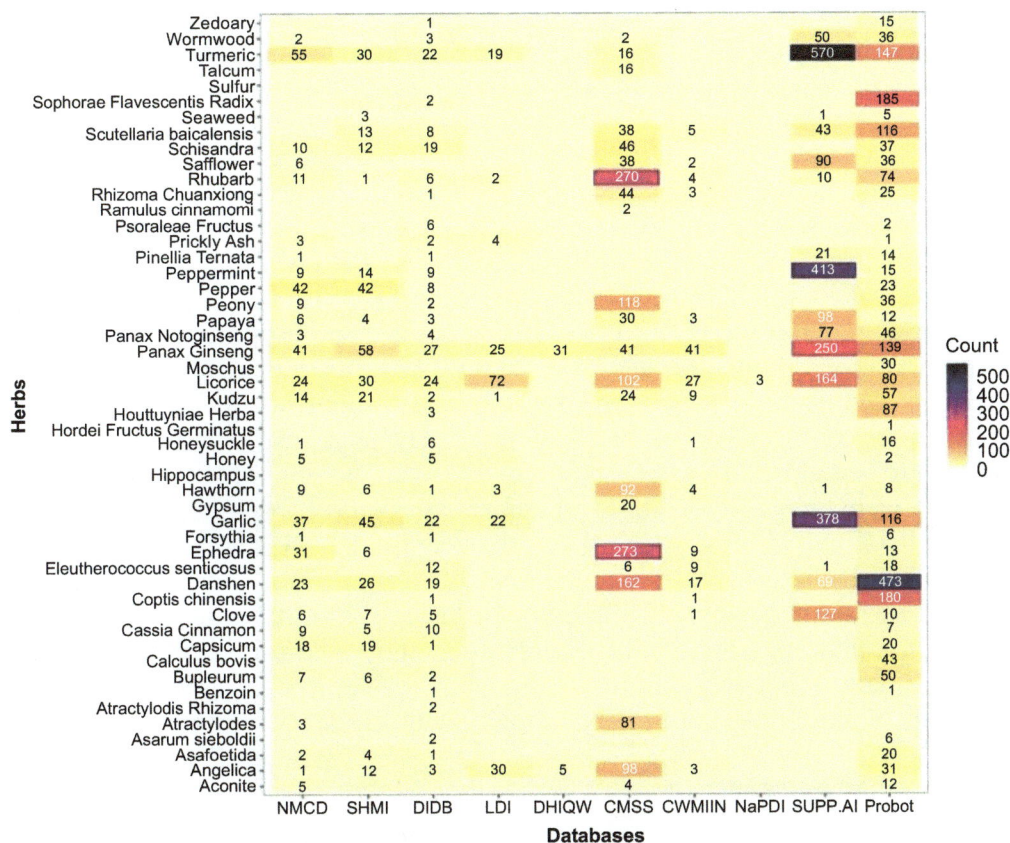

Herbs	NMCD	SHMI	DIDB	LDI	DHIQW	CMSS	CWMIIN	NaPDI	SUPP.AI	Probot
Zedoary			1							15
Wormwood	2		3			2			50	36
Turmeric	55	30	22	19		16			570	147
Talcum						16				
Sulfur										
Sophorae Flavescentis Radix			2							185
Seaweed		3							1	5
Scutellaria baicalensis		13	8			38	5		43	116
Schisandra	10	12	19			46				37
Safflower	6					38	2		90	36
Rhubarb	11	1	6	2		270	4		10	74
Rhizoma Chuanxiong			1			44	3			25
Ramulus cinnamomi						2				
Psoraleae Fructus			6							2
Prickly Ash	3		2	4						1
Pinellia Ternata	1		1						21	14
Peppermint	9	14	9						413	15
Pepper	42	42	8							23
Peony	9		2			118				36
Papaya	6	4	3			30	3		98	12
Panax Notoginseng	3		4						77	46
Panax Ginseng	41	58	27	25	31	41	41		250	139
Moschus										30
Licorice	24	30	24	72		102	27	3	164	80
Kudzu	14	21	2	1		24	9			57
Houttuyniae Herba			3							87
Hordei Fructus Germinatus										1
Honeysuckle	1		6				1			16
Honey	5		5							2
Hippocampus										
Hawthorn	9	6	1	3		92	4		1	8
Gypsum						20				
Garlic	37	45	22	22					378	116
Forsythia	1		1							6
Ephedra	31	6				273	9			13
Eleutherococcus senticosus			12			6	9		1	18
Danshen	23	26	19			162	17		69	473
Coptis chinensis			1				1			180
Clove	6	7	5				1		127	10
Cassia Cinnamon	9	5	10							7
Capsicum	18	19	1							20
Calculus bovis										43
Bupleurum	7	6	2							50
Benzoin			1							1
Atractylodis Rhizoma			2							
Atractylodes	3					81				
Asarum sieboldii			2							6
Asafoetida	2	4	1							20
Angelica	1	12	3	30	5	98	3			31
Aconite	5					4				12

Count: 500, 400, 300, 200, 100, 0

Fig. (4). Comparison of the 50 chosen herbs HDI database coverage. At the point where the databases for herbs and HDI collide, references with numbers greater than 0 were noted.

3.3. HDI Screening: Cutting-edge Intelligent *In silico* Methods

Using software and databases, xenobiotics may be virtually tested using the *in silico* method. *In vitro* and *in vivo* methods relate to experiments performed on cultured cells and living organisms, respectively [37].

3.3.1. In silico Forecasting HDIs

Herbal components will engage receptors and stimulate gene expression, changing DMEs or DTs as a result. The co-administration of herbs and medications may result in HDIs. The attempts to prevent toxic adverse effects might benefit from having reliable models for predicting metabolic HDIs [38, 39]. *In silico* methods for studying drug transport and interactions or metabolism have become more popular recently. In the future, HDIs are also anticipated to be predicted with the use of such a technique. The likelihood of ligand-CYP interactions has been widely studied and predicted using *in-silico* methods. Additionally, it decreases the number of tests, improves our comprehension of xenobiotic action, and forecasts potential HDIs [40]. To understand and foresee the potential effects of CYP inhibition, computational techniques, including pharmacokinetic modeling, pharmacophore modeling, QSAR investigations, and proteochemometric modeling are widely used. From the protein data library, the CYP isoform's crystal structures were obtained, and the protein production wizard of the Schrodinger suite was used to create 3D structures of proteins [41-43]. The Schrodinger suite's glide MD software will be used to do the MD of the ligand drug in the aforementioned CYP [44]. The glide docking programme was used to do computer-assisted docking analysis, which is useful to determine the binding affinity between each CYP isozyme and the ligand medication. To find possible mechanism-based CYP3A4 inhibitors from HMs, a mix of computational methods may be helpful. In research by He *et al.*, the herbal constituent picroside-III was utilized to predict HDI for cerebrovascular disorders using Autodock software [45-48]. By using MD and pharmacophore analyses with the help of AutoDock software, Wongrattanakamon *et al.* carried out computational modeling to examine several bioflavonoids that are important in HDIs [49, 50]. The modeling process used 25 flavonoids as ligands, and the findings showed a strong association between the docking scores. Table (**1**) provides the key compilation of herb-drug interactions stocking in clinical settings.

Table 1. Effects of herb-drug interactions in clinical settings.

Scientific Name of Herbs	Common Name	Key Interactions and Concerns	Avoid In
Withania somnifera (L.)	Ashwagandha	Enhancement of testosterone levels in men.	Prostate cancer
Curcuma longa	Turmeric	Increase in Oxalates CYP2C9 enzyme.	Renal disease
Inonotusobliquus (Ach. ex Pers) Pilat	Chaga	Increase in oxalates Anticoagulants Anti-platelets Antihyperlipidemic agents.	Renal disease Diabetic patients on treatment (acarbose).
Camellia sinensis	Green tea	Increased doses taken on an empty stomach can lead to liver toxicity.	Elevated liver function tests.

(Table 1) cont.....

Scientific Name of Herbs	Common Name	Key Interactions and Concerns	Avoid In
Taraxacummongolicum Hand-Mazz., *T. officinale* (L.) Weber ex F.H.Wigg	Dandelion	CYP1A2 enzyme Diuretic Antihyperglycemic agents Estrogenic action	Hormone-sensitive Breast cancer
Boswelliaserrata Roxb. exColebr.	Boswellia	Unknown	Contact dermatitis

3.3.2. Lexicomp Drug Interactions (LDI)

Using the Lexi-comp database's interaction module, one may retrieve details on HDI. This module is a component of UpToDate, a global source of expert knowledge for several industries and a division of Wolters Kluwer. The phrase "the most trusted evidence-based clinical decision support resource at the point of care" has been used to describe UpToDate [51]. Medical professionals who frequently need to give consumers medical advice are the intended users. An author and editor team pre-processed the clinical evidence. Data in LDI mainly correspond to those in DIDB for DDI. LDI detected only 85 herbs among 2,096 entities and a total of 295 drug-herb or herb-herb interaction pairs from 902 sources [52, 53].

3.3.3. Literature Excerpt

Health professionals and academics often handle the post-processing work in order to validate raw data and make complex technical information easier to grasp for general audiences. A notable exception is SUPP.AI, which extracts evidence phrases straight from the source references without additional rephrasing or expert confirmation. The advantage is that it requires less human work, but the accuracy of the HDI data collected without expert validation may be in doubt [54].

3.3.4. Frequency, Liability and Sustainability of Content Updates.

In order to offer users up-to-date and comprehensive information at all times, professional databases require frequent updates, expansions, and additions. Database knowledge that hasn't been updated in a timely manner might lead users to draw the wrong conclusions, endangering the patient's health [55, 56]. Additionally, SUPP.AI provides an automated mechanism for upgrading the database once every few months, which might be a workaround for database providers who lack the capacity to provide continuous updates [57-59]. The publisher is not liable for mistakes and omissions. The disclaimer on SUPP.AI further states that "the information contained herein should not be used as a substitute for the advice of a physician or other health care provider who is appropriately qualified and licensed" [60].

3.3.5. Interfaces for Users

Keyword-based searches are supported by all databases. The names of the medications, the herbs, or both are frequently required as keywords by users. For easier searching, certain databases (such as NMCD and SHMI) offer alphabetical indexes, whereas CWMIIN offers comparable indexes for Chinese-language herbs and medications [61]. Nearly none of the databases taken into consideration include a search capability that enables queries based purely on the HDI's properties without specifically naming the herbs and medications involved. One apparent exception is DIDB, which allows users to get study findings of HDI without necessarily supplying medication names. Graph databases, which allow searches based on connection descriptions, have recently gained increasing popularity [62-64]. An HDI graph database implies that users who are symptom experts but not herb or drug experts may nevertheless be able to utilize the database to uncover probable herb-drug pairings that produce the symptoms. The reason being that signs and symptoms depict interactions between medicines [65].To make it easier to filter out irrelevant search results, certain databases include advanced search options. For instance, NMCD lets users exclude certain criteria from a search. Users may only conduct searches in a single subset of either the NMCD or the SHMI because they are both collections of smaller, sub-categorized databases [66].

3.3.6. AI's Role in Creating HDI Databases

With the use of natural language processing (NLP), we can reduce the amount of time and effort required to construct and maintain these HDI databases by automatically classifying, extracting, translating, and interpreting this text data [67]. An HDI database can be built using natural language processing techniques such as article recognition, named entity recognition (such as herbs and drugs), relationship recognition (such as the interactions between herbs and drugs), and conclusion drawing from a piece (or large corpus) of text [68, 69].

Disease detection from relevant corpora has been aided by rule-based NLP methods. While Segura-Bedmar *et al.* sought to use a rule-based approach to extract DDIs from the biological literature, their efforts were unsuccessful. The authors reasoned that this was due to the idiomatic expressions seen in everyday speech [70, 71]. Extracting DDIs with an accuracy of over 80% was achieved by using a feature-based technique validated by Quoc-Chinh Bui and colleagues with an SVM classifier [72-74]. Although statistical models can be useful for unstructured data like articles covering a wide range of themes and writing styles, they can be difficult to debug and need a large amount of data, both of which can be time-consuming and resource-intensive to gather [75].

So far, SUPP.AI has been the most compelling example of how AI could be used with HDI data. The data was preprocessed by creating concept unique identifiers (CUIs) and grouping them according to the UMLS Metathesaurus, as well as by identifying and linking entities using the ScispaCy toolkit. At this stage, sentences from many abstracts are assembled in order to be further categorized. Using the classification that follows, you may figure out if a certain sentence involves an interaction or not [76]. The RoBERTa. and Bidirectional Encoder Representations from Transformers (BERT) model are used by SUPP.AI for DDI classification, and it has been improved using pre-trained embeddings [77, 78].

3.3.7. HDI Database Development Going Forward

The third stage of creating an HDI database, which involves uncovering and evaluating the hidden information in the literature to produce structured data, is the most time-consuming, as we said in the introduction. For example, extracting SDI terms in SUPP.AI and evaluating HDI-related studies in Probot are examples of how artificial intelligence appears to produce positive results when detecting plant and pharmaceutical names and when categorizing text. It may also significantly reduce the amount of manual effort required for database management.

CONCLUDING REMARKS

The use of herbal products will probably continue to rise, partly as a result of consumer efforts to cut down on medical expenditures by self-diagnosing and treating. Despite the growing incidence of HDIs, a standardized method for assessing the potential for medication interactions with herbal remedies is still lacking. Prospective evaluation of HDIs is more difficult than DDIs because of the considerable compositional variability inherent to herbal medicines, various perpetrator ingredients, lack of understanding about the pharmacokinetics of perpetrator elements, and different regulatory views. All things considered, there is a rare chance to create a system for enhancing HDI forecasts. The HDI performance ratings for all the tools were lower than the DDI performance values. In reality, the best tool available for spotting HDIs (Lexicomp: 0.54) underperformed the worst tool available for picking up on DDIs. (WebMD: 0.67). A significant barrier to the advancement of effective HDI screening tools is the absence of dependable information regarding potential pharmaceutical interactions with herbal supplements. These HDI ratings, on the other hand, consider how well these technologies compare to the state of the art in natural medicine. Natural Medicine may be a useful HDI screening tool if it does not require a subscription and saves data in raw format. The vast majority of the information we have on herbal remedies comes from the field of natural medicine.

ACKNOWLEDGEMENTS

The College of Pharmaceutical Sciences at DayanandaSagar University encouraged us to draft the current chapter, for which the writers are grateful.

REFERENCES

[1]   Zhang Y, Man Ip C, Lai YS, Zuo Z. Overview of current herb-drug interaction databases. Drug Metab Dispos 2022; 50(1): 86-94.
[http://dx.doi.org/10.1124/dmd.121.000420] [PMID: 34697080]

[2]   Glisson JK, Walker LA. How physicians should evaluate dietary supplements. Am J Med 2010; 123(7): 577-82.
[http://dx.doi.org/10.1016/j.amjmed.2009.10.017] [PMID: 20493463]

[3]   Izzo AA, Ernst E. Interactions between herbal medicines and prescribed drugs: A systematic review. Drugs 2001; 61(15): 2163-75.
[http://dx.doi.org/10.2165/00003495-200161150-00002] [PMID: 11772128]

[4]   Hu J, Chen G, Lo IMC. Removal and recovery of Cr(VI) from wastewater by maghemite nanoparticles. Water Res 2005; 39(18): 4528-36.
[http://dx.doi.org/10.1016/j.watres.2005.05.051] [PMID: 16146639]

[5]   Juurlink DN. Drug interactions with warfarin: What clinicians need to know. CMAJ 2007; 177(4): 369-71.
[http://dx.doi.org/10.1503/cmaj.070946] [PMID: 17698826]

[6]   Cheng C, Liye Z, Zhan RJ. Surface modification of polymer fibre by the new atmospheric pressure cold plasma jet. Surf Coat Tech 2006; 200(24): 6659-65.
[http://dx.doi.org/10.1016/j.surfcoat.2005.09.033]

[7]   Fugh-Berman A, Ernst E. Herb–drug interactions: Review and assessment of report reliability. Br J Clin Pharmacol 2001; 52(5): 587-95.
[http://dx.doi.org/10.1046/j.0306-5251.2001.01469.x] [PMID: 11736868]

[8]   Levine GN, Bates ER, Blankenship JC, *et al.* 2011 ACCF/AHA/SCAI guideline for percutaneous coronary intervention: A report of the american college of cardiology foundation/american heart association task force on practice guidelines and the society for cardiovascular angiography and interventions. Circulation 2011; 124(23): e574-651.
[PMID: 22064601]

[9]   Engels EA, Pfeiffer RM, Fraumeni JF Jr, *et al.* Spectrum of cancer risk among US solid organ transplant recipients. JAMA 2011; 306(17): 1891-901.
[http://dx.doi.org/10.1001/jama.2011.1592] [PMID: 22045767]

[10]   Vardell E. Natural medicines: A complementary and alternative medicines tool combining natural standard and the natural medicines comprehensive database. Med Ref Serv Q 2015; 34(4): 461-70.
[http://dx.doi.org/10.1080/02763869.2015.1082382] [PMID: 26496400]

[11]   Squires RW, Kaminsky LA, Porcari JP, Ruff JE, Savage PD, Williams MA. Progression of exercise training in early outpatient cardiac rehabilitation. J Cardiopulm Rehabil Prev 2018; 38(3): 139-46.
[http://dx.doi.org/10.1097/HCR.0000000000000337] [PMID: 29697494]

[12]   Wu Z, Shen C, van den Hengel A. Wider or deeper: Revisiting the resnet model for visual recognition. Pattern Recognit 2019; 90: 119-33.
[http://dx.doi.org/10.1016/j.patcog.2019.01.006]

[13]   Birer-Williams C, Gufford BT, Chou E, *et al.* A new data repository for pharmacokinetic natural product-drug interactions: From chemical characterization to clinical studies. Drug Metab Dispos 2020; 48(10): 1104-12.
[http://dx.doi.org/10.1124/dmd.120.000054] [PMID: 32601103]

[14]   Yang J, Zheng Y, Gou X, *et al.* Prevalence of comorbidities and its effects in patients infected with SARS-CoV-2: A systematic review and meta-analysis. Int J Infect Dis 2020; 94: 91-5.
[http://dx.doi.org/10.1016/j.ijid.2020.03.017] [PMID: 32173574]

[15]   Patwardhan B. Ayurveda: The designer medicine. Indian Drugs 2000; 37(5): 213-27.

[16]   Jiang X, Coffee M, Bari A, *et al.* Towards an artificial intelligence framework for data-driven prediction of coronavirus clinical severity. Comput Mater Continua 2020; 62(3): 537-51.
[http://dx.doi.org/10.32604/cmc.2020.010691]

[17]   Huang JZ, Huang M. How much of the corporate-treasury yield spread is due to credit risk? Rev Asset Pricing Stud 2012; 2(2): 153-202.
[http://dx.doi.org/10.1093/rapstu/ras011]

[18]   Huang HJ, Yu HW, Chen CY, *et al.* Current developments of computer-aided drug design. J Taiwan Inst Chem Eng 2010; 41(6): 623-35.
[http://dx.doi.org/10.1016/j.jtice.2010.03.017]

[19]   Kaur P, Khatik G. An overview of computer-aided drug design tools and recent applications in designing of anti-diabetic agents. Curr Drug Targets 2021; 22(10): 1158-82.
[http://dx.doi.org/10.2174/1389450121666201119141525] [PMID: 33213342]

[20]   Hachad H, Ragueneau-Majlessi I, Levy RH. A useful tool for drug interaction evaluation: The university of washington metabolism and transport drug interaction database. Hum Genomics 2010; 5(1): 61-72.
[http://dx.doi.org/10.1186/1479-7364-5-1-61] [PMID: 21106490]

[21]   Riechelmann RP, Tannock IF, Wang L, Saad ED, Taback NA, Krzyzanowska MK. Potential drug interactions and duplicate prescriptions among cancer patients. J Natl Cancer Inst 2007; 99(8): 592-600.
[http://dx.doi.org/10.1093/jnci/djk130] [PMID: 17440160]

[22]   Alqahtani S. *In silico* ADME-Tox modeling: Progress and prospects. Expert Opin Drug Metab Toxicol 2017; 13(11): 1147-58.
[http://dx.doi.org/10.1080/17425255.2017.1389897] [PMID: 28988506]

[23]   Kannan G, Rani VN, Alosh J, *et al.* A study of drug-drug interactions in cancer patients of a south Indian tertiary care teaching hospital. J Postgrad Med 2011; 57(3): 206-10.
[http://dx.doi.org/10.4103/0022-3859.85207] [PMID: 21941058]

[24]   Jiang L, Zhang X, Chen X, *et al.* Virtual screening and molecular dynamics study of potential negative allosteric modulators of mGluR1 from Chinese herbs. Molecules 2015; 20(7): 12769-86.
[http://dx.doi.org/10.3390/molecules200712769] [PMID: 26184151]

[25]   Mukhtar M, Arshad M, Ahmad M, Pomerantz RJ, Wigdahl B, Parveen Z. Antiviral potentials of medicinal plants. Virus Res 2008; 131(2): 111-20.
[http://dx.doi.org/10.1016/j.virusres.2007.09.008] [PMID: 17981353]

[26]   Sarma H, Upadhyaya M, Gogoi B, *et al.* Cardiovascular drugs: An insight of *in silico* drug design tools. J Pharm Innov 2021; 1-26.

[27]   Erlina L, Paramita RI, Kusuma WA, *et al.* Virtual screening of Indonesian herbal compounds as COVID-19 supportive therapy: Machine learning and pharmacophore modeling approaches. BMC Complementary Medicine and Therapies 2022; 22(1): 207.
[http://dx.doi.org/10.1186/s12906-022-03686-y] [PMID: 35922786]

[28]   Caccia S, Garattini S, Pasina L, Nobili A. Predicting the clinical relevance of drug interactions from

pre-approval studies. Drug Saf 2009; 32(11): 1017-39.
[http://dx.doi.org/10.2165/11316630-000000000-00000] [PMID: 19810775]

[29]   Chen J, See KC. Artificial intelligence for COVID-19: Rapid review. J Med Internet Res 2020; 22(10): e21476.
[http://dx.doi.org/10.2196/21476] [PMID: 32946413]

[30]   Seddon G, Lounnas V, McGuire R, *et al.* Drug design for ever, from hype to hope. J Comput Aided Mol Des 2012; 26(1): 137-50.
[http://dx.doi.org/10.1007/s10822-011-9519-9] [PMID: 22252446]

[31]   Fahmi OA, Hurst S, Plowchalk D, *et al.* Comparison of different algorithms for predicting clinical drug-drug interactions, based on the use of CYP3A4 *in vitro* data: predictions of compounds as precipitants of interaction. Drug Metab Dispos 2009; 37(8): 1658-66.
[http://dx.doi.org/10.1124/dmd.108.026252] [PMID: 19406954]

[32]   Fung KW, Kapusnik-Uner J, Cunningham J, Higby-Baker S, Bodenreider O. Comparison of three commercial knowledge bases for detection of drug-drug interactions in clinical decision support. J Am Med Inform Assoc 2017; 24(4): 806-12.
[http://dx.doi.org/10.1093/jamia/ocx010] [PMID: 28339701]

[33]   Nayarisseri A, Khandelwal R, Tanwar P, *et al.* Artificial intelligence, big data and machine learning approaches in precision medicine & drug discovery. Curr Drug Targets 2021; 22(6): 631-55.
[http://dx.doi.org/10.2174/18735592MTEzsMDMnz] [PMID: 33397265]

[34]   Muhammad J, Khan A, Ali A, Fang L, Yanjing W, Xu Q, Wei DQ. Network Pharmacology: Exploring the Resources and Methodologies. Curr Top Med Chem. 2018; 18(12): 949-964.
[http://dx.doi.org/10.2174/1568026618666180330141351] [PMID: 29600765]

[35]   Kibble M, Saarinen N, Tang J, Wennerberg K, Mäkelä S, Aittokallio T. Network pharmacology applications to map the unexplored target space and therapeutic potential of natural products. Nat Prod Rep 2015; 32(8): 1249-66.
[http://dx.doi.org/10.1039/C5NP00005J] [PMID: 26030402]

[36]   Chou TC. Theoretical basis, experimental design, and computerized simulation of synergism and antagonism in drug combination studies. Pharmacol Rev 2006; 58(3): 621-81.
[http://dx.doi.org/10.1124/pr.58.3.10] [PMID: 16968952]

[37]   Shi P, Lin X, Yao H. A comprehensive review of recent studies on pharmacokinetics of traditional Chinese medicines (2014–2017) and perspectives. Drug Metab Rev 2018; 50(2): 161-92.
[http://dx.doi.org/10.1080/03602532.2017.1417424] [PMID: 29258334]

[38]   Leung EL, Cao ZW, Jiang ZH, Zhou H, Liu L. Network-based drug discovery by integrating systems biology and computational technologies. Brief Bioinform 2013; 14(4): 491-505.
[http://dx.doi.org/10.1093/bib/bbs043] [PMID: 22877768]

[39]   Kharkar PS, Warrier S, Gaud RS. Reverse docking: A powerful tool for drug repositioning and drug rescue. Future Med Chem 2014; 6(3): 333-42.
[http://dx.doi.org/10.4155/fmc.13.207] [PMID: 24575968]

[40]   Ageno W, Gallus AS, Wittkowsky A, Crowther M, Hylek EM, Palareti G. Oral anticoagulant therapy: Antithrombotic therapy and prevention of thrombosis: American College of Chest Physicians evidence-based clinical practice guidelines. Chest 2012; 141(2) (Suppl.): e44S-88S.
[http://dx.doi.org/10.1378/chest.11-2292] [PMID: 22315269]

[41]   Banerjee S, Mitra A. Changing landscape of herbal medicine: Technology attributing renaissance. Int J Pharm Pharm Sci 2012; 4(1): 47-52.

[42]   Mehmood MA, Sehar U, Ahmad N. Use of bioinformatics tools in different spheres of life sciences. J Data Mining Genomics Proteomics 2014; 5(2): 1.

[43]   Won CS, Oberlies NH, Paine MF. Mechanisms underlying food–drug interactions: Inhibition of intestinal metabolism and transport. Pharmacol Ther 2012; 136(2): 186-201.

[http://dx.doi.org/10.1016/j.pharmthera.2012.08.001] [PMID: 22884524]

[44]   Gasteiger J. Chemoinformatics: Achievements and challenges, a personal view. Molecules 2016; 21(2): 151.
[http://dx.doi.org/10.3390/molecules21020151] [PMID: 26828468]

[45]   Zhang A, Fang H, Wang Y, *et al.* Discovery and verification of the potential targets from bioactive molecules by network pharmacology-based target prediction combined with high-throughput metabolomics. RSC Advances 2017; 7(81): 51069-78.
[http://dx.doi.org/10.1039/C7RA09522H]

[46]   Santana K, do Nascimento LD, Lima e Lima A, *et al.* Applications of virtual screening in bioprospecting: Facts, shifts, and perspectives to explore the chemo-structural diversity of natural products. Front Chem 2021; 9: 662688.
[http://dx.doi.org/10.3389/fchem.2021.662688] [PMID: 33996755]

[47]   Seden K, Khoo SH, Back D, *et al.* Global patient safety and antiretroviral drug-drug interactions in the resource-limited setting. J Antimicrob Chemother 2013; 68(1): 1-3.
[http://dx.doi.org/10.1093/jac/dks346] [PMID: 22915459]

[48]   Chandran U, Mehendale NE, Tillu GI, Patwardhan BH. Network pharmacology: An emerging technique for natural product drug discovery and scientific research on ayurveda. Proc Indian Natn Sci Acad 2015; 81(3): 561-8.

[49]   Rathod V, Jain S, Nandekar P, Sangamwar AT. Human pregnane X receptor: A novel target for anticancer drug development. Drug Discov Today 2014; 19(1): 63-70.
[http://dx.doi.org/10.1016/j.drudis.2013.08.009] [PMID: 23974067]

[50]   Li WH, Han JR, Ren PP, Xie Y, Jiang DY. Exploration of the mechanism of Zisheng Shenqi decoction against gout arthritis using network pharmacology. Comput Biol Chem 2021; 90: 107358.
[http://dx.doi.org/10.1016/j.compbiolchem.2020.107358] [PMID: 33243703]

[51]   Das AP, Agarwal SM. Recent advances in the area of plant-based anti-cancer drug discovery using computational approaches. Mol Divers 2023; 1-25.
[http://dx.doi.org/10.1007/s11030-022-10590-7] [PMID: 36670282]

[52]   Koulouridi E, Valli M, Ntie-Kang F, Bolzani VS. A primer on natural product-based virtual screening. Physical Sciences Reviews 2019; 4(6): 20180105.
[http://dx.doi.org/10.1515/psr-2018-0105]

[53]   Mathew T, Sree RA, Aishwarya S, *et al.* Graphene-based functional nanomaterials for biomedical and bioanalysis applications. FlatChem 2020; 23: 100184.
[http://dx.doi.org/10.1016/j.flatc.2020.100184]

[54]   Singh N, Decroly E, Khatib AM, Villoutreix BO. Structure-based drug repositioning over the human TMPRSS2 protease domain: Search for chemical probes able to repress SARS-CoV-2 Spike protein cleavages. Eur J Pharm Sci 2020; 153: 105495.
[http://dx.doi.org/10.1016/j.ejps.2020.105495] [PMID: 32730844]

[55]   Choi S, Oh DS, Jerng UM. A systematic review of the pharmacokinetic and pharmacodynamic interactions of herbal medicine with warfarin. PLoS One 2017; 12(8): e0182794.
[http://dx.doi.org/10.1371/journal.pone.0182794] [PMID: 28797065]

[56]   Kiguba R, Karamagi C, Bird SM. Incidence, risk factors and risk prediction of hospital-acquired suspected adverse drug reactions: A prospective cohort of Ugandan inpatients. BMJ Open 2017; 7(1): e010568.
[http://dx.doi.org/10.1136/bmjopen-2015-010568] [PMID: 28110281]

[57]   Santana Azevedo L, Pretto Moraes F, Morrone Xavier M, *et al.* Recent progress of molecular docking simulations applied to development of drugs. Curr Bioinform 2012; 7(4): 352-65.
[http://dx.doi.org/10.2174/157489312803901063]

[58]   Priest J, Sanchez J. Product development and design for manufacturing: a collaborative approach to

producibility and reliability. CRC Press 2001.

[59]   Chobanian AV, Bakris GL, Black HR, *et al.* Seventh report of the joint national committee on prevention, detection, evaluation, and treatment of high blood pressure. Hypertension 2003; 42(6): 1206-52.
[http://dx.doi.org/10.1161/01.HYP.0000107251.49515.c2] [PMID: 14656957]

[60]   Li JJ, Corey EJ, Eds. Drug discovery: Practices, processes, and perspectives. John Wiley & Sons 2013.
[http://dx.doi.org/10.1002/9781118354483]

[61]   Ayres LB, Gomez FJV, Linton JR, Silva MF, Garcia CD. Taking the leap between analytical chemistry and artificial intelligence: A tutorial review. Anal Chim Acta 2021; 1161: 338403.
[http://dx.doi.org/10.1016/j.aca.2021.338403] [PMID: 33896558]

[62]   Rodríguez-Mazahua L, Rodríguez-Enríquez CA, Sánchez-Cervantes JL, Cervantes J, García-Alcaraz JL, Alor-Hernández G. A general perspective of Big Data: Applications, tools, challenges and trends. J Supercomput 2016; 72(8): 3073-113.
[http://dx.doi.org/10.1007/s11227-015-1501-1]

[63]   Bansal T, Jaggi M, Khar R, Talegaonkar S. Emerging significance of flavonoids as P-glycoprotein inhibitors in cancer chemotherapy. J Pharm Pharm Sci 2009; 12(1): 46-78.
[http://dx.doi.org/10.18433/J3RC77] [PMID: 19470292]

[64]   Zhang QR, Zhong ZF, Sang W, *et al.* Comparative comprehension on the anti-rheumatic Chinese herbal medicine Siegesbeckiae Herba: Combined computational predictions and experimental investigations. J Ethnopharmacol 2019; 228: 200-9.
[http://dx.doi.org/10.1016/j.jep.2018.09.023] [PMID: 30240786]

[65]   Issa NT, Stathias V, Schürer S, Dakshanamurthy S. Machine and deep learning approaches for cancer drug repurposing. Semin Cancer Biol 2021; 68: 132-42.
[http://dx.doi.org/10.1016/j.semcancer.2019.12.011] [PMID: 31904426]

[66]   Pfisterer PH, Wolber G, Efferth T, Rollinger JM, Stuppner H. Natural products in structure-assisted design of molecular cancer therapeutics. Curr Pharm Des 2010; 16(15): 1718-41.
[http://dx.doi.org/10.2174/138161210791164027] [PMID: 20222854]

[67]   Arora G, Joshi J, Mandal RS, Shrivastava N, Virmani R, Sethi T. Artificial intelligence in surveillance, diagnosis, drug discovery and vaccine development against COVID-19. Pathogens 2021; 10(8): 1048.
[http://dx.doi.org/10.3390/pathogens10081048] [PMID: 34451513]

[68]   Saha S, Nandi R, Vishwakarma P, Prakash A, Kumar D. Discovering potential RNA dependent RNA polymerase inhibitors as prospective drugs against COVID-19: An *in silico* approach. Front Pharmacol 2021; 12: 634047.
[http://dx.doi.org/10.3389/fphar.2021.634047] [PMID: 33716752]

[69]   Parikesit AA, Ratnasari NRP, Anurogo D. Application of artificial intelligence-based computation in the health sciences to ward off the COVID-19 pandemic. International Journal of Human and Health Sciences (IJHHS) 2020; 5(2): 177-84. [IJHHS].
[http://dx.doi.org/10.31344/ijhhs.v5i2.256]

[70]   Brents LK, Prather PL. The K2/Spice Phenomenon: Emergence, identification, legislation and metabolic characterization of synthetic cannabinoids in herbal incense products. Drug Metab Rev 2014; 46(1): 72-85.
[http://dx.doi.org/10.3109/03602532.2013.839700] [PMID: 24063277]

[71]   Makhouri FR, Ghasemi JB. *In silico* studies in drug research against neurodegenerative diseases. Curr Neuropharmacol 2018; 16(6): 664-725.
[http://dx.doi.org/10.2174/1570159X15666170823095628] [PMID: 28831921]

[72]   Vora J, Patel S, Sinha S, *et al.* Molecular docking, QSAR and ADMET based mining of natural compounds against prime targets of HIV. J Biomol Struct Dyn 2019; 37(1): 131-46.
[http://dx.doi.org/10.1080/07391102.2017.1420489] [PMID: 29268664]

[73]   Nagpal K, Singh SK, Mishra DN. Drug targeting to brain: A systematic approach to study the factors, parameters and approaches for prediction of permeability of drugs across BBB. Expert Opin Drug Deliv 2013; 10(7): 927-55.
[http://dx.doi.org/10.1517/17425247.2013.762354] [PMID: 23330786]

[74]   Protti M, Mandrioli R, Marasca C, Cavalli A, Serretti A, Mercolini L. New-generation, non-SSRI antidepressants: Drug-drug interactions and therapeutic drug monitoring. Part 2: NaSSAs, NRIs, SNDRIs, MASSAs, NDRIs, and others. Med Res Rev 2020; 40(5): 1794-832.
[http://dx.doi.org/10.1002/med.21671] [PMID: 32285503]

[75]   Molassiotis A, Xu M. Quality and safety issues of web-based information about herbal medicines in the treatment of cancer. Complement Ther Med 2004; 12(4): 217-27.
[http://dx.doi.org/10.1016/j.ctim.2004.09.005] [PMID: 15649835]

[76]   Dunkel M, Fullbeck M, Neumann S, Preissner R. SuperNatural: A searchable database of available natural compounds. Nucleic Acids Res 2006; 34(90001) (Suppl. 1): D678-83.
[http://dx.doi.org/10.1093/nar/gkj132] [PMID: 16381957]

[77]   Pandey RK, Narula A, Naskar M, *et al.* Exploring dual inhibitory role of febrifugine analogues against *Plasmodium* utilizing structure-based virtual screening and molecular dynamic simulation. J Biomol Struct Dyn 2017; 35(4): 791-804.
[http://dx.doi.org/10.1080/07391102.2016.1161560] [PMID: 26984239]

[78]   Guleria A, Kumar A, Kumar U, Raj R, Kumar D. NMR based metabolomics: An exquisite and facile method for evaluating therapeutic efficacy and screening drug toxicity. Curr Top Med Chem 2018; 18(20): 1827-49.
[http://dx.doi.org/10.2174/1568026619666181120141603] [PMID: 30465509]

CHAPTER 10

Deliberations and Considerations of Mesodyn Simulations in Pharmaceuticals

Manisha Yadav[1], Dhriti Mahajan[1], Om Silakari[1] and Bharti Sapra[1,*]

[1] *Department of Pharmaceutical Sciences & Drug Research, Punjabi University, Patiala, Punjab, India*

Abstract: The main aim of this chapter is the detailed analysis of the Mesodyn module and how it is beneficial in the pharmaceuticals or drug delivery systems. These models are the generalization of a coarse-grained model in mesoscopic dynamics which is used for the field-based simulations of complex systems. A set of functional Langevin equations characterize the system's behavior. These computer-based simulation tools have been proven effective for providing information at molecular and mesoscopic scales and also for overcoming the limitations of wet lab experiments. So, this chapter will discuss the potential use of Mesodyn simulations in pre-formulations and various other applications for the rational designing of drug delivery systems after providing a brief theoretical background.

Keywords: Coarse-grained model, Drug delivery, Formulation, Investigations, Material studio, Molecular dynamics, Mesodyn, Pharmaceutics, Simulations.

1. INTRODUCTION

Drug delivery has been constrained due to biological systems' mechanical and chemical fragility [1]. To produce a precise and intended drug release profile; *in silico* simulations can assist in determining the needed composition of the drug delivery system and manufacturing process. Thus, the creation of therapeutic products may speed up and costly and time-consuming sets of trials can be substituted [2]. Recently, computer-assisted simulations have proven to be useful tools for delivering extra information at the molecular or mesoscopic scale, to address the limitations of the practical experimental investigations [3]. Mesoscopic dynamics models are gaining popularity as a link between slow macroscale thermodynamic relaxation and fast molecular kinetic features [4]. Mesoscale simulations offer important insights into the precise physico-chemical

* **Corresponding author Bharti Sapra:** Department of Pharmaceutical Sciences and Drug Research, Punjabi University, Patiala, Punjab, India; E-mail: Bhartijatin2000@yahoo.co.in.

Dilpreet Singh and Prashant Tiwari (Eds.)

behavior of molecules and materials. Numerous properties, such as shape, solidity action, heterogeneity solubility, adherence adsorption, dispersion, spectral information, and structural properties can be directly determined [5]. "Biovia Material Studio 2022," suite of software consists of a Mesoscale simulation tool [6] which consists of two edge-cutting techniques that successfully address the physics and chemistry of mesoscale modeling Mesodyn and Dissipative Particle Dynamics (DPD) [1]. To achieve precise and targeted drug release profiles, *in silico* simulations can assist in determining the necessary composition of drug delivery systems and manufacturing processes. Consequently, the development of therapeutic products could be expedited, and resource-intensive and time-consuming sets of trials might be replaced [2]. Recently, computer-assisted simulations have demonstrated their utility as tools for providing additional insights at both the molecular and mesoscopic scales, aiming to address the limitations inherent in practical experimental investigations [3].

Mesoscopic dynamic models are increasingly gaining attraction as a bridge between the gradual thermodynamic relaxation observed at the macroscale and the rapid molecular kinetic features [4]. Mesoscale simulations offer valuable insights into the precise physicochemical behavior of molecules and materials. Various properties, such as shape, rigidity, mobility, heterogeneity, solubility, adhesion, adsorption, dispersion, spectral information, and structural properties, can be directly ascertained [5]. The suite of software known as "Biovia Material Studio 2022" includes a Mesoscale simulation tool [6], incorporating two state-of-the-art techniques that effectively address the physics and chemistry of mesoscale modeling: MesoDyn and Dissipative Particle Dynamics (DPD) [7]. One of the first applications of the Mesodyn method was for the study of the Microphase separation kinetics of real polymer systems, in which the water-based solution of the triblock polymer surfactant *i.e.* [(Ethylene oxide)$_{13}$ (Propylene oxide)$_{13}$ (Ethylene oxide)$_{13}$] was used [8]. Early in the 1990s, Akzo Nobel created Mesodyn to address a challenge, related to the stability of water-borne coatings [9]. Mesodyn is reported to have numerous applications in the field of drug delivery including active release patterns and formulation stability, compatibilization, the impact of hydrophobic medicines on the size of micelle in a Pluronic solution, as well as excipients' function [5]. According to classical mechanics, Molecular dynamics (MD) is a technique that simulates a complex system as a collection of interacting particles represented as atoms [10]. It is possible to resolve dynamic problems in MD without utilizing any approximations, where the intervals between completely elastic collisions, and particles move at a constant speed. Rahman made the first successful attempt in 1964 to use MD to model the way Lennard-Jones particles behave in phases in a realistically promising molecular system [11]. Drug delivery studies use MD simulations because they can monitor system behavior changes across large spatial-sequential domain durations with atomic accuracy and high resolutions [12]. Fig. (**1**) depicts the process of molecular dynamics (MD).

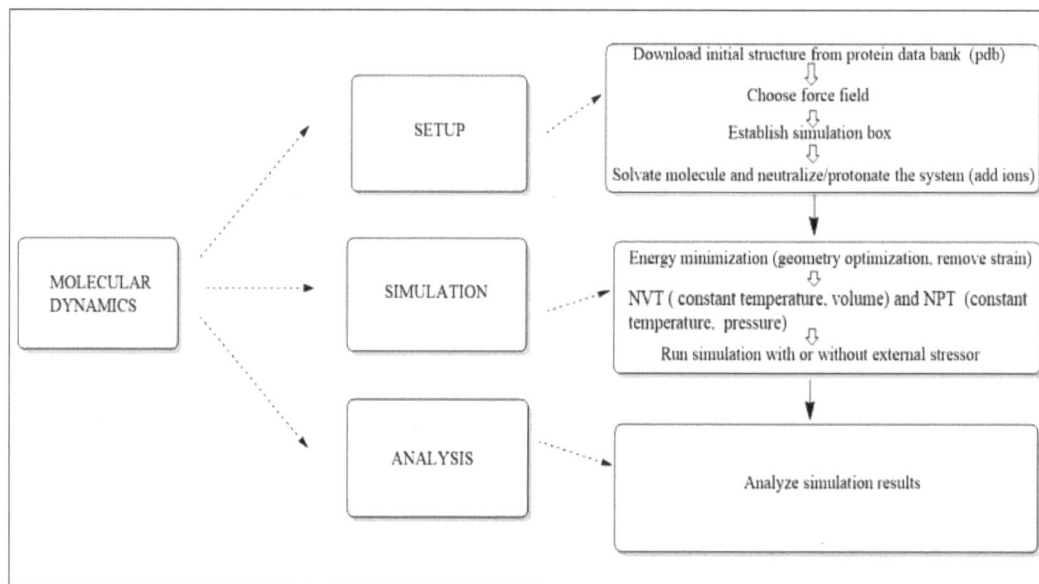

Fig. (1). Flow chart describing the process of molecular dynamics (MD).

2. MESOSCALE SIMULATIONS IN PHARMACEUTICS

Before development and manufacture, simulations predict the performance of innovative materials. Monte Carlo (MC) and MD simulations provide precise molecular information. The 0.1-10 nm molecule structure may be examined using atomistic simulation methods. A product's mesoscale range in formulation design is too large for atomistic simulations since it is 100–1000 nm. In such a scenario, coarse-grained simulations, such as dissipative particle dynamics (DPD), self-consistent field theory (SCFT), and dynamic mean-field density functional theory (Mesodyn) can provide a bridge between the atomistic scale and the continuum by providing direct input, such as property data into coarse-grained models [13]. The idea of a coarse-grained model has become an attractive alternative to atomistic models in building complex mesostructures like surfactants and various polymers. Considering the time and length scales at which these phenomena may occur, the utilization of standard molecular dynamics (MD) simulations at atomic precision is currently impractical. However, through the application of such potentials, the simulation of substantially larger systems over extended periods becomes feasible

[14]. Specific attributes of drug delivery systems (DDSs), such as their phase behaviors and the distribution of drugs within nanomaterials [15, 16], are characterized on the mesoscopic scale. These simulations offer the potential to determine these properties at early stages.

2.1. Theoretical Background of Mesodyn

Mesodyn is a multipurpose program that can be used for field-based simulations of complex systems [9]. Mesodyn offers a broad framework for computing the dynamics of mesoscale patterns created on a coarse-grained scale of 1-1000 nm, in various block polymer combinations and complicated mixtures, including charged systems and systems with hard particles like colloids or surfaces. Mesoscale structures play an important part in the manufacturing of polymer blends, block-copolymer systems, and in surfactant aggregates systems that are used in various drug delivery systems [4]. Performance in the mesoscale domain is determined in the real world. Clear comprehension is provided by mathematical models based on computer simulations of these structures' growth and stability. Mesodyn utilizes the dynamic mean-field density functional theory to compute mesoscale pattern creation in intricate fluid systems and block polymer mixtures. It is a combination of two traditional theories in which polymer phase diagrams utilize the Flory-Huggins theory and dynamic pattern analysis using the Ginzburg-Landau model for creation. The interactions between the several types of beads that make up the Mesodyn model are specified by a mean-field potential for all other interactions and harmonic oscillator potentials for intra-molecular contacts (Gaussian chains) [17]. The simulation objective is to forecast how density patterns of the different beads present in the system. In this model, each bead corresponds to covalently bonded groups of atoms, typically comprising one or a few chains of structural units. These beads symbolize distinct components present within the system's composition. The approach employed is coarse-grained, where molecules are described as Gaussian chains formed by these beads. The Mesodyn technique entails inputs such as self-diffusion coefficients, the noise parameter, and the bond length of the bead constituents, among others [18, 19]. This method encompasses dynamic and numerical aspects, as well as order parameters. As illustrated in Fig. (**2**), the configuration involves a box populated with these beads.

2.1.1. Dynamical Considerations

Mesodyn assumes that all thermodynamic functions arise from free energy and act as a function of its local density. It expands MD spatial investigation and further investigates mesoscopic space evolution. Mesodyn converts a polymer or a molecule into a coarse-grained Gaussian chain model and characterizes its movement with the Langevin equation, which is important for studying complex fluid and polymer blend systems [20]. Fig. (**3**) represents the basic protocol of mesoDyn simulation.

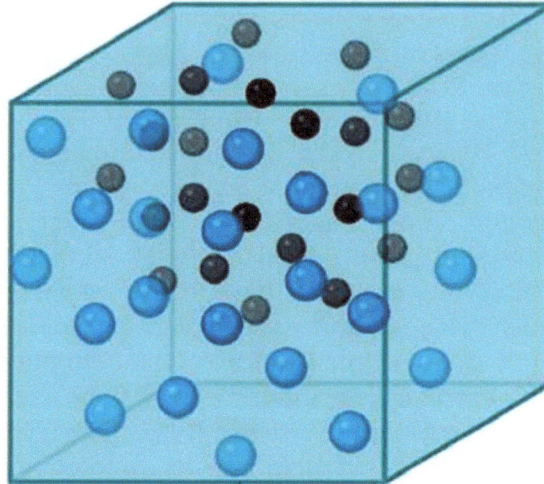

Fig. (2). Box depicting beads (changed).

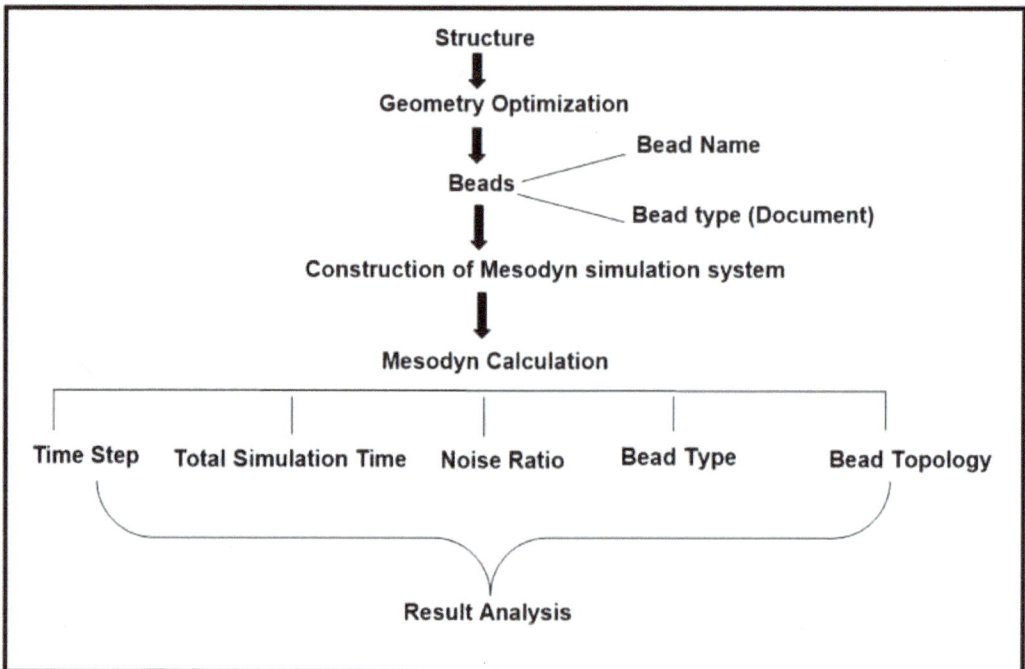

Fig. (3). Basic protocol of mesoDyn simulation.

Functional Langevin equations characterize the system dynamics. They represent noise-accounting diffusion equation in component densities:

$$\frac{\partial \rho i}{\partial t} = Mvj\nabla\rho i\rho j\nabla[\mu j - \mu i] + n \tag{1}$$

Where,

Mvj = The mobility of the bead

$\blacktriangledown \rho i\rho j$ = Kinetic coefficient

η = Blank Gaussian noise

μ = Diffusion coefficient

Integrating these mathematical equations generates Boltzmann density fields.

2.1.2. Numerical Considerations

For each kind of bead, the areas of density and external potential are related in a one-to-one manner by the Gaussian chain density function. The density fields and the external potentials both have an impact on the intrinsic chemical potentials (μ_j). Between the time derivatives and the intrinsic chemical potentials there is a relationship that is formed by the coupled Langevin equations. The fluctuation-dissipation theorem links the noise source to the exchange kinetic coefficients in the final step. Altogether, such mathematical expressions make up a finished set that a Crank-Nicholson technique can effectively integrate on a cubic mesh. Because there are a lot of equations to solve (approximately 10^6 nested Fredholm integrals every time step) and because systems with a mesh with 10 beads, about 100^3 demand a lot of memory, for parallel systems with distributed memory, using the industry standard MPI (Message Passing Interface), a domain decomposition approach has been utilized and implemented [20].

2.1.3. Order parameters

Mesodyn defines the free energy of a mixed system which causes phase separation as a time-dependent function of local density. To measure the degree of phase separation, order parameters (P) are determined [19]. The equation which is utilized to express the P is:

$$P_I = 1/V \int_v [n_i^2(r) - n_i^2] \, dr \tag{2}$$

where,

P_I= Time-based order parameter of a particle

V = Cell volume

$n_i^2(r)$= The particle's local density distributions at a sampling distance of r

n_i^2= The component beads' system-wide average density distribution P explains how the local bead distribution deviates from the anticipated homogeneous mean distribution. A rise in P indicates phase dissociation has occurred, and it can be consequently noticed that distinct P for every particle over time.

3. PROPERTIES OF MESODYN

3.1. Aggregation and Coagulation

Computer simulation techniques like DPD and Mesodyn are employed widely. These mesoscopic models connect slow thermodynamic moderation of macro-scale features with rapid molecular kinetics. Zheng and colleagues conducted a study on the phase behavior of the amphiphilic polymer poly(methyl methacrylate-co-methacrylic acid)-b-poly(poly(ethylene glycol) methyl ether monomethacrylate), denoted as P(MMA-co-MAA)-b-PPEGMA, within a water environment. This investigation was facilitated by employing Mesodyn modeling. The study delved into the creation of drug-loaded micelles and their impact on aggregate morphology concerning factors such as composition, hydrophobicity, and pH levels. The Mesodyn simulations provided insights indicating that generating polymer chain-ordered aggregates using less hydrophobic MMA beads is more challenging. Furthermore, it was observed that order parameters require a longer period to reach equilibrium, as revealed by the Mesodyn data.

The aggregate process is also influenced by pH. The polymer was found to be capable of forming conventional micelles. Yet, at pH levels above 5 micelles shape was noted abnormal as well as baggy for the release of the drug. MAA clustered externally rather than internally. In addition to complementing experiments and more effectively guiding the exploratory fabrication of drug delivery systems with covet features, computer simulation could shed light on the method through which drug-loaded polymeric micelles have mesoscopic structures [21].

3.2. Phase Morphology

With the aid of phase morphology, it is possible to manipulate the block copolymer thin film's shape to create an ordered phase-separated microdomain. Amphiphilic graft and block copolymers composed of PMMA poly (methyl methacrylate) and PEO (polyethylene oxide) blocks have drawn more interest because of the possibility of using them to modify keratoprostheses, and create drug carriers, and biomedical materials. The phase morphologies of plain polystyrene-block-poly (methyl methacrylate) copolymers (PS- b- PMMA) were investigated. Mesoscopic modeling of 12 miktoarm polystyrene-block-poly (methyl methacrylate) copolymers was done at 383k, 413k, and 443 K.

Order parameters (P) were almost the same at 413 and 443 K. Changes in temperature had a noticeable impact on long and PMMA-rich chains. It was observed that the microscopic phase of copolymers infused with nanoparticles was influenced not only by the composition and architecture of the copolymers but also by the characteristics of the nanoparticles, such as size, number, and density. The iso-surface images revealed that regardless of the nanoparticles' arrangement, the 13214-type copolymers had lamellar phase morphologies at 443 K [22].

3.3. Effect of Confinement on Miscibility

When it comes to material modeling and technology development of pharmaceuticals, molecular simulations have been crucial due to the increasing computational capability and assets. These simulations can offer insightful microscopic and mesoscopic information about immiscible compounds' interfacial behaviors that significantly impact the rheological and mechanical characteristics of materials. To verify the miscibility and immiscibility, as well as the mesoscopic morphology of the polypropylene (PP)/polyamide-11 (PA11) PP/PA11 polymer blends, Fu and their co-researchers, conducted mesoDyn simulations, which revealed that the phase difference between PP and PA11 polymers could be attributed to the inclusion of 10% wt. of PA11 within the polymer mixture. The presence of this specific PA11 concentration was identified as the cause of the observed phase disparity. Moreover, the order parameter, registering a value of 0.1, served to validate the notion of a moderate level of polymer mixing between PP and PA11 blends containing less than 10% PA11. This observation underscored the significance of the PA11 content in influencing the phase behavior and the degree of polymer amalgamation in the blends [23].

3.4. Effect of Shear on Morphology

Mesoscopic simulation can be thought of as a technique for explaining how mesoscale morphological forms and provides mesoscale data for investigations. The Mesodyn approach is used to model the phase behaviors of Pluronic (ethylene oxide)$_{19}$ (propylene oxide)$_{29}$ (ethylene oxide)$_{19}$, (P65) solution in the presence and absence of shear. Phase separation was studied about the time evolution under the effects of concentration and shear. Three morphologies emerge in an aqueous solution when there is no shear. The prior type of micelle is spherical and has a worm-like structure, face-centered cubic micellar structure, and hexagonal packing. The latter is the bi-continuous phase, which is a part of the connectivity like a type of gyroid shape, and the 3rd one is the lamellar phase. Only hexagonal and lamellar phases, which line up with the direction of flow, are created in all concentrations when shear is applied. It was observed that the order parameters fluctuate in the presence of weak shears, whereas under strong shears they go to constant values after shorter oscillations. Moreover, it was observed that the morphologies of systems alter similarly under various strong shears. Hence, the study of order parameters in the case of formulation that undergoes shear conditions can help to explain the stability of their morphologies [24].

3.5. Compositional Order Parameters

The Mesodyn simulation analyses the microphase separation of block copolymers using the dynamic mean field density functional theory. The temporal integration of functional Langevin equations is one of MesoDyn's key advantages. Because the average screening time is less than that of the thermodynamic moderation period, these irregular states significantly affect how the finished materials behave. Using the Mesodyn approach, the phase behaviors of the (polyethylene oxide- polypropylene oxide) PEO27-PPO61-PEO27 (P104) solution were simulated. The order parameter's value increases with the degree of phase separation. Better compatibility or miscibility was indicated by a drop in P, and the polymer got mixed more randomly. Phase separation occurs over a period measured in milliseconds. Spherical micelle's phase separation was rapidly generated, according to the order parameters, however micellar structures or disk-shaped micelles took longer to reach a stable equilibrium [25].

3.6. Free Energy and Entropy Evolution

Molecular Dynamics Simulation (MDS) is a powerful tool that allows us to predict and understand the behaviors of molecules and atoms on the nanometer scale. In the realm of colloidal systems, MDS provides valuable insights into characteristics like fluidity, interfacial stability, and thermodynamics. It fills gaps in knowledge that traditional physical laboratory experiments might not fully address. By utilizing the mesoDyn approach, researchers have delved into the phenomenon of Pickering stabilization. They achieved this by employing Nanocellulose (NC) as a particle stabilizer within a mixture of poly-1-butene (PB) and water. When dealing with a system where an amphiphilic polymer like NC exists at low concentrations in a continuous water phase, a random movement pattern is observed. This pattern indicates a homogeneous phase structure. However, as the concentrations of both NC and PB increase, the hydrophobic effect becomes the dominant force. This transition causes NC to adhere to interfaces between oil and water, eventually forming droplets. This process leads to a reduction in the system's free energy. As the oil-water interface expands, the system's spontaneity decreases, which results in a decline in Gibbs free energy [26].

3.7. Density Histograms

The Kinetics of phase separation can be determined utilizing density profiles resolved from the Mesodyn approach to inspect the miscibility and immiscibility characteristics of the blends. To understand the compatibility and incompatibility of poly (L-lactide) and poly (vinyl alcohol) blends, studies using molecular modeling simulation techniques have been conducted by different researchers. To confirm the contribution of certain polymer atoms to a blend's miscibility or immiscibility, an RDF plot can be created. Understanding their significance, the thermodynamic interactions between PLL Poly (L-lactide) and PVA poly (vinyl alcohol), and the complete spectrum of mix compositions to confirm experimental results on their compatibility/incompatibility characteristics have been analyzed. To forecast blend miscibility, the Flory-Huggins interaction parameter, c, could be calculated for various blends. It was discovered that while miscibility was seen at a 1:9 blend of PLL/PVA, increasing immiscibility was seen as more common at higher PLL component compositions. When simulations were conducted at larger time steps, equilibrium was obtained from the calculated free energy from the mesoscopic simulation of blends, suggesting the consistency of the mixture at specific compositions. The hydrogen bonding effect was found to be responsible for the miscibility of PLL and PVA polymers. Mesoscopic density slices revealed that the phase separations between PLL and PVA caused because of the concentration of polymers above 25% PLL of the blend [27]. Table **1** represents some research investigations reported in the literature using mesoDyn.

4. APPLICATIONS OF MESODYN SIMULATIONS IN FORMULATION DEVELOPMENT

Research using simulations is evolving in various fields whether it is related to physics, chemistry, or biology. Applications of simulations in every aspect will lead to better innovations, and ideas, and the research will be less time-consuming experimentally. Here are some applications of Mesodyn simulations in the field of pharmaceuticals. Fig. (**4**) represents the applications of mesoDyn simulations in pre-formulations.

Table 1. Summary of various investigations carried out using Mesodyn simulations.

Sr. No.	Research	Inference	References
1	Phase behaviors and phase structure of Pluronic L64, water, and *p*-xylene systems were investigated using the mesoDyn simulation. Pluronic L64 with varied concentrations of solvents was examined.	• Water was considered to be the most important component in determining whether or not micelles develop. • As no micelles were created in *p*-xylene alone or with a very small amount of water. • When the water hydration level was increased, micelles of varying forms were formed in vast numbers. • With rising temperatures, reverse micelle formation becomes more challenging and time-consuming. At 348 K, pluronic L64 did not produce reverse micelles. • Reverse micelle production was an exothermic process.	[28]
2	Mesodyn simulation and binding isotherm measurements were used to study β-hydroxyl trimethylammonium (C_9phNBr) and xanthan (XC) aggregation in aqueous solution.	• XC was kept constant at 10% whereas, the C_9phNBr concentrations varied between 5% and 30% in the solution. At 20% concentration, small molecules are aggregated on the XC chain, and at 30% larger aggregates were observed on the XC chain. • With a rise in the concentration of C_9phNBr, the rate of aggregate development was found to decrease which was found to be evident from the order parameter. • With an increase in temperature, production of C_9phNBr-XC aggregates becomes challenging. • The process was observed to be exothermic. • With a rise in volume fraction of XC (20%), the formation of aggregates decreased significantly.	[17]
3	The aggregation behavior between carboxymethyl chitosan (CMCHS) and cetyltrimethylammonium bromide (CTAB) was studied by Mesodyn simulation, and results of the viscosity, hydrodynamic radius of the aggregates were compared with their aggregation behavior.	• CMCHS was kept at 10% while CTAB ranged from 35% to 60%. They clustered out at low CTAB concentrations. • Increasing CTAB concentration formed a network with CMCHS. Excellent network structure was observed at 35% CTAB volume. • When CTAB volume was raised to 60%, a network architecture disappeared and was displaced by an ellipsoidal aggregate structure.	[29]

(Table 1) cont.....

Sr. No.	Research	Inference	References
		• The interaction between CMCHS and CTAB strengthened with increasing CTAB concentration. • As simulation time increased, the order parameter also increased, suggesting that CTAB molecules were progressively accumulating on the CMCHS chain. • After simulating the (10%) CMCHS and (40%) CTAB system at three different temperatures (52°C, 25°C, 37°C), it was found that the higher the temperature, the lower the order parameter values, indicating that it is slightly more difficult to form CMCHS-CTAB aggregates.	
4	The aggregation behavior of the linear polyethers Ethylene oxide and propylene oxide (EO) 60 (PO) 40 (EO) 60 (Polyether B) and the branched block polyether T1107 (Polyether A) in an aqueous solution were compared.	• Polyether A has a lower aggregation number and can form micellar structures at a lower concentration than polyether B. • The polyether showed time-dependent growth behavior. Both block polyether micelles may change from sphere to rod when the shear rate is 1×10^5 s$-$1, which was found after simulating the systems.	[30]
5	The aggregation behavior of weakly water-soluble baicalin with glycyrrhizin micelles was studied to demonstrate the solubilization mechanism of glycyrrhizin.	• A core/shell structured micelle was created by the (hydrophilic) group forming a peak in the shell position and the B group (hydrophobic) being more concentrated in the center of the aggregate. • Due to hydrophobic interactions, a hydrophobic core formed for entrapping the poorly water-soluble medicines, while a hydrophilic shell created a stabilizing interface between the core and water. • According to DPD, when concentration was increased, glycyrrhizin molecules formed core/shell-structured spherical, cylindrical, and lamellae aggregates.	[31]
6	To explore the phase behavior of dye-polyether derivatives in aqueous solutions. The influence of dye-polyether derivative concentration on micelle morphology and micelle production process was investigated.	• At a concentration of 10%, micelles that are round and those that resemble worms were produced in dye-polyether derivative solutions. • At 5% concentration, no micelles were produced. The asymmetric micelles were seen at concentrations of 20 and 30% by volume. The development of micelles can be explained using thermodynamics; A balance between entropy and enthalpy can cause micelles to spontaneously develop.	[32]

(Table 1) cont.....

Sr. No.	Research	Inference	References
		• Monomers were present in the solution at very low concentrations. The negative entropy implications from the hydrophobic end of the molecules became more prominent as the polymer concentration grew.	
7	Simulation of the mesoscopic phase separation behavior of Polyvinyl alcohol (PVA) /polyvinyl pyrrolidone (PVP) blends.	• At a temperature of 298 K, morphologies that were developed using Mesodyn provided more evidence that PVA and PVP could be miscible in varying proportions due to the enthalpy contributions of H-bonds.	[33]
8	To examine the phase behavior and mesoscopic structure of the Brij97 (Non-ionic surfactant), Isopropanol (Cosurfactant), and Isoamyl acetate (Oily phase) and water system.	• Microemulsion (ME) was observed to go through a change from bicontinuous to O/W as the proportion of water in its composition increased from 60 to 80%. • Order parameters revealed that the rate of formation of ME was dependent on the parts of oil and surfactant used for formulating as well as on the ratio of surfactant and cosurfactant mixture used. • Different chain lengths of alcohol like isopropanol, ethanol, and 1,2-propanediol also exhibited an impact on the particle size and morphology of the system.	[34]
9	Molecular and mesoscale simulations (MesoDyn and DPD) were carried out to determine the key groups involved in solubilizing and dispersing of flavonoids in various solvents.	• Flory Huggins parameter assessed flavonoid solubility in different organic mediums and helped to identify key groups. • The B2 component of the three investigated flavonoids interacted favorably (χ_{ij} near 0.5) with tertiary-amyl alcohol and isopropanol. The quercetin and tertiary-amyl alcohol system's order parameter displayed phase separation, demonstrating that quercetin clustered in tertiary-amyl alcohol.	[19]
10	The aggregation processes of poly-lactic-co-glycolic acid (PLGA), polyvinyl alcohol (PVA), quercetin, and water systems were investigated using a mesoscale modeling approach.	• The outcomes showed that the investigated systems enabled the production of spherical nanoparticles that were evenly dispersed in the dispersion medium that is water. • The particle size and rate of aggregation increase as PLGA and quercetin concentrations increase. • Compared to quercetin, PLGA's impact was more noticeable in aggregation behavior.	[35]

(Table 1) cont.....

Sr. No.	Research	Inference	References
		• The order parameter of water was found to be 0.018. Creating this large-size nanoparticle is encouraged by the rise in poly-lactic--o-glycolic acid and quercetin concentrations as well as the lactic acid of poly-lactic--o-glycolic acid.	
11	MesoDyn simulation was used to examine the clustering behavior of a Pluronic copolymer with a surfactant.	• According to simulation data, sodium dodecyl sulfate (SDS) stimulates the formation of $EO_{13}PO_{30}EO_{13}$ (L64) spherical micelles at substantially lower L64 concentrations than in a system without SDS. • When SDS concentration is increased but L64 concentration remains constant, the dimensions of the micelles grow while the number of micelles decreases. • The process of forming a cluster is exothermic and as the temperature rises, the process becomes more challenging, and the rate of formation declines.	[36]

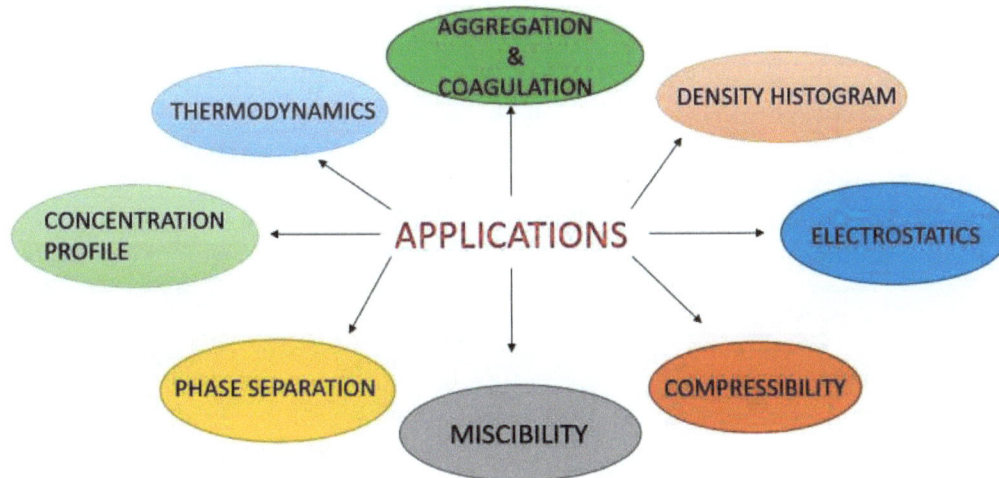

Fig. (4). Applications of mesoDyn simulations in pre-formulations.

4.1. Precipitation Membrane Formation

The mesoDyn module has been used to explore the solvent exchange aspects with the nano-porous membrane. The pathway dependence of the structures and the complex kinetic phenomena can be examined by the mesoDyn module with variable concentration. The membrane precipitation and the overall structural development were simulated with the interaction of the solvent and non-solvent. This method has proven useful for forecasting advanced and higher systems. Example- Phase diagrams become more complicated when there are several solvents, non-solvents, and polymers [37].

4.2. Nanoscale Drug Delivery Systems

Mesodyn module can be applied for simulating the impact of hydrophobic drug particles *e.g.* (Haloperidol) on the structure of aqueous solutions of amphiphilic block copolymer, Poly (ethylene oxide)–poly (propylene oxide) (PEO and PPO). It is based on the dynamic mean field density functional theory. It was observed that with the increase in the number of hydrophobic molecules the micellar size increases and a rod-like structure forms. The width of size distribution and aggregation number of the micelles also increased in this investigation [38].

4.3. Development of thin Films from Block Copolymers

In the mesodyn framework, a mean-field model was used for field-based simulations such as to determine the microphase diagram of a thin, neat, three-block polymer film made of polystyrene and polybutadiene that was placed on a solid substrate. The comparison was made between the lamellar, cylindrical, and micellar microphases of diblock copolymers and the gyroid phase, which lies between the lamellar and cylindrical phases. One component frequently having lower interfacial energy than the other is another driving element for structure creation at interfaces and in thin films. This occurrence is exclusive to the specified filmmaking system and or procedure and is a member of the class of modulated phase interfaces. Results indicated that in a single system and under the same experimental circumstances, all phases can be manifested [9].

4.4. Solubility of Menthol by Platycodin D

Mesodyn simulation has been used to investigate the solubilizing property of platycodin D in menthol and also the effect of temperature on the solubility of menthol by platycodin D was studied. The isodensity surface and density profile stated the structural properties whereas, the order parameter indicated the various effects of phase separation. During the process of solubilization of menthol by platycodin D, the order parameters revealed that the solubilization occurred in four stages *i.e.* I, II, III-a, III-b, and IV which stated that platycodin D acts as a solubilizer at low concentration and the solubilization occurred by the formation of micelles of platycodin D in which menthol was diffused. Also, the temperature showed a positive effect on the diffusion of menthol in platycodin D micelles [39].

4.5. Prediction of Chitosan and Poly(e-caprolactone) Binary Systems, Miscibility and Phase Separation Behaviour

Molecular Dynamic and Mesodyn simulations were used to investigate the compatibility of chitosan (CS) and poly (e-caprolactone) (PCL) as binary blends and further compared with experimental results. In MD simulations, the solubility parameters of CS and PCL were computed using various repeating units to provide the best possible monomer units at room temperature. Also, the COMPASS force field can be used to simulate the polymer blends at different ratios. Using the Flory-Huggins interaction (χ) parameter and (ΔH_m) enthalpy of mixing the compatibility of the polymer blends to a certain extent at specific compositions were calculated. Through the Mesodyn simulations, the phase separation behavior of the polymer blends at coarse grain level can be calculated. The phase separation behavior at different ratios is represented by the order parameter and density profile. The order parameter was found to be less than 0.1 which indicates the congeniality of blends. These types of studies can be used to demonstrate the compatibility of blends with different components [40].

4.6. Nanotube Self-forming Process of Amphiphilic Copolymer

Utilizing Mesodyn modeling, the self-forming process of an amphiphilic AB block copolymer contained in well-aligned cylindrical nanopores was investigated. Models are created for various patterned nanotubes with a few tens of nanometers. As a result of quenching of a homogeneous solution of polymer surfactant, several patterned structures were produced. Order parameter was used to calculate the phase separation and compressibility of the system. Chain topology and interaction parameters can be used to describe the system's chemical composition. The results of the study showed that hexagonal nanotubes were formed by the AB block polymers, as carbon nanotubes and were acquired at an average ratio of amphiphilic and amphophobic solvent blocks. The morphological patterns can be correlated with the polymer content, polymer-solvent interaction parameter, and nanotube diameter, *etc* [41].

4.7. Utilizing a Macrocyclic Terbium Compound with Hetero-ligands to Create an Effective Luminous Soft Media in a Lyomesophase

A novel method for the manufacturing of luminous soft media was established which utilizes a complex of terbium β-diketonate and substituted *p*-ter--butylthiacalix arene 1 and introduces it into the lyotropic mesophase of Pluronic (P123) —Ethylene glycol ($C_2H_6O_2$). Utilizing the mesoDyn approach, the phase behavior of the P123—H_2O system was modeled. It was demonstrated that only the lamellar mesophase forms in the range of 80-84 Vol% Pluronic content, while there was a region of mixed phases which is (hexagonal and lamellar) at a

Pluronic concentration above 65 Vol%. Unique characteristics of the produced soft composites include the creation of controllably ordered supramolecular structures and the presence of strong luminescence [42].

4.8. Design and Development of Polymersome Chimaera

Indeed, Mesodyn simulation has been employed to create two-component polymersome chimaeras, a type of structure that comprises at least two phase-separated components. These chimaeras are formed by combining two distinct diblock polymers. These systems can also be referred to as polymer particles or polymersome composites. In these structures, the hydrophilic blocks exhibit a preference for specific curvatures, while the two hydrophobic blocks undergo a process of demixing. The design and behavior of these chimaeras are governed by Flory-Huggins (χ) parameters, which play a role in considering the ratios between block sizes and the tendency for demixing. These parameters influence the ability of the hydrophilic and hydrophobic segments to form distinct domains within the polymer structure. By employing Mesodyn simulations, researchers can gain insights into the self-assembly and behavior of these two-component polymersome chimaeras. This approach enables a deeper understanding of how different components within the chimaera interact and organize themselves, ultimately leading to the formation of specific structural features. This research holds significance in fields such as nanotechnology and materials science, as it contributes to the development of tailored nanostructures with unique properties and functionalities.

Diblock polymer surfactants in solvents with great selectivity and strong segregation are the major components of simulation parameters. They are the two polymer structures, A13H7 and B13H17, in which every chain resembles a necklace of beads. Barriers in the world of free energy can be overcome with the aid of white noise. The bending moduli of the structures are affected by both block length (entropic effect) and chemical composition (Flory-Hugging's parameter). The fundamental finding is that the trick to manipulating hierarchical structures on a mesoscopic scale is the counterintuitive co-assembly of demixing block copolymers. Block copolymers that are longer than a certain length inevitably divide the assembly, thus polymer-based chimera is inherently unstable without additional safeguards. Hence, these types of studies can help in designing and developing scaffolds of different polymers and copolymers [43].

4.9. Mesoscopic Simulation for the Phase Separation Behavior of the Pluronic Aqueous Mixture

The equivalent chain technique of the mesoDyn module was utilized to simulate the Pluronic water solutions and to explain the effects of temperature, the effect of the polymer surfactant solution's relative block size and the dynamic changes in the Pluronic water mixes. Three poly (ethylene oxide) -b- poly (propylene oxide) -b- poly (ethylene oxide) (PEO-PPO-PEO) block copolymers of different compositions, *i.e.* $(EO)_6 (PO)_{34} (EO)_6$ (L62), $(EO)_{13} (PO)_{30} (EO)_{13}$ (L64) and $(EO)_{37} (PO)_{58} (EO)_{37}$ (P105) have been studied for their phase-related studies using the Mesodyn simulation. To find a Gaussian chain that acts like the Pluronic molecule, the equivalent chain approach was applied. The Flory-Huggins expression is a thermodynamic model often used to estimate the interactions between a solvent and a polymer in a solution. Further, the Flory-Huggins expression was used to estimate the interactions between the solvent (water) and the polymer components (pluronic) within different types of pluronic-water mixtures. The three pluronic water mixes were simulated at 373 K, 298 K, and 323 K to explore the impact of temperature on the temporal development of phase separation. The simulation findings demonstrate that the L62, L64, and P105 solutions generated gel, transition, and micelles morphologies respectively at the same volume concentration of polymer (0.50). The findings of the Fourier transform show that when the temperature rises, the frequency of the density field rises as well, also the phase separation process slows down as the temperature rises. The copolymer and water solvent interact more slowly, and the morphology's periodicity becomes more frequent. Hence, these types of studies can help determine the stability of mixtures containing different components or in studying the phase separation kinetics [44].

4.10. Quercetin's Solubility and Release Characteristics

In this investigation, the flavanol quercetin solid dispersions in polyethylene glycol at various compositions showed a rise in solubility, but with time, the concentration of dissolved flavanols decreased. Mesodyn describes the free energy of a variety of systems that might result in phase separation. The average order parameter (P), of the component can be used to characterize how the various beads separate into phases. It was illustrated that a rise in P was demonstrated in sample systems, which suggests that phase difference (or an increase in local bead order) occurred. As a result, one may examine the various P continuations for each component throughout time [45].

CONCLUSION

Mesodyn is based on a novel fusion of well-known ideas taken from soft condensed matter physics and polymer research. Modern computational and mathematical techniques are used in its execution. These simulations carry out an effective and logical *in silico* chemical space exploration, which is sometimes impractical from an experimental standpoint when several steps and poor yielding synthetic pathways are involved. Mesodyn can expand the scope of computer simulations due to the vast number of potentially industrially relevant applications that are available.

ACKNOWLEDGEMENT

We gratefully acknowledge Bioinformatics Centre (BIC) Sanctioned by DBT, New Delhi, Sanction No: BT/PR39876/BTIS/137/7/2021.

REFERENCES

[1]   Zakaria J, Abd Shukor SR, Abd Razak K. Intermolecular interaction of tween 80, water and butanol in micelles formation *via* molecular dynamics simulation. IOP Conf Ser Mater Sci Eng 2020; 778(1): 012091.

[2]   Moghtaderi M, Sedaghatnia K, Bourbour M, *et al.* Niosomes: A novel targeted drug delivery system for cancer. Med Oncol 2022; 39(12): 240.
[http://dx.doi.org/10.1007/s12032-022-01836-3] [PMID: 36175809]

[3]   Ashwini T, Narayan R, Shenoy PA, Nayak UY. Computational modeling for the design and development of nano-based drug delivery systems. J Mol Liq 2022; 120596.

[4]   Altevogt P, Evers OA, Fraaije JG, Maurits NM, van Vlimmeren BA. The MesoDyn project: Software for mesoscale chemical engineering. J Mol Struct THEOCHEM 1999; 463(1-2): 139-43.
[http://dx.doi.org/10.1016/S0166-1280(98)00403-5]

[5]   McGrother S, Goldbeck-Wood G, Lam YM. Integration of modelling at various lengths and time scales. Comput Mater Sci 2004; 223-33.

[6]   Shankar U, Gogoi R, Sethi SK, Verma A. Introduction to materials studio software for the atomistic-scale simulations. InForcefields for Atomistic-Scale Simulations: Materials and Applications 2022; 299-13.
[http://dx.doi.org/10.1007/978-981-19-3092-8_15]

[7]   Nicolaides D. Mesoscale modelling. Mol Simul 2001; 26(1): 51-72.
[http://dx.doi.org/10.1080/08927020108024200]

[8]   van Vlimmeren BA. Mesoscopic dynamics: Simulation of microphase separation dynamics in complex liquids. Thesis fully internal. University of Groningen 1998.

[9]   Fraaije H, Sevink A, Zvelindovsky A. 7 Dynamical Microphase Modelling with Mesodyn. Dev. Block Copolymer Sci. Technol 2004; pp. 245-64.

[10]   Bunker A, Róg T. Mechanistic understanding from molecular dynamics simulation in pharmaceutical research 1: Drug delivery. Front Mol Biosci 2020; 7: 604770.
[http://dx.doi.org/10.3389/fmolb.2020.604770] [PMID: 33330633]

[11]   Kumar H, Maiti PK. Introduction to molecular dynamics simulation. Comput Stat Phys 2011; 161-97.
[http://dx.doi.org/10.1007/978-93-86279-50-7_6]

[12]   Omolo CA, Kalhapure RS, Agrawal N, Rambharose S, Mocktar C, Govender T. Formulation and molecular dynamics simulations of a fusidic acid nanosuspension for simultaneously enhancing solubility and antibacterial activity. Mol Pharm 2018; 15(8): 3512-26.
[http://dx.doi.org/10.1021/acs.molpharmaceut.8b00505] [PMID: 29953816]

[13]   Guo XD, Zhang LJ, Qian Y. Systematic multiscale method for studying the structure–performance relationship of drug-delivery systems. Ind Eng Chem Res 2012; 51(12): 4719-30.
[http://dx.doi.org/10.1021/ie2014668]

[14]   Ryjkina E, Kuhn H, Rehage H, Müller F, Peggau J. Molecular dynamic computer simulations of phase

behavior of non-ionic surfactants. Angew Chem Int Ed 2002; 41(6): 983-6.
[http://dx.doi.org/10.1002/1521-3773(20020315)41:6<983::AID-ANIE983>3.0.CO;2-Y] [PMID: 12491288]

[15] Fermeglia M, Pricl S. Multiscale molecular modeling in nanostructured material design and process system engineering. Comput Chem Eng 2009; 33(10): 1701-10.
[http://dx.doi.org/10.1016/j.compchemeng.2009.04.006]

[16] Gates BD, Xu Q, Stewart M, Ryan D, Willson CG, Whitesides GM. New approaches to nanofabrication: Molding, printing, and other techniques. Chem Rev 2005; 105(4): 1171-96.
[http://dx.doi.org/10.1021/cr030076o] [PMID: 15826012]

[17] Li YM, Xu GY, Chen AM, Yuan SL, Cao XR. Aggregation between xanthan and nonyphenyloxypropyl β-hydroxyltrimethylammonium bromide in aqueous solution: MesoDyn simulation and binding isotherm measurement. J Phys Chem B 2005; 109(47): 22290-5.
[http://dx.doi.org/10.1021/jp0528414] [PMID: 16853902]

[18] Fraaije JGEM, van Vlimmeren BAC, Maurits NM, *et al.* The dynamic mean-field density functional method and its application to the mesoscopic dynamics of quenched block copolymer melts. J Chem Phys 1997; 106(10): 4260-9.
[http://dx.doi.org/10.1063/1.473129]

[19] Slimane M, Ghoul M, Chebil L. Mesoscale modeling approach to study the dispersion and the solubility of flavonoids in organic solvents. Ind Eng Chem Res 2018; 57(37): 12519-30.
[http://dx.doi.org/10.1021/acs.iecr.8b02321]

[20] Yuan SL, Zhang XQ, Chan KY. Effects of shear and charge on the microphase formation of P123 polymer in the SBA-15 synthesis investigated by mesoscale simulations. Langmuir 2009; 25(4): 2034-45.
[http://dx.doi.org/10.1021/la8035133] [PMID: 19161270]

[21] Zheng LS, Yang YQ, Guo XD, Sun Y, Qian Y, Zhang LJ. Mesoscopic simulations on the aggregation behavior of pH-responsive polymeric micelles for drug delivery. J Colloid Interface Sci 2011; 363(1): 114-21.
[http://dx.doi.org/10.1016/j.jcis.2011.07.040] [PMID: 21824624]

[22] Mu D, Li JQ, Wang S. MesoDyn simulation study on the phase morphologies of miktoarm PS- *b* - PMMA copolymer doped by nanoparticles. J Appl Polym Sci 2013; 127(3): 1561-8.
[http://dx.doi.org/10.1002/app.37510]

[23] Fu Y, Liao L, Lan Y, *et al.* Molecular dynamics and mesoscopic dynamics simulations for prediction of miscibility in polypropylene/polyamide-11 blends. J Mol Struct 2012; 1012: 113-8.
[http://dx.doi.org/10.1016/j.molstruc.2011.12.026]

[24] Zhang X, Yuan S, Wu J. Mesoscopic simulation on the phase behavior of ternary copolymeric solution in the absence and presence of shear. Macromolecules 2006; 39(19): 6631-42.
[http://dx.doi.org/10.1021/ma061201b]

[25] Zhang X, Yuan S, Xu G, Liu C. Mesoscopic simulation of the phase separation on triblock copolymer in aqueous solution. Wuli Huaxue Xuebao 2007; 23(2): 139-44.

[26] Lee KK, Low DYS, Foo ML, *et al.* Molecular dynamics simulation of nanocellulose-stabilized pickering emulsions. Polymers 2021; 13(4): 668.
[http://dx.doi.org/10.3390/polym13040668] [PMID: 33672331]

[27] Jawalkar SS, Aminabhavi TM. Molecular modeling simulations and thermodynamic approaches to investigate compatibility/incompatibility of poly(l-lactide) and poly(vinyl alcohol) blends. Polymer 2006; 47(23): 8061-71.
[http://dx.doi.org/10.1016/j.polymer.2006.09.030]

[28] Guo SL, Hou TJ, Xu XJ. Simulation of the phase behavior of the (EO)13(PO)30(EO)13(pluronic L64)/water/p-xylene system using MesoDyn. J Phys Chem B 2002; 106(43): 11397-403.

[http://dx.doi.org/10.1021/jp026314l]

[29]   Li Y, Xu G, Wu D, Sui W. The aggregation behavior between anionic carboxymethylchitosan and cetyltrimethylammonium bromide: MesoDyn simulation and experiments. Eur Polym J 2007; 43(6): 2690-8.
[http://dx.doi.org/10.1016/j.eurpolymj.2007.03.003]

[30]   Gong H, Xu G, Shi X, Liu T, Sun Z. Comparison of aggregation behaviors between branched and linear block polyethers: MesoDyn simulation study. Colloid Polym Sci 2010; 288(16-17): 1581-92.
[http://dx.doi.org/10.1007/s00396-010-2294-7]

[31]   Wang Y, Dai X, Shi X, Qiao Y. Mesoscopic simulation studies on the aggregation behavior of glycyrrhizin micelles for drug solubilization. Computer Modelling & New Technologies 2014; 18(10): 452-6.

[32]   Zhang B, Liu R, Zhang J, Liu B, He J. MesoDyn simulation study of phase behavior for dye–polyether derivatives in aqueous solutions. Comput Theor Chem 2016; 1091: 8-17.
[http://dx.doi.org/10.1016/j.comptc.2016.06.028]

[33]   Wu H, Xin Y. Molecular dynamics and MesoDyn simulations for the miscibility of polyvinyl alcohol/polyvinyl pyrrolidone blends. Plast Rubber Compos 2017; 46(2): 69-76.
[http://dx.doi.org/10.1080/14658011.2017.1280642]

[34]   Zhao X, Wang Z, Yuan S, Lu J, Wang Z. MesoDyn prediction of a pharmaceutical microemulsion self-assembly consistent with experimental measurements. RSC Advances 2017; 7(33): 20293-9.
[http://dx.doi.org/10.1039/C7RA01541K]

[35]   Slimane M, Gaye I, Ghoul M, Chebil L. Mesoscale modeling and experimental study of quercetin organization as nanoparticles in the poly-lactic-co-glycolic acid/water system under different conditions. Ind Eng Chem Res 2020; 59(10): 4809-16.
[http://dx.doi.org/10.1021/acs.iecr.9b06630]

[36]   Li Y, Xu G, Zhu Y, Wang Y, Gong H. Aggregation behavior of Pluronic copolymer in the presence of surfactant: Mesoscopic simulation. Colloids Surf A Physicochem Eng Asp 2009; 334(1-3): 124-30.
[http://dx.doi.org/10.1016/j.colsurfa.2008.10.029]

[37]   Maiti A, Wescott J, Wood GG. Mesoscale modelling: Recent developments and applications to nanocomposites, drug delivery and precipitation membranes. Int J Nanotechnol 2005; 2(3): 198-214.
[http://dx.doi.org/10.1504/IJNT.2005.008059]

[38]   Lam YM, Goldbeck-Wood G, Boothroyd C. Mesoscale simulation and cryo-TEM of nanoscale drug delivery systems. Mol Simul 2004; 30(4): 239-47.
[http://dx.doi.org/10.1080/0892702031000165911 5]

[39]   Ding H, Shi X, Dai X, Yin Q, Qiao Y. A mesoscopic simulation study on the solubilization of menthol by platycodin D. Journal of Engineering Science and Technology Review 2013; 6(2): 125-9.
[http://dx.doi.org/10.25103/jestr.062.26]

[40]   Gu C, Gu H, Lang M. Molecular simulation to predict miscibility and phase separation behavior of chitosan/poly(ϵ-caprolactone) binary blends: A comparison with experiments. Macromol Theory Simul 2013; 22(7): 377-84.
[http://dx.doi.org/10.1002/mats.201300109]

[41]   Yang SH, Cheng YK, Yuan SL. Mesoscale simulation on patterned nanotube model for amphiphilic block copolymer. J Mol Model 2010; 16(12): 1819-24.
[http://dx.doi.org/10.1007/s00894-010-0673-0] [PMID: 20217162]

[42]   Selivanova NM, Zimina MV, Padnya PL, Stoikov II, Gubaidullin AT, Galyametdinov YG. Development of efficient luminescent soft media by incorporation of a hetero-ligand macrocyclic terbium complex into a lyomesophase. Russ Chem Bull 2020; 69(9): 1763-70.
[http://dx.doi.org/10.1007/s11172-020-2960-y]

[43]   Fraaije JGEM, van Sluis CA, Kros A, Zvelindovsky AV, Sevink GJA. Design of chimaeric

polymersomes. Faraday Discuss 2005; 128: 355-61.
[http://dx.doi.org/10.1039/b403187c] [PMID: 15658783]

[44] Li Y, Hou T, Guo S, Wang K, Xu X. The Mesodyn simulation of pluronic water mixtures using the 'equivalent chain' method. Phys Chem Chem Phys 2000; 2(12): 2749-53.
[http://dx.doi.org/10.1039/b002060p]

[45] Otto DP, Otto A, de Villiers MM. Experimental and mesoscale computational dynamics studies of the relationship between solubility and release of quercetin from PEG solid dispersions. Int J Pharm 2013; 456(2): 282-92.
[http://dx.doi.org/10.1016/j.ijpharm.2013.08.039] [PMID: 24004565]

CHAPTER 11

Computational Tools to Predict Drug Release Kinetics in Solid Oral Dosage Forms

Devendra S. Shirode[1,*], **Vaibhav R. Vaidya**[1] and **Shilpa P. Chaudhari**[1]

¹ Department of Pharmacology, Dr. D. Y. Patil College of Pharmacy, Akurdi, Pune, Maharashtra, India

Abstract: Dissolution is the concentration of a drug that goes into solution per unit of time under standard conditions of solid-liquid interface, temperature, and composition of solvent. In the pharmaceutical industry, *in vitro* dissolution testing has been established as a preferred method to evaluate the development potential of new APIs and drug formulations and to select the most appropriate solid form for further development. Dissolution allows the measurement of some important physical parameters, like drug diffusion coefficient, and is also used in model fitting on experimental release data. Kinetic modeling of drug release in dosage forms has served as a promising alternative to reduce bio studies in the development stage of pharmaceutical formulations. Qualitative as well as quantitative changes in a formulation that influence the performance of formulations can be predicted with the help of different computational tools. The present chapter plans to highlight various computational tools available online as well as offline such as PCP disso, DD solver, Kinetds, *etc.* along with these software, the effective use of Microsoft Office Excel tool for calculating drug kinetic studies is also discussed here.

Keywords: Computational tools, Dissolution, Drug release, Integrated software, Kinetics.

1. INTRODUCTION

The field of pharmaceutical sciences has witnessed remarkable advancements over the years, particularly in the design and development of solid oral dosage forms [1]. Among the critical attributes influencing the efficacy and safety of these dosage forms is the drug release kinetics, which governs the rate and extent at which a therapeutic agent is released from its matrix and becomes available for absorption in the body. Accurate prediction and control of drug release kinetics are pivotal for optimizing drug performance, ensuring patient compliance, and achieving desired therapeutic outcomes [2]. Traditionally, empirical approaches

* **Corresponding author Devendra S. Shirode:** Department of Pharmacology, Dr. D. Y. Patil College of Pharmacy, Akurdi, Pune, Maharashtra, India; E-mail: devendrashirode@dyppharmaakurdi.ac.in

Dilpreet Singh and Prashant Tiwari (Eds.)
All rights reserved-© 2024 Bentham Science Publishers

and trial-and-error methods have been employed to understand and manipulate drug release from solid oral dosage forms. However, these conventional techniques are often time-consuming, resource-intensive, and may not provide a comprehensive understanding of the underlying release mechanisms. In response to these challenges, computational tools have emerged as indispensable assets, offering a more systematic and efficient means of predicting and analyzing drug release kinetics [3].

This paper delves into the exciting realm of computational tools for predicting drug release kinetics in solid oral dosage forms. By harnessing the power of mathematical modeling, molecular simulations, and data-driven algorithms, researchers and pharmaceutical scientists can gain deeper insights into the intricate interplay of formulation parameters, physicochemical properties, and release mechanisms [4]. These computational tools not only expedite the drug development process but also enable a more rational and personalized approach to dosage form design. In the subsequent sections, we will explore a spectrum of computational techniques that have gained prominence in predicting drug release kinetics. From mechanistic models that elucidate diffusion and dissolution processes to advanced simulations that capture molecular interactions within the matrix, each approach contributes to unraveling the complexities of drug release behavior [5]. Moreover, we will discuss the potential implications of these tools in optimizing therapeutic efficacy, minimizing side effects, and shaping the future landscape of pharmaceutical research.

As the pharmaceutical industry continues its trajectory towards precision medicine and individualized therapies, the integration of computational tools into the drug development paradigm holds immense promise. By augmenting our understanding of drug release kinetics in solid oral dosage forms, these tools empower scientists to expedite innovation, enhance product quality, and ultimately improve patient outcomes [6]. This paper serves as a guide to navigating this burgeoning field, shedding light on the transformative potential of computational methodologies in shaping the future of pharmaceutical science.

2. CALCULATIONS AND VALIDATION OF DISSOLUTION MODELS USING MS EXCEL

In vitro dissolution testing has emerged as a promising method to ensure the quality of pharmaceutical formulations [6]. Dissolution studies are not only used to measure the amount of released drug but are also employed to ensure batch-to-batch reproducibility and analyze the effects of changes made to the formulation or manufacturing process on the formulation's performance [7]. They are also used to establish pharmaceutical equivalence for new formulations with similar marketed formulations [8]. An appropriate dissolution model and its statistical analysis assist formulation scientists in predicting the drug's performance within a given formulation [9]. The different methods used to investigate the kinetics of drug release are classified into three types:

- Statistical methods like exploratory data analysis method, repeated measures design, and multivariate approach (MANOVA: Multivariate analysis of variance) [9, 10]
- Model dependent methods (zero order, first order, Higuchi, Korsmeyer-Peppas model, Hixson Crowell, Baker-Lonsdale model, Weibull model, *etc.*) [11 - 14]
- Model independent methods [difference factor (f1), similarity factor (f2)] [15 - 19]

Different mathematical functions of model-dependent methods can be used to describe the dissolution profile. Various model-dependent methods are zero order, Higuchi, first order, Hixson-Crowell, and Korsmeyer-Peppas. [20]

2.1. Zero-order Drug Release Model

In this, the equation used as per Kinetics to predict the dissolution profile is:

$$Co\text{-}Ct\text{=}Kot \tag{1}$$

Rearranging the equation number 1,

$$Ct\text{=}Co\text{+}Kot \tag{2}$$

where Ct is the amount of drug dissolved in time *t*, $C0$ is the initial amount of drug in the solution (most times, $C0 = 0$) and $K0$ is the zero-order release constant which is expressed in units of concentration/time [21 - 23]. Using this equation, we can calculate ko value. If it is constant then it follows Zero order release kinetics. The calculation of the percentage of drug released using zero-order and first-order dissolution methods it is given in Table **1**.

Table 1. Calculation of Percentage drug released using zero order and first order dissolution model.

Time (t)	Initial Amount Dissolved (Q0Q0)	Zero-Order Rate Constant (k0k0)	Amount Dissolved (QQ)
0	100	0.5	100
1	-	-	99.5
2	-	-	99.0

(Table 1) cont.....

Time (t)	Initial Amount Dissolved (Q0Q0)	Zero-Order Rate Constant (k0k0)	Amount Dissolved (QQ)
3	-	-	98.5
4	-	-	98.0
5	-	-	97.5
0	100	0.2	100
1	-	-	80.877
2	-	-	66.686
3	-	-	54.598
4	-	-	44.429
5	-	-	35.937

The plot of cumulative amount of drug released *versus* time gives the Release kinetics data which produces a straight line which graphically depicts that the order kinetics of drug follows zero order process in a unit dosage form (Fig. **1**).

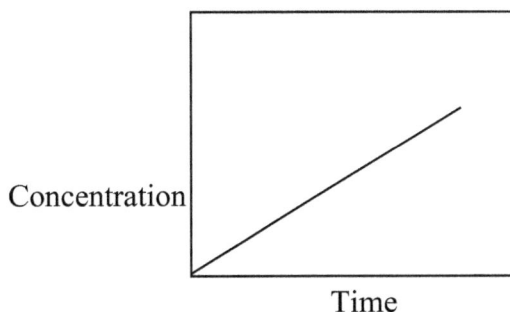

Fig. (1). Concentration *v/s* time curve for zero order drug release.

This order is followed by slow-release drug, modified release, osmotic release and some transdermal dosage forms.

2.2. First Order Drug Release Model

Here the equation used for calculation is:

$$\log C = \log C_0 - K_1 t / 2.303 \tag{3}$$

Where;

C_0 is the initial concentration of the drug, k is the first order rate constant, and t is the time, as the initial amount of drug in the solution (most times, $C_0 = 0$). We calculate the K value using eq. (3) if it is constant then it follows first-order kinetics. During the observation gathered, the log of concentration is graphically ploted against the log cumulative percentage of drug remaining *vs*. time, t gives a straight line with a slope of $-K1/2.303$ (Fig. **2**). This kinetics is usually followed by a water-soluble porous matrix [24 - 28].

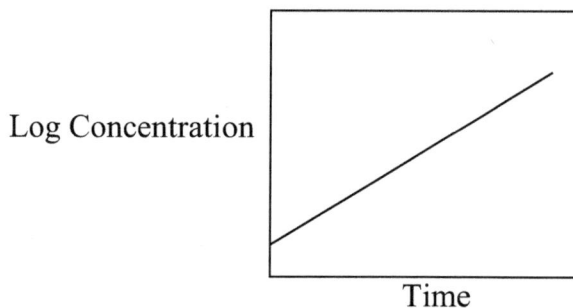

Fig. (2). Log Concentration *v/s* time curve for first order drug release.

2.3. Higuchi Model

This simplified Higuchi equation can be used for the calculation,

$$C = KH * t1/2 \tag{4}$$

where, KH is the Higuchi dissolution constant.

Concentration is calculated by taking the Concentration /Square root of time. You will get KH if it is constant, it follows the Higuchi equation. Graphically, it can be calculated by plotting Concentration *versus* the square root of time (Fig. **3**).

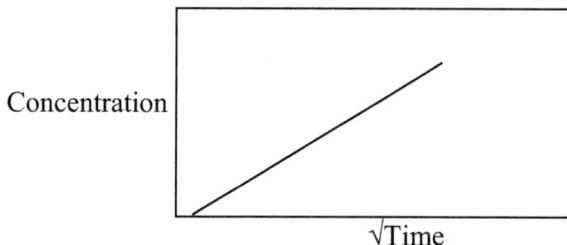

Fig. (3). Concentration *v/s* $\sqrt{}$ time curve for Higuchi model.

This model was proposed by Higuchi to describe drug release from a matrix system. Higuchi assumed that (i) the initial drug concentration in the matrix is higher than the solubility of a drug; (ii) drug diffusion is unidirectional and takes place through one facet; (iii) drug particles are much smaller than system thickness; (iv) matrix swelling and dissolution are negligible; (v) drug diffusivity is constant; and (vi) perfect sink conditions are always attained in the release environment [29 - 31].

2.4. Hixson Crowell Model

For the calculation of the Hixon Crowell model, the equation used is

$$C01/3 - Ct1/3 = \kappa\ t \tag{5}$$

Where;

C0 is the initial amount of drug formulation, Ct is the remainder drug amount in the formulation at time t and κ (kappa) is a constant incorporating the surface volume relation.

The calculation of the amount of drug to be released is provided. The total amount of drug is subtracted from the drug released at given point of time and its cube root is calculated. The % of the drug remaining to be released is divided by the time to give Kappa value which should be constant and the cube root of % drug remaining *versus* time I is plotted. It yields a straight line, whose slope value represents K (Fig. **4**) [32 - 36].

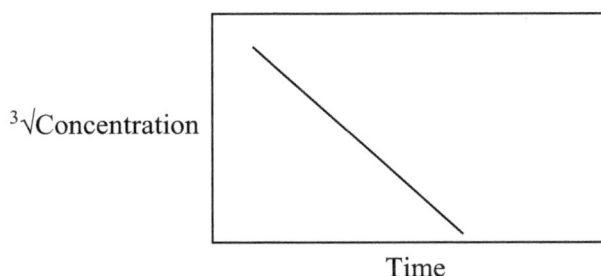

Fig. (4). $^3\sqrt{}$concentration *v/s* time curve for hixson crowell model.

2.5. Korsmeyer-Peppas Model

Kosemer derived a simple equation for determining the drug release from the polymeric dosage form:

$$Ct / C\infty = Ktn \tag{6}$$

Where Ct / C∞ is a fraction of the drug released at time t, k is the drug release rate constant and n is the release exponent. Here the n value is used to set apart different releases for cylindrical-shaped matrices [7].

In mathematical calculations, C∞ is considered when 100% drug release takes place. The relative drug release at different time intervals is divided by the complete drug release. The plot of relative concentration *v/s* time gives the slope value and intercept K (Fig. **5**). This calculation can be made simple by taking a log of relative drug release. It is represented by the following equation,

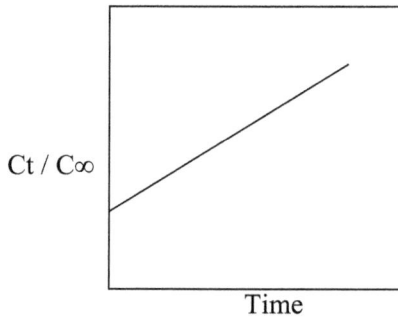

Fig. (5). Ct / C∞ *v/s* Time curve for Korsmeyer-Peppas model.

$$\text{Log C rel} = \log K + n \text{ Log t} \tag{7}$$

If we consider the case of cylindrical shape tablet, if the value of n is $0.45 \leq n$ it indicates Fickian diffusion mechanism, whereas $0.45 < n < 0.89$, indicates the release is non-Fickian, n = 0.89 to Case II (relaxational) transport, and n > 0.89 to super case II transport [37]. To find out the exponent of *n* the portion of the release curve, where Ct / C∞ < 0.6 should only be used. Here first 60% drug release data must be fitted in Korsmeyer-Peppas model [38, 39].

2.6. Similarity Factor Calculations Using Excel

Similarity Factor (f2) calculations in pharmaceutical sciences are used to compare the dissolution profiles of two drug products, often a test formulation and a reference formulation (such as a brand-name product) [40]. The f2 value assesses the similarity between these dissolution profiles, indicating whether the test formulation releases the drug in a manner comparable to the reference formulation [41]. This is an essential step in ensuring generic drug products are equivalent to their branded counterparts in terms of their *in vivo* performance. The f2 calculation is based on the difference in cumulative percent dissolved between the test and reference formulations at multiple time points. The formula for calculating the f2 value is:

$$f2 = 50 \times \log_{10}\left(\frac{1}{N}\sum_{i=1}^{N}[1 + (Ri - Ti)2Ri + Ti]\right) f2 = 50 \times \log_{10}\left(N\frac{1}{\sum_{i=1}^{N}}[1 + Ri + Ti(Ri - Ti)2]\right) \tag{8}$$

Where:

- NN = number of time points
- RiRi = percent dissolved of the reference formulation at time point ii
- TiTi = percent dissolved of the test formulation at time point ii

2.6.1. Using Excel for f2 Calculation

2.6.1.1. Data Collection

Dissolution data is gathered for both the test and reference formulations at various time points. This data should include the percentage of drug dissolved at each time point.

2.6.1.2. Data Entry in Excel

A spreadsheet is created where you can enter the time points, percentage dissolved for the reference formulation, and percentage dissolved for the test formulation.

2.6.1.3. Calculation

In a new column, the squared differences $(Ri-Ti)2(Ri-Ti)2$ between the reference and test formulation percentages for each time point are calculated. Then, in another column, we calculate the sum of $(Ri+Ti)(Ri+Ti)$ for each time point.

2.6.1.4. Average Calculation

The average of the squared differences and the sum of (reference + test percentages) across all time points is calculated.

2.6.1.5. Final Calculation

The f2 formula mentioned earlier is used to calculate the f2 value using the averaged values [42].

2.6.1.6. Interpretation

The calculated f2 value is compared with the interpretation guidelines mentioned above to assess the similarity between the dissolution profiles.

2.7. Difference Factor (f1)

The Difference Factor (f1) is another parameter used in pharmaceutical sciences to compare dissolution profiles of two drug products, typically a test formulation and a reference formulation [43]. The f1 value quantifies the overall difference between the dissolution profiles and is often used in conjunction with the

Similarity Factor (f2) to provide a comprehensive assessment of similarity or dissimilarity [44]. The f1 calculation involves calculating the absolute differences between the cumulative percent dissolved of the test and reference formulations at multiple time points. The formula for calculating the f1 value is:

$$f1 = \frac{1}{N}\sum_{i=1}^{N}|R_i - T_i| \qquad (9)$$

Where:

- NN = number of time points

- R_iR_i = percent dissolved of the reference formulation at time point ii

- T_iT_i = percent dissolved of the test formulation at time point ii

2.7.1. Interpretation

The f1 value ranges from 0 to 100. A lower f1 value indicates greater similarity between the two dissolution profiles, while a higher f1 value suggests greater difference. The general interpretations are as follows:

- $f1 < 15f1 < 15$: The test and reference formulations are considered similar or interchangeable.

- $f1 \geq 15f1 \geq 15$: The test formulation may exhibit differences from the reference formulation, indicating potential dissimilarity.

2.7.2. Using Excel for f1 Calculation

2.7.2.1. Data Collection

Dissolution data is gathered for both the test and reference formulations at various time points. This data should include the percentage of drug dissolved at each time point [45].

2.7.2.2. Data Entry in Excel

A spreadsheet is created where one can enter the time points, percentage dissolved for the reference formulation, and percentage dissolved for the test formulation.

2.7.2.3. Calculation

In a new column, the absolute differences $(R_i-T_i)(R_i-T_i)$ between the reference and test formulation percentages for each time point are calculated.

2.7.2.4. Average Calculation

The average of the absolute differences across all time points is calculated [46].

2.7.2.5. Interpretation

The calculated f1 value is compared with the interpretation guidelines mentioned above to assess the difference between the dissolution profiles. Table **2** provides the calculation of similarity and difference factor [47].

Table 2. Calculation of similarity and difference factor.

Similarity Factor f2		-	Difference Factor f1	
A	**B**	**C**	**D**	**E**
Number of Time Points, n = 4				
10	45	55	10	100
15	65	75	10	100
20	80	90	10	100
30	90	100	10	100
45	-	-	0	0
60	-	-	0	0
75	-	-	0	0
120	-	-	0	0
-	-	sum (Rt-Tt)		40
-	-	sum (Rt-Tt)^2		400
-	-	sum Rt	-	280
-	-	Similarity factor f2		50
-	-	Difference factor f1		14

3. INTEGRATED TOOLS FOR DISSOLUTION MODELING OF VARIOUS DOSAGE FORMS

3.1. DD Solver: Modeling and Comparison of Drug Dissolution Profiles

DD Solver is a menu-driven add-in program for Microsoft Excel written in Visual Basic for Applications. Calculation using Excel offers a number of advantages over other software packages, the most attractive of which is the ease of use. Most scientists are already familiar with Excel because of its wide availability and high flexibility [58]. For each module, DD Solver offers a number of customizability choices, including convergence and the maximum number of nonlinear optimization algorithm rounds, starting parameter estimations, the number of decimal places in the computed results, chart output, and Microsoft Word report generating [59]. The DD Solver is used for various purposes in the pharmaceutical industry. The *in-vitro* dissolution, algorithms, comparisons, and estimation of various parameters can be done by this add-in, in Excel [60]. Table (**3**) provides the drug dissolution models assembled using DD solver software.

Table 3. Drug dissolution models.

Zero-order	Zero-order with T_{lag}	Zero-order with FO First-order
First-order with T_{lag} First-order with F_{max}	First-order with T_{lag} and F_{max}	Higuchi
Higuchi with Tlag	Higuchi with FO	Korsmeyer-Peppas with
Tlag Korsmeyer-Peppas with FO	Hixson-Crowell with T_{lag}	Hixson-Crowell
Hopfenberg with T_{lag}	Hopfenberg	Baker-Lonsdale

3.2. Key Features and Capabilities

3.2.1. Mechanistic Modeling

DD Solver provides a platform for applying mechanistic models to drug dissolution profiles. These models are based on fundamental principles of diffusion, dissolution, and other relevant physical and chemical processes. By inputting key parameters such as formulation characteristics, drug properties, and release kinetics, researchers can simulate how the drug is released over time and gain insights into its behaviour [60].

3.2.2. Curve Fitting and Prediction

The software offers curve-fitting algorithms that enable users to fit experimental dissolution data to various mathematical models. This aids in understanding the kinetics of drug release and helps in predicting the dissolution behavior of formulations under different conditions. By comparing experimental data with model predictions, researchers can validate and refine their hypotheses [61].

3.2.3. Dissimilarity and Similarity Assessment

DD Solver incorporates algorithms to calculate parameters such as the Difference Factor (f1) and Similarity Factor (f2) for comparing dissolution profiles. This capability allows researchers to quantitatively evaluate the likeness or disparity between two or more formulations, aiding in regulatory submissions, generic product development, and ensuring bioequivalence [62].

3.2.4. Visualization Tools

The software offers visualization tools that generate graphical representations of dissolution profiles, model fits, and comparison parameters. These visual aids assist researchers in communicating their findings effectively and identifying trends or anomalies in the data [63].

3.2.5. Predictive Insights

Through simulation and modeling, DD Solver can provide predictive insights into how changes in formulation parameters, manufacturing processes, or environmental conditions might impact the dissolution behavior. This empowers researchers and formulators to make informed decisions early in the development process, saving time and resources [64].

3.2.6. User-Friendly Interface

DD Solver is designed with a user-friendly interface that enables researchers with varying levels of expertise to navigate through the software's features. Its intuitive design and guided workflows make it accessible for both novice and experienced users [65].

3.3. Procedure for Using DD Solver for Drug Dissolution Profile Analysis and Comparison

Using DD Solver for drug dissolution profile analysis and comparison involves a systematic process to model, simulate, and assess dissolution behavior across various dosage forms. The following procedure outlines the key steps to effectively utilize the software:

3.3.1. Data Collection and Input Preparation

This begins by collecting experimental dissolution data for the formulations you intend to analyze. Organize the data to include time points and corresponding percentages of drug dissolved. Input this data into the DD Solver software either manually or by importing relevant files.

3.3.2. Model Selection and Curve Fitting

Choose an appropriate mathematical model that represents the expected release kinetics of the drug from your dosage forms. DD Solver offers a selection of mechanistic models, including diffusion-based, dissolution-controlled, and hybrid models. Curve-fitting algorithms are applied to the collected data to determine the best-fit parameters for the selected model [66].

3.3.3. Simulation and Prediction

The fitted model parameters are utilized to simulate drug dissolution profiles under various conditions, such as changes in formulation composition or manufacturing processes. This predictive capability allows you to assess the impact of these changes on dissolution behavior without requiring additional experimental work.

3.3.4. Dissolution Profile Comparison

DD Solver enables the calculation of parameters like the Difference Factor (f1) and Similarity Factor (f2) to compare dissolution profiles. The experimental data is input for both the reference and test formulations, and the software will provide quantitative measures of their similarity or dissimilarity. This step aids in assessing bioequivalence and regulatory compliance.

3.3.5. Visualization and Reporting

Visualize the modeled and experimental dissolution profiles using graphical representations generated by DD Solver. These visual aids help in understanding trends, discrepancies, and changes in the dissolution behavior. Comprehensive reports are created summarizing the modeling results, curve fitting details, comparison parameters, and any insights gained from the analysis [67].

3.3.6. Interpretation and Decision-Making

The results obtained from DD Solver's simulations and comparisons are analyzed. The f1 and f2 values are interpreted alongside the graphical representations to make informed decisions about the equivalence or disparity of dissolution profiles. The clinical relevance of any differences and the potential implications on therapeutic efficacy are considered. The screenshots of software used to describe multi-step proceudre for analysis is depicted in Fig. (**6a-f**).

A)

	A	B	C	D	E	F	G	H	I
1		Time	No. 1	No. 2	No. 3	No. 4	No. 5	No. 6	
2		(min)	F(%)	F(%)	F(%)	F(%)	F(%)	F(%)	
3		5	12.03	13.41	13.67	12.68	13.07	13.71	
4		10	20.63	25.92	24.08	25.61	24.71	26.06	
5		15	30.61	31.12	30.05	31.78	31.05	32.81	
6		20	33.02	33.63	34.68	35.55	34.42	36.77	
7		30	40.13	42.99	40.57	41.74	39.54	44.15	
8		45	42.55	44.61	44.86	44.62	43.66	46.78	
9		60	45.86	46.71	46.06	46.06	47.96	47.61	

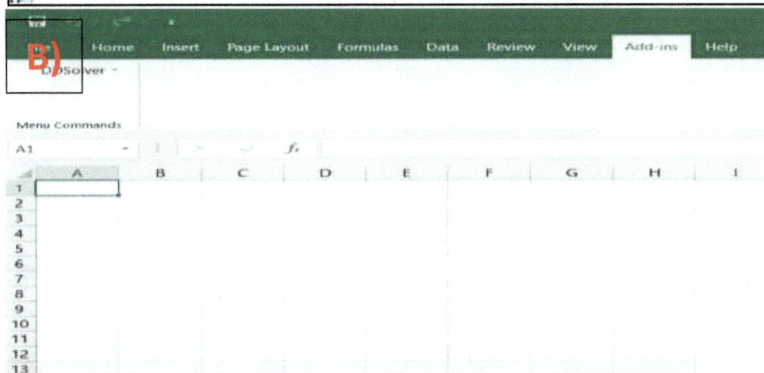

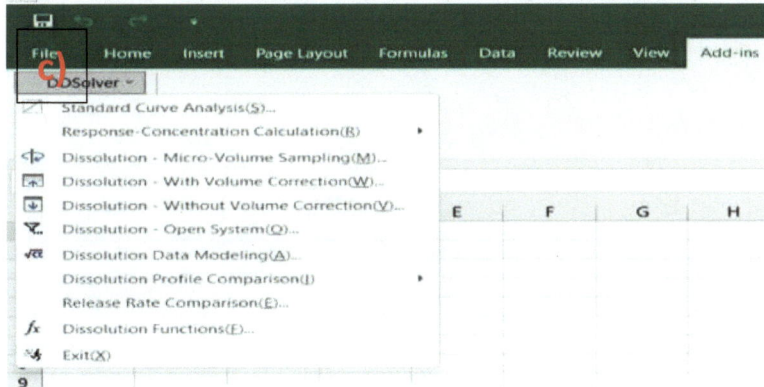

B)

DDSolver ▾

Menu Commands

C)

DDSolver ▾

Standard Curve Analysis(S)...
Response-Concentration Calculation(R)   ▸
Dissolution - Micro-Volume Sampling(M)...
Dissolution - With Volume Correction(W)...
Dissolution - Without Volume Correction(V)...
Dissolution - Open System(O)...
Dissolution Data Modeling(A)...
Dissolution Profile Comparison(I)   ▸
Release Rate Comparison(E)...
fx Dissolution Functions(F)...
Exit(X)

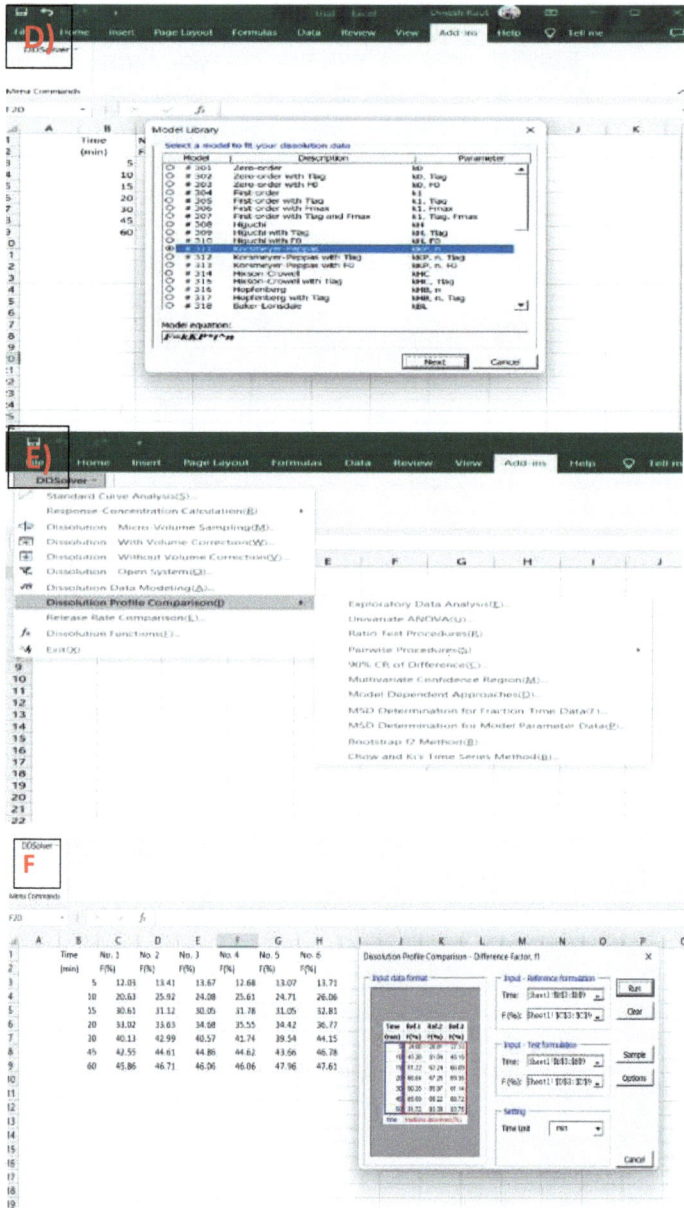

Fig. (6). Various steps involved in DD solver *i.e.* **a**) Excel sheet of DD solver; **b**) menu command; **c**) selecting dissolution models; **d**) data analysis; **e**) dissolution profile comparison and **f**) prediction of kinetic models.

3.4. Applications of DD Solver

The DD Solver software finds applications in a range of pharmaceutical areas, including:

3.4.1. Generic Product Development

Assessing the similarity of generic drug products to their reference counterparts, ensuring bioequivalence, and meeting regulatory requirements.

3.4.2. Formulation Optimization

Identifying the most suitable formulation parameters to achieve desired drug release profiles.

3.4.3. Comparative Studies

Conducting comparative dissolution studies for various dosage forms, polymorphs, or formulations.

3.4.4. Risk Assessment

Evaluating the impact of formulation or process changes on dissolution behavior and subsequent clinical outcomes.

4. PCPDISSO (V3I)

PCP Disso Dissolution Software is a sophisticated and extensively utilized software tool that plays a pivotal role in the realm of pharmaceutical research and development. Specifically tailored for managing and dissecting dissolution testing data, this software has emerged as an indispensable asset within the pharmaceutical industry [67]. The primary function of PCP Disso Dissolution Software revolves around automating, monitoring, and orchestrating the complex processes inherent to dissolution testing. One of its standout features lies in its ability to seamlessly collect data from various dissolution testers, expediting the otherwise intricate data collection procedure. In conjunction with real-time monitoring capabilities, the software enables users to observe dissolution profiles as they evolve, providing invaluable insights into the behavior of pharmaceutical formulations [68]. Through intuitive graphical representations, the software generates dissolution curves, aiding researchers and scientists in comprehending the dissolution kinetics with visual clarity.

A critical aspect of PCP Disso Software is its capability to calculate a spectrum of crucial dissolution parameters. These calculations encompass a diverse range of metrics, such as dissolution rates and percentages dissolved, offering comprehensive insights into the performance of pharmaceutical products. Moreover, the software provides robust tools for statistical analysis, further enhancing the depth of data interpretation. In the context of regulatory compliance, PCP Disso Software emerges as an invaluable asset. The software diligently generates reports that align with stringent regulatory standards, an essential attribute for pharmaceutical quality control and assurance. This capacity to seamlessly create compliant reports simplifies the arduous task of documentation and ensures that data-driven decisions are founded upon accurate, reliable information [69].

PCP Disso Software caters to a multitude of applications within the pharmaceutical landscape [70 - 72]. It facilitates rigorous research and development endeavors, assists in quality control and batch release testing, and is integral to the conduction of bioavailability and bioequivalence studies. Its versatility spans formulation optimization and regulatory adherence, thereby encapsulating the entire spectrum of pharmaceutical testing requirements. With PharmaTest as its manufacturer, this software carries a seal of credibility and competence. PharmaTest is a recognized name in the field of pharmaceutical testing equipment and software solutions, underlining the reliability and proficiency associated with PCP Disso Software [73, 74]. The equations for this software is written as follows [75]. The home screen of PCP disso software is depicted in Fig. (**7**).

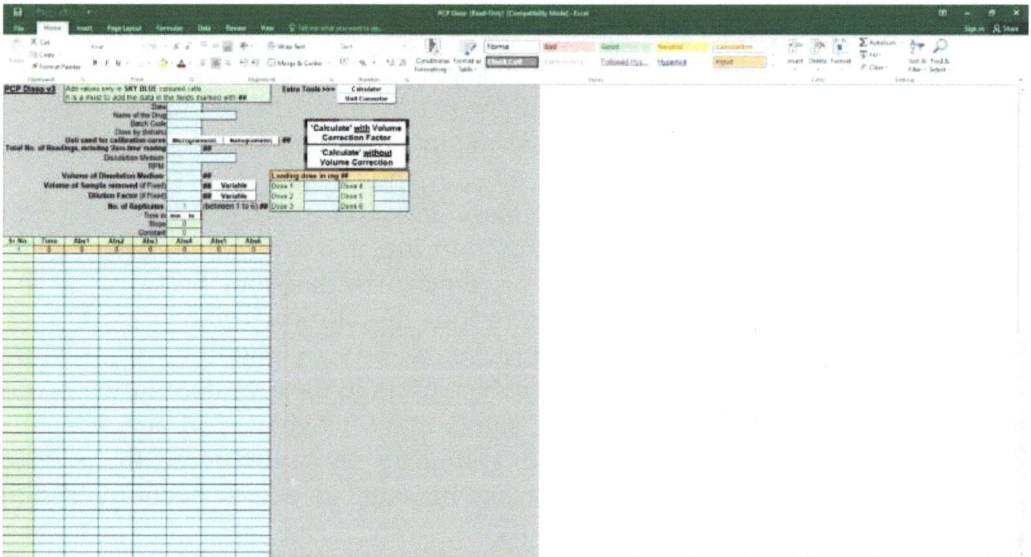

Fig. (7). Home screen of PCP disso software.

This model is widely used, whether the release mechanism is not well known or when more than one type of release phenomena could be involved [76]. The 'n' value could be used to characterize different release mechanisms as depicted in Table **4**.

Table 4. Release kinetic models in PCP disso.

'n'	Mechanism
0.5	Fickian Diffusion (HiguchiMatrix)0.5<n<1 Anomalous Transport
1	Case-IITransport(ZeroOrderRelease)
n>1	Super Case-II transport

4.1. Hypothetical Release Models for PCP Disso

Hypothetical release models that PharmaTest PTWS Dissolution Software might offer enhance its capabilities in dissolution testing data analysis:

4.1.1. Zero-Order Release Model

PCP Disso Software could incorporate a zero-order release model that allows users to fit and analyze data based on the zero-order kinetics of dissolution. This model assumes a constant release rate of the drug over time, irrespective of the concentration of the drug remaining in the dosage form. By providing insights into how well a drug maintains a consistent release rate, this model could assist in predicting the drug's long-term behaviour [77].

4.1.2. First-order Release Model

A first-order release model integrated into the software could enable researchers to analyze dissolution data based on first-order kinetics. This model assumes that the rate of drug release is proportional to the remaining drug concentration. By fitting data to this model, scientists can assess the overall dissolution rate and better understand how the drug's concentration changes over time [78].

4.1.3. Higuchi Release Model

PCP Disso Software might offer a Higuchi release model, which focuses on drug release from a solid dosage form based on Fickian diffusion principles. By fitting data to this model, researchers can gain insights into the drug's diffusion behavior within the matrix of the dosage form. This is particularly useful for understanding drug release from systems where diffusion plays a significant role [79].

4.1.4. Weibull Release Model

A Weibull release model within the software could cater to more complex release profiles. This model can account for various release mechanisms, including a burst release at the initial stages followed by more controlled release behavior. It is particularly useful when the dissolution behavior is influenced by multiple factors and stages.

4.1.5. Fractional Release Model

The software could also incorporate a fractional release model that allows users to estimate the fraction of drug released at different time points during the dissolution process. This model can provide insights into the immediate release portion, sustained release phase, and other distinct phases of drug release [80]. Various flow charts for working behaviour of PCP disso are depicted in Fig. (8a-c).

5. KINETDS 3.0 SOFTWARE

KinetDS is a software application designed for pharmacokinetic and pharmacodynamic modeling and simulation. It provides researchers, pharma-cologists, and pharmaceutical scientists with a platform to analyze drug behavior in biological systems, predict drug concentrations over time, and understand the relationships between drug doses, concentrations, and their effects [81 - 83]. *KinetDS* software is developed by Olenzor, based on curve fitting method, which comments on the cumulative dissolution curve by a simple equation or a set of equations. Other curves, which are derived from different data sources, also can be analyzed if their dependent-variables are in a range of 0 and 100. Different mechanistic and empirical models like zero to second order, Higuchi, Korsmeyer–Peppas, Hill and Hixon-Crowell can be applied to the drug dissolution curve description [84 - 87]. Fig. (**9**) shows the various models for KinetDS software.

Model-independent measures of dissolution process are also available in the software like Dissolution efficiency (DE) and Mean Dissolution Time (MDT), which are expressed by following equations:

$$DE = \frac{\int_0^t Q \, dt}{Q_{max} \times t} \times 100 \qquad \qquad \textbf{(10)}$$

$$MDT = \frac{\sum_{j=1}^{n} t_j^{AV} \times \Delta Q_j}{\sum_{j=1}^{n} \Delta Q_j} \qquad \qquad \textbf{(11)}$$
$$\Delta Q = Q_{(t)} - Q_{(t-1)}$$
$$t_j^{AV} = (t_i + t_{i-1})/2$$

where

Q: Percent of drug substance released at time t

Qmax: Maximum value of Q

Fig. (8a). Flow chart (integrated steps and initialization) of PCP disso.

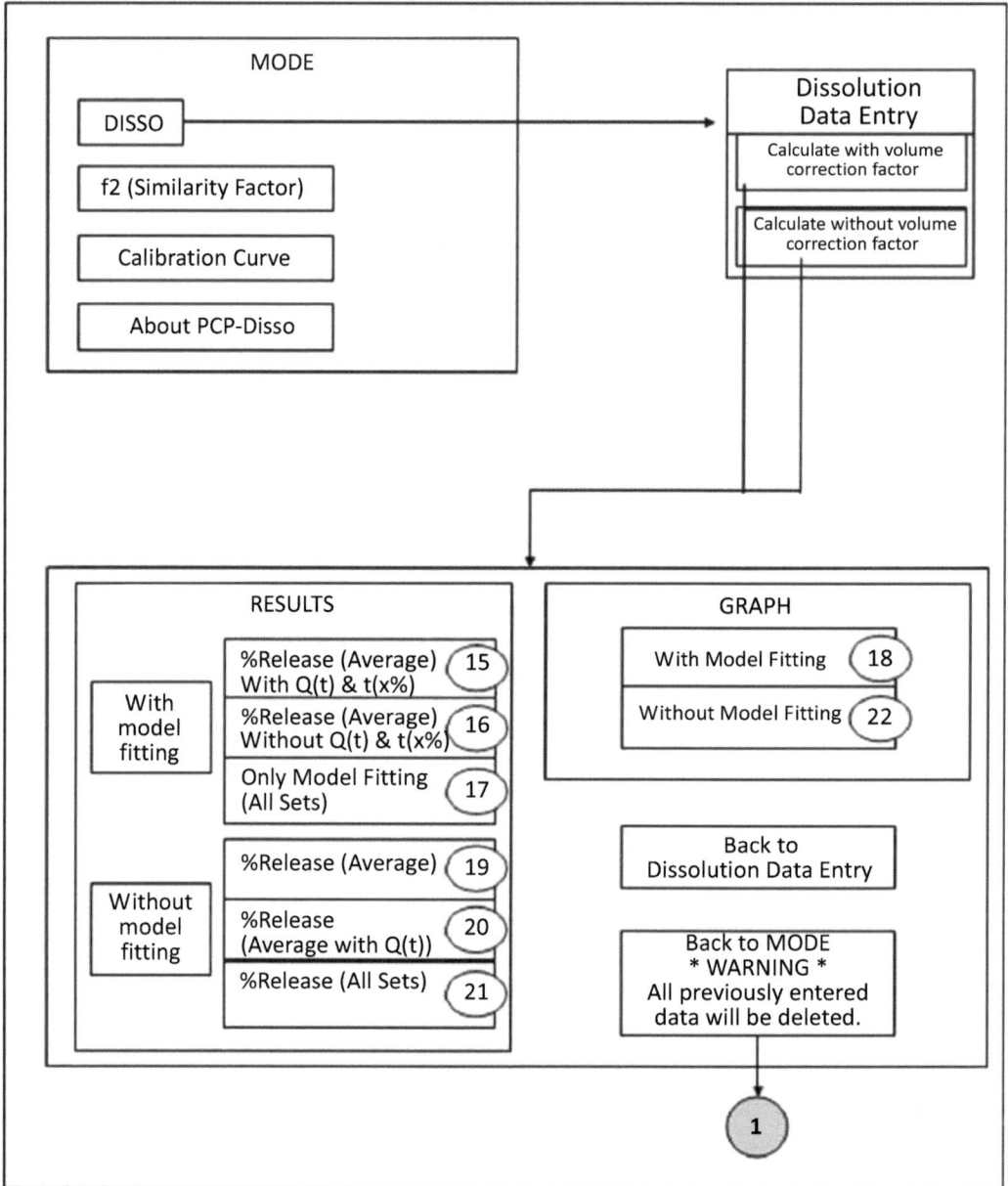

Fig. (8b). Flow chart (working operations) of PCP disso.

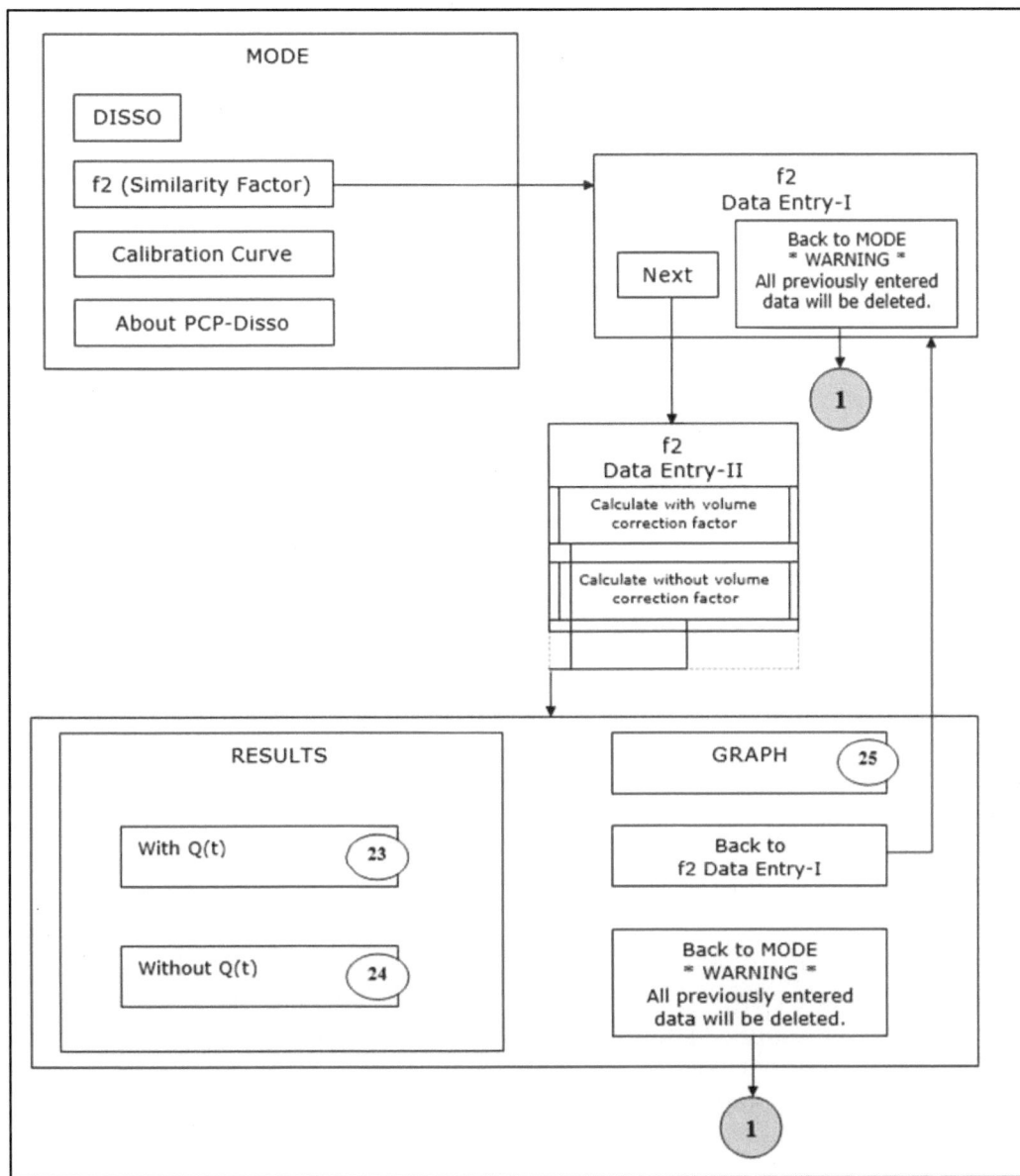

Fig. (8c). Flow chart (Interpretation of results) of PCP disso.

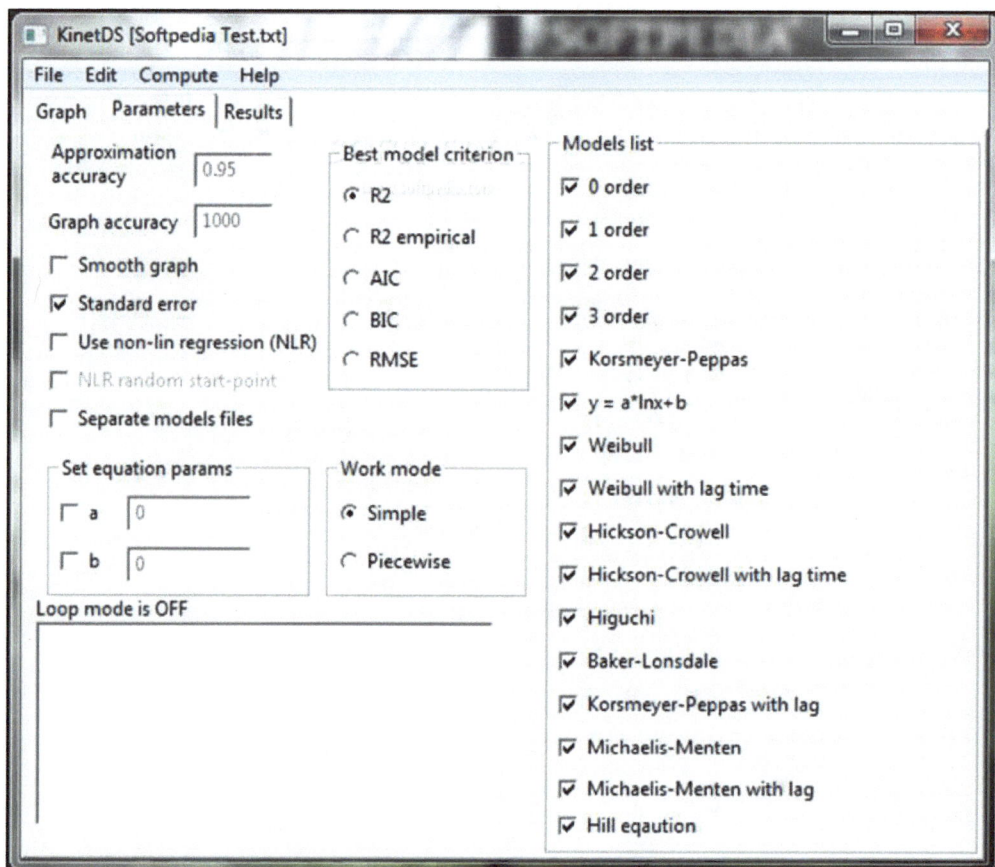

Fig. (9). Menu showing different models available in kinetDS.

Apart from this, the available data can be treated with different model fitting measures such as coefficient of determination, $R2$, and empirical coefficient of determination, $R2$emp; Akaike Information Criterion; Bayesian Information Criterion or Schwarz criterion; and root-mean-squared error, RMSE. Additional diagnostics of the linear regression expressed in the percent of the relevant parameter value (relative standard error of the regression coefficient). Model parameters are established by linear and nonlinear regression which is represented by the simplex method. So far, all the models included in KinetDS 3.0 are based on linearization. Therefore, they might be fit by both types of regression analyses [88 - 90].

6. OTHER DISSOLUTION WORKSTATIONS/TOOLS

Agilent Dissolution Workstation offers a comprehensive suite of features for automated dissolution testing, encompassing real-time monitoring, data acquisition, analysis, and reporting [91]. Its integration with Agilent's dissolution testing equipment provides a seamless workflow for researchers. LabWare Dissolution, integrated within LabWare's Laboratory Information Management System (LIMS), serves as a holistic solution for managing dissolution processes. It simplifies sample tracking, data management, and compliance adherence through its dedicated module. Thermo Fisher Scientific's UV Dissolution Software focuses on UV-visible spectroscopic data acquisition during dissolution testing. This real-time analysis capability allows researchers to gain immediate insights into drug release behavior [92]. Varian's Dissolution 700 software, now part of Agilent, caters to the control and management of dissolution testing instruments, ensuring precision data collection and compliance with regulatory standards. SOTAX DISI-II, provided by SOTAX, offers automation and management of dissolution testing [93]. It enables control over SOTAX's dissolution testing equipment, data collection, and comprehensive report generation. Pharma Test PTWS Dissolution Software, renowned for its versatility, enables automated data collection, real-time monitoring, and graphical representation of dissolution curves. This software is a reliable companion for researchers seeking accurate dissolution insights [94].

Hanson Research's VISION Dissolution Software is designed for compatibility with their dissolution testing instruments, empowering users with control, monitoring, data acquisition, and analysis capabilities. Pion Inc.'s DissolvIt Software provides real-time monitoring, data acquisition, and modeling for dissolution profiles, aiding in the prediction of drug release behavior [95]. Agilent's Dissolution 2000 Software serves as another option for comprehensive dissolution testing management, covering data acquisition, reporting, and regulatory compliance. DISOCS by SciPharma stands as a solution for dissolution testing management, featuring method development, data analysis, and compatibility with various dissolution instruments [96]. Table **5** provides key information on various secondary softwares for dissolution data analysis.

Table 5. List of some other online dissolution softwares.

Tool	Link
Performance verification test calculation tool	https://apps.usp.org/app/USPNF/pvtCalculationTool/results.html
Dissolution Performance Verification Tool **v. 1.4.9**	https://mypharmatools.com/pharmtools/pvt_[

(Table 5) cont.....

Tool	Link
Dissolution Profile Comparison Tool **v. 1.0 (beta)**	https://mypharmatools.com/pharmtools/dpct_[
Drug Dissolution Analysis- Data Analysis	https://www.originlab.com/fileExchange/details.aspx?fid=661

CONCLUSION

In conclusion, the realm of dissolution testing software has evolved into an indispensable aspect of pharmaceutical research and development. These software tools offer an array of functionalities that streamline data acquisition, analysis, and interpretation, fostering efficiency and accuracy in understanding drug release behaviors from various dosage forms. From automated data collection to real-time monitoring, graphical visualization, and comprehensive reporting, these software solutions cater to the diverse needs of researchers, quality control professionals, and regulatory compliance experts. In an era where data-driven insights are paramount, dissolution software stands as an enabler of innovation and precision in pharmaceutical sciences. Its ability to transform intricate dissolution profiles into comprehensible data points empowers researchers to unlock deeper understandings of drug behavior within biological systems. As technology continues to advance, these software solutions are poised to play an increasingly pivotal role in revolutionizing pharmaceutical research, promoting efficient regulatory compliance, and advancing the forefront of medical advancements for the benefit of patients worldwide.

ACKNOWLEDGEMENTS

We are thankful to all software developers of the mentioned software *i.e.* PCP disso, DD solver, Kinetds, *etc.* for calculating the kinetics of *in vitro* drug release studies.

REFERENCES

[1]   U.S. FDA/CDER. Guidance for Industry, "Dissolution testing of immediate release solid oral dosage forms. 1997.

[2]   Wang Q, Fotaki N, Mao Y. Biorelevant dissolution: Methodology and application in drug development. Dissolut Technol 2009; 16(3): 6-12.
[http://dx.doi.org/10.14227/DT160309P6]

[3]   Martin G, Gray V. Valid Technol 2011; 17(1): 8-11.

[4]   Lu X. Filling in the gaps for *in-vitro* release and dissolution testing of solid oral dosage formulations. AAPS Blog 2016.

[5]   Kochling J. Approaches to the investigation of dissolution testing changes and failures. AAPS Webinar 2013.

[6]   Costa P, Sousa Lobo JM. Modeling and comparison of dissolution profiles. Eur J Pharm Sci 2001; 13(2): 123-33.

[http://dx.doi.org/10.1016/S0928-0987(01)00095-1] [PMID: 11297896]

[7] Podczeck F. Comparison of *in vitro* dissolution profiles by calculating mean dissolution time (MDT) or mean residence time (MRT). Int J Pharm 1993; 97(1-3): 93-100.
[http://dx.doi.org/10.1016/0378-5173(93)90129-4]

[8] Gibaldi M, Feldman S. Establishment of sink conditions in dissolution rate determinations. Theoretical considerations and application to nondisintegrating dosage forms. J Pharm Sci 1967; 56(10): 1238-42.
[http://dx.doi.org/10.1002/jps.2600561005] [PMID: 6059440]

[9] O'Hara T, Dunne A, Butler J, Devane J. A review of methods used to compare dissolution profile data. Pharm Sci Technol Today 1998; 1(5): 214-23.
[http://dx.doi.org/10.1016/S1461-5347(98)00053-4]

[10] Yuksel N. Kan k AE, Baykara T. Int J Pharm 2000; 209(1-2): 57-67.
[http://dx.doi.org/10.1016/S0378-5173(00)00554-8] [PMID: 11084246]

[11] Crank J. The mathematics of diffusion. Oxford: Clarendon Press 1975; p. 23.

[12] Arhewoh MI, Okhamafe OA. J Med Biomed Res 2004; 3(2): 7-17.

[13] Ritger PL, Peppas NA. A simple equation for description of solute release I. Fickian and non-fickian release from non-swellable devices in the form of slabs, spheres, cylinders or discs. J Control Rel 1987; 5(1): 23-36.

[14] Ritger PL, Peppas NA. A simple equation for description of solute release I. Fickian and non-fickian release from non-swellable devices in the form of slabs, spheres, cylinders or discs. Journal of Controlled Release. 1987 Jun 1; 5(1): 23-36.

[15] Sathe PM, Tsong Y, Shah VP. *In-vitro* dissolution profile comparison: statistics and analysis, model dependent approach. Pharmaceut Res 1996; 13: 1799-803.
[http://dx.doi.org/10.1023/A:1016020822093]

[16] Moore JW, Flanner HH. Pharm Technol 1996; 20: 64.

[17] Anonymous. Guideline for Industry. US Department of Health and Human Services, Food and Drug Administration 1995.

[18] Letícia S, Koester GG, Ortega PM, Bassani VL. Eur J Pharm Biopharm 2004; 58: 177.
[http://dx.doi.org/10.1016/j.ejpb.2004.03.017] [PMID: 15207552]

[19] Kalam MA, Humayun M, Parvez N, *et al.* Cont. J Pharm Sci 2007; 1(1): 30-5.

[20] Siepmann J, Siepmann F. Mathematical modeling of drug dissolution. Int J Pharm 2013; 453(1): 12-24.
[http://dx.doi.org/10.1016/j.ijpharm.2013.04.044] [PMID: 23618956]

[21] Raslan HK, Maswadeh H. *In vitro* dissolution kinetic study of theophylline from mixed controlled release matrix tablets containing hydroxypropylmethyl cellulose and glycerylbehenate. Indian J Pharm Sci 2006; 68(3)

[22] Loisios-Konstantinidis I, Dressman J. Physiologically based pharmacokinetic/pharmacodynamic modeling to support waivers of *in vivo* clinical studies: Current status, challenges, and opportunities. Mol Pharm 2021; 18(1): 1-17.
[http://dx.doi.org/10.1021/acs.molpharmaceut.0c00903] [PMID: 33320002]

[23] Zhang X, Duan J, Kesisoglou F, *et al.* Mechanistic oral absorption modeling and simulation for formulation development and bioequivalence evaluation: Report of an FDA public workshop. CPT Pharmacometrics Syst Pharmacol 2017; 6(8): 492-5.
[http://dx.doi.org/10.1002/psp4.12204] [PMID: 28571121]

[24] Kesisoglou F, Chung J, van Asperen J, Heimbach T. Physiologically based absorption modeling to impact biopharmaceutics and formulation strategies in drug development-industry case studies. J Pharm Sci 2016; 105(9): 2723-34.

[http://dx.doi.org/10.1016/j.xphs.2015.11.034] [PMID: 26886317]

[25] Heimbach T, Suarez-Sharp S, Kakhi M, *et al.* Dissolution and translational modeling strategies toward establishing an *in vitro-in vivo* link-a workshop summary report. AAPS J 2019; 21(2): 29.
[http://dx.doi.org/10.1208/s12248-019-0298-x] [PMID: 30746576]

[26] Pepin XJ, Parrott N, Dressman J, *et al.* Current state and future expectations of translational modeling strategies to support drug product development, manufacturing changes and controls: A workshop summary report. J Pharm Sci 2020.
[PMID: 32380182]

[27] Shebley M, Sandhu P, Emami Riedmaier A, *et al.*

[28] Gurny R, Doelker E, Peppas NA. Modelling of sustained release of water-soluble drugs from porous, hydrophobic polymers. Biomaterials 1982; 3(1): 27-32.
[http://dx.doi.org/10.1016/0142-9612(82)90057-6] [PMID: 7066463]

[29] Higuchi T. Rate of release of medicaments from ointment bases containing drugs in suspension. J Pharm Sci 1961; 50(10): 874-5.
[http://dx.doi.org/10.1002/jps.2600501018] [PMID: 13907269]

[30] Higuchi T. Mechanism of sustained-action medication. Theoretical analysis of rate of release of solid drugs dispersed in solid matrices. J Pharm Sci 1963; 52(12): 1145-9.
[http://dx.doi.org/10.1002/jps.2600521210] [PMID: 14088963]

[31] Shoaib MH, Tazeen J, Merchant HA, Yousuf RI. Evaluation of drug release kinetics from ibuprofen matrix tablets using HPMC. Pak J Pharm Sci 2006; 19(2): 119-24.
[PMID: 16751122]

[32] Hixson AW, Crowell JH. Dependence of reaction velocity upon surface and agitation. Ind Eng Chem 1931; 23(8): 923-31.
[http://dx.doi.org/10.1021/ie50260a018]

[33] Baishya H. Application of mathematical models in drug release kinetics of carbidopa and levodopa er tablets. J Dev Drugs 2017; 6(2): 1-8.
[http://dx.doi.org/10.4172/2329-6631.1000171]

[34] Karasulu H, Ertan G, Köse T. Modeling of theophylline release from different geometrical erodible tablets. Eur J Pharm Biopharm 2000; 49(2): 177-82.
[http://dx.doi.org/10.1016/S0939-6411(99)00082-X] [PMID: 10704902]

[35] Chen S, Zhu J, Cheng J. Preparation and *in vitro* evaluation of a novel combined multiparticulate delayed-onset sustained-release formulation of diltiazem hydrochloride. Pharmazie 2007; 62(12): 907-13.
[PMID: 18214341]

[36] Korsmeyer RW, Gurny R, Doelker E, Buri P, Peppas NA. Mechanisms of solute release from porous hydrophilic polymers. Int J Pharm 1983; 15(1): 25-35.
[http://dx.doi.org/10.1016/0378-5173(83)90064-9]

[37] Shivhare UD, Pardhi DM. Effect of non-volatile solvent on dissolution profile of Carvedilol Liquisolid compact. Res J Pharm Technol 2011; 4(4): 537-44.

[38] Korsmeyer RW, Peppas NA. Solute and penetrant diffusion in swellable polymers. III. Drug release from glassy poly(HEMA-co-NVP) copolymers. J Control Release 1984; 1(2): 89-98.
[http://dx.doi.org/10.1016/0168-3659(84)90001-4]

[39] Peppas NA, Narasimhan B. Mathematical models in drug delivery: How modeling has shaped the way we design new drug delivery systems. J Control Release 2014; 190: 75-81.
[http://dx.doi.org/10.1016/j.jconrel.2014.06.041] [PMID: 24998939]

[40] Brooke D. Dissolution profile of log-normal powders II: Dissolution before critical time. J Pharm Sci 1974; 63(3): 344-7.

[http://dx.doi.org/10.1002/jps.2600630306] [PMID: 4820361]

[41]   Brahmankar DM, Jaiswal SB. Biopharmaceutics and pharmacokinetics: A treatise. Vallabh Publications 2009; p. 20.

[42]   Hopfenberg HB. Controlled release polymeric formulations. In: Paul DR, Haris FW, Eds. American Chemical Society. ACS Symp Ser 1976.

[43]   Wu F, Mousa Y, Raines K, *et al.* Regulatory utility of physiologically based pharmacokinetic modeling to support alternative bioequivalence approaches and risk assessment: A workshop summary report. CPT Pharmacomet Syst Pharmacol 2022.

[44]   Dressman JB, Krämer J, Eds. Pharmaceutical dissolution testing. Boca Raton, FL: Taylor & Francis 2005.
[http://dx.doi.org/10.1201/9780849359170]

[45]   Meineke I, Brockmöller J. Simulation of complex pharmacokinetic models in Microsoft EXCEL. Comput Methods Programs Biomed 2007; 88(3): 239-45.
[http://dx.doi.org/10.1016/j.cmpb.2007.09.007] [PMID: 17981357]

[46]   Sato H, Sato S, Wang YM, Horikoshi I. Add-in macros for rapid and versatile calculation of non-compartmental pharmacokinetic parameters on Microsoft Excel spreadsheets. Comput Methods Programs Biomed 1996; 50(1): 43-52.
[http://dx.doi.org/10.1016/0169-2607(96)01730-0] [PMID: 8835839]

[47]   Haddad S, Pelekis M, Krishnan K. A methodology for solving physiologically based pharmacokinetic models without the use of simulation softwares. Toxicol Lett 1996; 85(2): 113-26.
[http://dx.doi.org/10.1016/0378-4274(96)03648-X] [PMID: 8650694]

[48]   Delboy H. A non-linear fitting program in pharmacokinetics with Microsoft® Excel spreadsheet. Int J Biomed Comput 1994; 37(1): 1-14.
[http://dx.doi.org/10.1016/0020-7101(94)90066-3] [PMID: 7896432]

[49]   Hussain M, Sahudin S, Yussof I. Exploring the use of computer-aided learning modules (CAL) to enhance the teaching and learning of pharmacokinetics to pharmacy students. J Young Pharm 2020; 12(4): 354-9.
[http://dx.doi.org/10.5530/jyp.2020.12.91]

[50]   Dash S, Murthy PN, Nath L, Chowdhury P. Kinetic modeling on drug release from controlled drug delivery systems. Acta Pol Pharm 2010; 67(3): 217-23.
[PMID: 20524422]

[51]   Ruiz-Garcia A, Bermejo M, Moss A, Casabo VG. Pharmacokinetics in drug discovery. J Pharm Sci 2008; 97(2): 654-90.
[http://dx.doi.org/10.1002/jps.21009] [PMID: 17630642]

[52]   Chhabra G, Chuttani K, Mishra AK, Pathak K. Design and development of nanoemulsion drug delivery system of amlodipine besilate for improvement of oral bioavailability. Drug Dev Ind Pharm 2011; 37(8): 907-16.
[http://dx.doi.org/10.3109/03639045.2010.550050] [PMID: 21401341]

[53]   Jain SK, Agrawal GP, Jain NK. Evaluation of porous carrier-based floating orlistat microspheres for gastric delivery. AAPS PharmSciTech 2006; 7(4): E54-62.
[http://dx.doi.org/10.1208/pt070490] [PMID: 17233542]

[54]   Jagdale SC, Patil S, Kuchekar BS. Application of design of experiment for floating drug delivery of tapentadol hydrochloride. Comput Math Methods Med 2013.
[http://dx.doi.org/10.1155/2013/625729]

[55]   Thube MW, Shahi SR, Gulecha B. Formulation development and evaluation of extended release tablets of pentoxifylline. Int J Pharm Res Dev 2010; 2(4): 1-1.

[56]   Ayyoubi S, Cerda JR, Fernández-García R, *et al.* 3D printed spherical mini-tablets: Geometry *versus*

composition effects in controlling dissolution from personalised solid dosage forms. Int J Pharm 2021; 597: 120336.
[http://dx.doi.org/10.1016/j.ijpharm.2021.120336] [PMID: 33545280]

[57]   Ekenna IC, Abali SO. Comparison of the Use of Kinetic Model Plots and DD Solver Software to Evaluate the Drug Release from Griseofulvin Tablets. J Drug Deliv Ther 2022; 12(2-S): 5-13.
[http://dx.doi.org/10.22270/jddt.v12i2-S.5402]

[58]   Shaikh HK, Kshirsagar R, Patil S. Mathematical models for drug release characterization: A review. World J Pharm Pharm Sci 2015; 4(4): 324-38.

[59]   British Pharmacopoeia Commission. British Pharmacopoeia. London: The Stationery Office 2009; Vol. III.

[60]   Akhtar M, Ahmad M, Khan SA, Murtaza G. Novel modified release tableted microspheres of ibuprofen and misoprostol in a combined formulation: Use of software DDSolver. Afr J Pharm Pharmacol 2012; 6(36): 2613-20.
[http://dx.doi.org/10.5897/AJPP12.349]

[61]   Siswanto A, Fudholi A, Nugroho AK, Martono S. *In vitro* release modeling of aspirin floating tablets using DDSolver. Indones J Pharm 2015; 26(2): 94.
[http://dx.doi.org/10.14499/indonesianjpharm26iss2pp94]

[62]   Colombo P, Santi P, Bettini R, Brazel CS, Peppas NA. Drug release from swelling-controlled system. In: Wise DL, Ed. Handbook of Pharmaceutical Controlled Release Technology. 2nd ed. New York: Marcel Dekker 2000; pp. 183-205.

[63]   Dave BS, Amin AF, Patel MM. Gastroretentive drug delivery system of ranitidine hydrochloride: Formulation and *in vitro* evaluation. AAPS PharmSciTech 2004; 5(2): 77-82.
[http://dx.doi.org/10.1208/pt050234] [PMID: 15760092]

[64]   Davies P. Oral solid dosage form. In: Gibson M, Ed. Pharmaceutical Preformulation and Formulation. 2nd ed. New York: Informa Healthcare 2009; p. 389.

[65]   Fudholi A. Methodology formulation in direct compression. Medika 1983; 7(9): 586-93.

[66]   Sabitha Ananthi D. Formulation and evaluation of modified release oral solid dosage form (mr-osdf) of vildagliptin using carbopol 71g-nf polymer for the treatment of type-2 diabetes mellitus.

[67]   Rajput A, Sevalkar G, Pardeshi K, Pingale P. Computational nanoscience and technology. OpenNano 2023; 100147.

[68]   Zhang Y, Huo M, Zhou J, *et al.* DDSolver: An add-in program for modeling and comparison of drug dissolution profiles. AAPS J 2010; 12(3): 263-71.
[http://dx.doi.org/10.1208/s12248-010-9185-1] [PMID: 20373062]

[69]   Zuo J, Gao Y, Bou-Chacra N, Löbenberg R. Evaluation of the DDSolver software applications. BioMed Res Int 2014; 1-9.
[http://dx.doi.org/10.1155/2014/204925] [PMID: 24877067]

[70]   Johnson KC. Dissolution and absorption modeling: Model expansion to simulate the effects of precipitation, water absorption, longitudinally changing intestinal permeability, and controlled release on drug absorption. Drug Developm Indust Pharma 2003; 29(8): 833-42.
[http://dx.doi.org/10.1081/DDC-120024179]

[71]   Kharia AA, Hiremath SN, Singhai AK, Omray LK, Jain SK. Design and optimization of floating drug delivery system of acyclovir. Indian J Pharm Sci 2010; 72(5): 599-606.
[http://dx.doi.org/10.4103/0250-474X.78527] [PMID: 21694992]

[72]   Tiwari SB, Murthy TK, Raveendra Pai M, Mehta PR, Chowdary PB. Controlled release formulation of tramadol hydrochloride using hydrophilic and hydrophobic matrix system. AAPS PharmSciTech 2003; 4(3): 18-23.
[http://dx.doi.org/10.1208/pt040331] [PMID: 14621963]

[73]   Reddy KR, Mutalik S, Reddy S, Raju PS, Diwan PV. Once-daily sustained-release matrix tablets of nicorandil: Formulation and *in vitro* evaluation. AAPS PharmSciTech 2003; 4(4): 480-8.
        [http://dx.doi.org/10.1208/pt040461] [PMID: 15198556]

[74]   Pharmaceutical Preformulation Services information from Ricerca Chemical Development. Available from : https://olonricerca.com/wp-content/uploads/2020/01/Pharmaceutical-Preformulation-Servi-es-Olon-RB.pdf

[75]   Lachman L, Lieberman L, Kanig J. The theory and practice of industrial pharmacy. Philadelphia, PA: Lea and Febiger. 1986; pp. 66-99.

[76]   Longer MN, Ching HS, Robinson JR. A mechanistic investigation of the controlled release of macromolecules from polyanhydride matrices. J Pharm Sci 1985; 74(4): 406-11.
        [http://dx.doi.org/10.1002/jps.2600740408] [PMID: 3999000]

[77]   Park K, Robinson JR. Bioadhesive polymers as platforms for oral-controlled drug delivery: Method to study bioadhesion. Int J Pharm 1984; 19(2): 107-27.
        [http://dx.doi.org/10.1016/0378-5173(84)90154-6]

[78]   Manwar J, Kumbhar DD, Bakal R, Baviskar S, Manmode R. Response surface based co-optimization of release kinetics and mucoadhesive strength for an oral mucoadhesive tablet of cefixime trihydrate. Bull Fac Pharm Cairo Univ 2016; 54(2): 227-35.
        [http://dx.doi.org/10.1016/j.bfopcu.2016.06.004]

[79]   Mauger JW, Chilko D, Howard S. On the analysis of dissolution data. Drug Dev Ind Pharm 1986; 12(7): 969-92.
        [http://dx.doi.org/10.3109/03639048609048052]

[80]   Polli JE, Rekhi GS, Augsburger LL, Shah VP. Methods to compare dissolution profiles and a rationale for wide dissolution specifications for metoprolol tartrate tablets. J Pharm Sci 1997; 86(6): 690-700.
        [http://dx.doi.org/10.1021/js960473x] [PMID: 9188051]

[81]   Shah VP, Lesko LJ, Fan J, *et al.* FDA guidance for industry 1 dissolution testing of immediate release solid oral dosage forms. Dissolut Technol 1997; 4(4): 15-22.
        [http://dx.doi.org/10.14227/DT040497P15]

[82]   Costa P. An alternative method to the evaluation of similarity factor in dissolution testing. Int J Pharm 2001; 220(1-2): 77-83.
        [http://dx.doi.org/10.1016/S0378-5173(01)00651-2] [PMID: 11376969]

[83]   Rohrs BR. Dissolution method development for poorly soluble compounds. Dissol Technol 2001; 8(3): 6-12.

[84]   Mendyk A, Jachowicz R, Fijorek K, Dorożyński P, Kulinowski P, Polak S. KinetDS: An open source software for dissolution test data analysis. Dissolut Technol 2012; 19(1): 6-11.
        [http://dx.doi.org/10.14227/DT190112P6]

[85]   Patel N, Chotai N, Patel J, Soni T, Desai J, Patel R. Comparison of *in vitro* dissolution profiles of oxcarbazepine-HP b-CD tablet formulations with marketed oxcarbazepine tablets. Dissolut Technol 2008; 15(4): 28-34.
        [http://dx.doi.org/10.14227/DT150408P28]

[86]   Kulinowski P, Malczewski P, Pesta E, *et al.* Selective laser sintering (SLS) technique for pharmaceutical applications—Development of high dose controlled release printlets. Addit Manuf 2021; 38: 101761.
        [http://dx.doi.org/10.1016/j.addma.2020.101761]

[87]   Kulinowski P, Hudy W, Mendyk A, *et al.* The relationship between the evolution of an internal structure and drug dissolution from controlled-release matrix tablets. AAPS PharmSciTech 2016; 17(3): 735-42.
        [http://dx.doi.org/10.1208/s12249-015-0402-1] [PMID: 26335419]

[88]   Huanbutta K, Sangnim T. Design and development of zero-order drug release gastroretentive floating tablets fabricated by 3D printing technology. J Drug Deliv Sci Technol 2019; 52: 831-7.
[http://dx.doi.org/10.1016/j.jddst.2019.06.004]

[89]   Endashaw E, Tatiparthi R, Mohammed T, Tefera YM, Teshome H, Duguma M. Dissolution profile evaluation of seven brands of amoxicillin-clavulanate potassium 625 mg tablets retailed in hawassa town, sidama regional state, ethiopia. Pharmacology 2021; 107(9): 431-40.

[90]   Antosik-Rogóż A, Szafraniec-Szczęsny J, Knapik-Kowalczuk J, *et al.* How does long-term storage influence the physical stability and dissolution of bicalutamide from solid dispersions and minitablets? Processes 2022; 10(5): 1002.
[http://dx.doi.org/10.3390/pr10051002]

[91]   Brouers F, Al-Musawib TJ. The use of the fractal Brouers-Sotolongo formalism to analyze the kinetics of drug release. arXiv:190701540 2019.

[92]   Shah VA, Rathod DN, Basuri T, Modi VS, Parmar IJ. Applications of bioinformatics in pharmaceutical product designing: A review. World J Pharm Pharm Sci 2015; 4: 477-93.

[93]   Plock N, Buerger C, Joukhadar C, Kljucar S, Kloft C. Does linezolid inhibit its own metabolism? Population pharmacokinetics as a tool to explain the observed nonlinearity in both healthy volunteers and septic patients. Drug Metab Dispos 2007; 35(10): 1816-23.
[http://dx.doi.org/10.1124/dmd.106.013755] [PMID: 17639029]

[94]   Moore JW, Flanner HH. Mathematical comparison of dissolution profiles. Pharm Technol 1996; 20: 64-74.

[95]   Brown W, Perivilli S, Podolsky D, Stippler ES, Walfish S. The critical role of the USP Performance Verification Test in dissolution testing and qualification of the paddle apparatus. Dissolut Technol 2019; 26(1): 6-12.
[http://dx.doi.org/10.14227/DT260119P6]

[96]   Elsayed I, Abdelbary AA, Elshafeey AH. Nanosizing of a poorly soluble drug: Technique optimization, factorial analysis, and pharmacokinetic study in healthy human volunteers. Int J Nanomedicine 2014; 9: 2943-53.
[PMID: 24971006]

CHAPTER 12

Warp and Woof of Drug Designing and Development: An In-Silico Approach

Monika Chauhan[1], Vikas Gupta[2], Anchal Arora[3], Gunpreet Kaur[2], Parveen Bansal[2] and Ravinder Sharma[4,*]

[1] *School of Health Sciences and Technology, UPES, Dehradun, India*

[2] *University Center of Excellence in Research, BFUHS, Faridkot, India*

[3] *All India Institute of Medical Sciences, Bathinda, India*

[4] *University Institute of Pharmaceutical Sciences and Research, BFUHS, Faridkot, India*

Abstract: Designing and developing a novel therapeutic drug candidate remains a daunting task and requires a long time with an investment of approximately ~USD 2-3 billion. Owing to the subpar pharmacokinetic or toxicity profiles of the therapeutic candidates, only one molecule enters the market over a period of 12 to 24 years. So, the reduction of cost, time, high attrition rate in the clinical phase, or drug failure has become a challenging and dire question in front of the pharmaceutical industry. In the last few decades, steep advancements in artificial intelligence, especially computer-aided drug design have emerged with robust and swift drug-designing tools. Existing reports have clearly indicated an imperative and successful adoption of virtual screening in drug design and optimization. In parallel, advanced bioinformatics integrated into genomics and proteomics discovering molecular signatures of disease based on target identification or signaling cascades has directly or indirectly smoothened the roadmap of the clinical trial. Integrated genomics, proteomics, and bioinformatics have produced potent new strategies for addressing several biochemical challenges and generating new approaches that define new biological products. Therefore, it is fruitful to utilize the computational-based high throughput screening methods to overcome the hurdles in drug discovery and characterize ventures. Besides that, bioinformatic analysis speed up drug target selection, drug candidate screening, and refinement, but it can also assist in characterizing side effects and predicting drug resistance. In this chapter, the authors have discussed a snapshot of State-of-the-Art technologies in drug designing and development.

Keywords: Artificial intelligence, Computer aided drug designing, Genomics, Pharmacokinetics, Proteomics.

---

[*] **Corresponding author Ravinder Sharma:** University Institute of Pharmaceutical Sciences and Research, BFUHS, Faridkot, India; E-mail: ravindersharma7@yahoo.com

Dilpreet Singh and Prashant Tiwari (Eds.)

1. INTRODUCTION

Drug discovery process, it is not shocking that the period from lead to the clinical candidate is sometimes called as "valley of death" [1]. The process of drug discovery is an expensive, time-taking process and the success rate is too low [2]. However, the lack of artificial intelligence limits the drug discovery and development process, by making it a tedious, costly, time-consuming, and challenging task in front of the pharmaceutical industry [3, 4]. Despite the implementation of successful strategies, 90% of drugs fail in the later stages of the drug development process due to a lack of optimal pharmacokinetic profile [5]. This urged and suggested the need to gear up with advanced *in silico* approaches to expedite the drug discovery process [6].

Computer-based applications can recognize hit to lead molecules, genomics, proteomics, and new biology involved in the genesis of disease. Further, it allows the improved and quicker validation and optimization of drug candidates and receptors [7]. Today, the complete adoption of *in silico* approaches in R&D has shifted the paradigm in drug discovery and, ultimately clinical development [8]. Herein, in this chapter, authors have tried to highlight commonly used software for new drug development along with their applications.

1.1. Drug Design

Inventing a new drug molecule remains a daunting task and requires a long time with an investment of ~USD 2-3 billion [9]. Integration of artificial intelligence with high throughput experimentation (Fig. **1**) has fastened up the research process by serving advanced applications such as PubChem (millions of small molecules) [10].

ZINC database (annotating millions of commercially available molecules) [11]; WOMBAT (claiming bioactivity information for molecules reported in medicinal chemistry journals) [12]; MDDR (candidate molecules under development) [13]; 3D MIND (Ligand-receptor interaction & cancer cell line screen data) *etc* [14, 15].

2. STRUCTURE-BASED DRUG DESIGN

Structure-based drug design (SBDD) relies on the information and availability of target structure and is based on the hypothesis that a molecule should interact with a receptor or target protein to exert a desirable therapeutic effect [16]. Since 1932, advancements in X-ray diffraction unveiled the chemical composition and 3D structure of organic molecules [17]. During the same time, immense growth in X-ray and NMR technology led to the determination of crystallographic structures of the proteins and successfully elucidated the first (1st) structure of protein *i.e.* myoglobin [18]. Further, the establishment of the National Institute of Health US, National Library in the 1980s, attracted considerable attention toward the target structure determination. Crystallographic structures of the proteins are being reported in the databases like protein data bank (PDB) and in the past few decades, immense growth has been observed with a plethora (≈203084) of known proteins structure starting from seven (7) proteins in 1971 [19, 20]. The availability of high-quality three-dimensional (3D) structures of protein not only allows the researcher to understand the physiology and pathology of disease, but also fastened up the process of screening, and opens the door for rational-based drug discovery approaches. Captopril an angiotensin-converting enzyme inhibitor, in the 1980s, was the first successful outcome of target-based drug discovery [21]. Further, Viracept (nelfinavir mesylate; in 1997), the first ever completely designed HIV protease inhibitor using its known target structure [22, 23], sparked and facilitated the young minds in adopting these technologies for the development of new drug molecules. Some of the drug-target repositories and prediction tools are shown in Fig. (**2**).

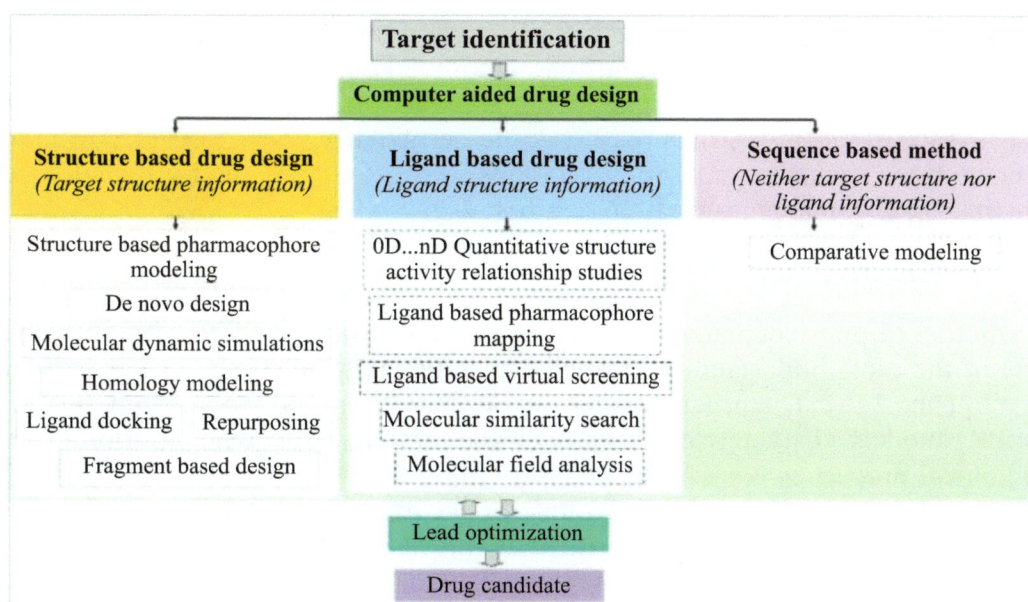

Fig. (1). High throughput computing tools in drug design.

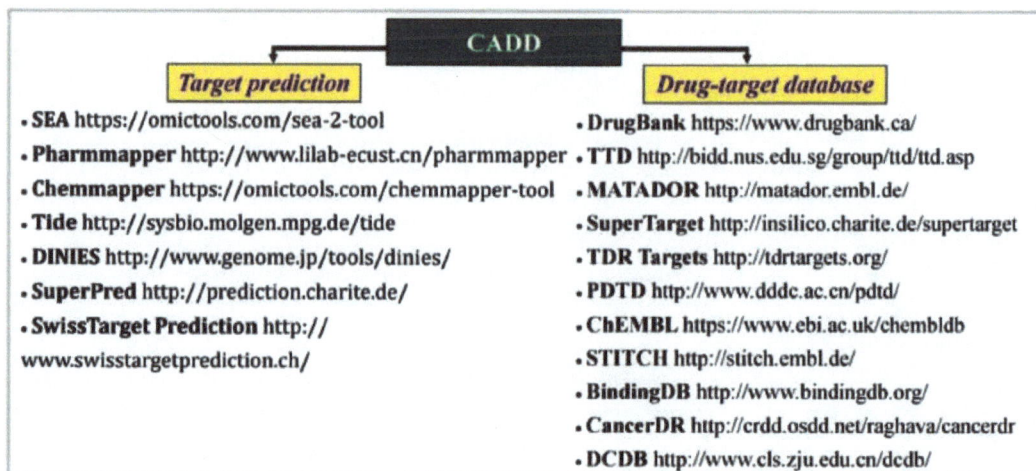

Fig. (2). Drug-target repositories and prediction tools.

3. LIGAND-BASED DRUG DESIGN

On the other hand, ligand-based drug design (LBDD) uses the existing information of the known bioactive molecule against the particular target, to predict the behaviour of compounds with improved biological attributes [24]. This method has become a key computational approach to promote drug designing in the absence of receptor structure [25]. More particularly, this approach covers the quantitative structure–activity relationships (QSAR) and ligand-based pharmacophore mapping as discussed below.

3.1. Quantitative Structure-activity Relationship

Prediction of physicochemical properties or chemical bioactivities has remained one of the most important applications of statistically relational-based artificial intelligence [26]. Quantitative structure-activity/property relationship (QSAR/QSPR) models characterize the associations among molecular descriptors and biological properties [27]. More precisely, models are obtained in the form of a mathematical equation on an account of a correlation between the chemical properties of the dataset and experimental values. In the 1980s, the First (1st) 3D structural information was introduced into the QSAR study, since then continuous development from 0D to nD QSAR has swiftly secured its valuable position in drug designing [28]. These models play a significant role in the prediction, and design of a new molecule that can be skipped or prioritized during lead identification or optimization [29].

3.2. Pharmacophore Modelling

The International Union of Pure and Applied Chemistry (IUPAC) defined pharmacophore as an "ensemble of steric and electronic features that is necessary to ensure the optimal supramolecular interactions with a specific biological target and to trigger its biological response" [30]. On the interaction of a ligand molecule with its receptor, it produces geometrically, energetically matched active and stable conformation with the binding cavity to produce pharmacological effect [31].

Medicinal chemist strategies and observations found that different descriptors or functional groups in drug molecules have a different biological response, and their replacement with some other groups or bio-isosteres have a great influence on binding interaction with receptor [32, 33]. Therefore, in 1909, Ehrlich was credited with the concept of "privileged structures" *i.e.* pharmacophore. Gund *et al.* 1977 first updated the view of pharmacophore and developed the first graph theory-based software for pharmacophore *i.e.* 2D structure searching method that could be transformed into a 3D structure [34]. Two principle methods including structure-based pharmacophore modeling (receptor target is known) and ligand-based pharmacophore modeling (target is unknown) are used for the identification and determination of pharmacophores [35]. Structure-based pharmacophore modelling has become a convenient approach when target receptor is known. Hypothetically designed molecules or possible pharmacophore structures can be screened to identify the hit molecules, by analysing their best fit with the binding site of receptor. The advent of docking software and the availability of large number of databases or huge chemical libraries have opened the door to perform pharmacophore modelling (*e.g.* MODELER, SWISS-MODEL) and facilitated the search at a reasonable cost and time [36, 37].

On the other hand, ligand information-based pharmacophore modelling, a key computational approach can be used to promote and guide drug development when the target receptor structure is not available [38]. This approach is highly influenced by the selection of dataset, and the protocol is summarized by means of low energy conformer generation, model development and the validation. With this advancing technology, several software *e.g.* ChemAxon"s PF, CCG"s GpiDAPH3 fingerprints are able to screen million compounds from the database and have become one of the important means to drug discovery [39, 40].

4. VIRTUAL SCREENING

Virtual screening, (machine learning technique) uses *in silico* based algorithmic programming to search potential hits from chemical databases; this offers a precedence advantage over physical screening [41 - 43]. Further, a brief overview is exemplified in Fig. (**3**).

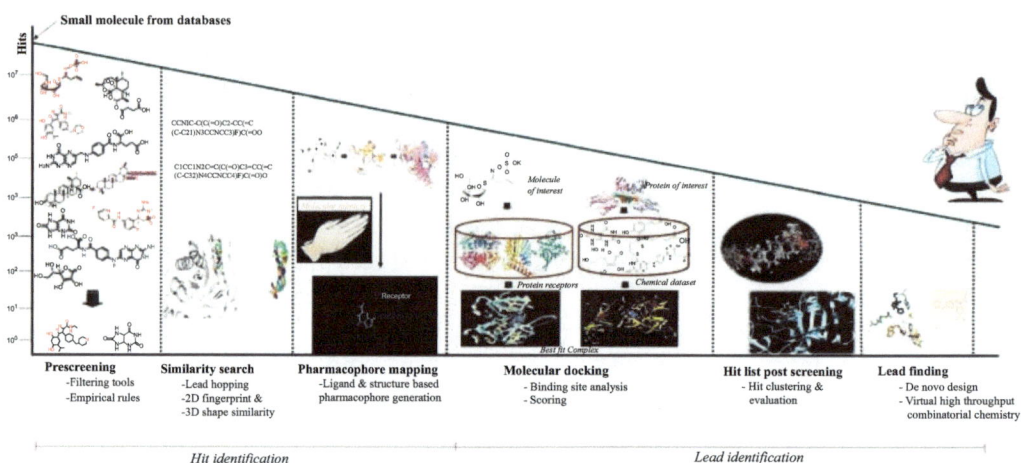

Fig. (3). Virtual screening in hit to lead identification.

4.1. Molecular Docking

Hunting a lead molecule is a long, tedious and arduous task and fortunately, *in silico* tools with advanced algorithms playing a pivotal role in rationalizing the drug discovery path [44]. Of all the techniques, undoubtedly molecular docking remains one of the holy grails in the modern scenario of drug discovery campaign. For molecular docking, simulation-based algorithms serve to identify potential ligands by predicting and evaluating the preferred binding mode and finally putative energetic binding affinities between receptor and small molecules or proteins in order to define a stable complex [45, 46]. Ligand receptor interaction studies are based on the hypothesis of "induced fit theory" and rely on the spatial conformation of ligands, energy matching and stability of the complex [47]. In docking analysis determination of correct binding conformation *i.e.* ligand pose of a molecule and receptor complex remain the key point for understating their affinity and mechanism of action [48, 49]. Binding sites on the surface of proteins using a multi-scale algorithm was successfully determined and reported for the 1st time by Collins *et al.* and greatly promoted the development of molecular docking [50].

Molecular docking could be roughly categorized into rigid docking, semiflexible docking, and flexible docking. In the rigid docking, the structure of molecules remains static and does not undergo any change, and the method mainly involves the study of the degree of conformation with relatively simple calculations. Binding modes and interactions are further considered on the basis of static geometry, physical and chemical complementarity between the molecules. This fast and highly effective approach is used for studying macromolecular systems including protein–protein, and protein–nucleic acid systems or to screen a larger number of datasets from the libraries [51, 52].

On the other hand, semi-flexible docking, wherein the conformation of the ligand molecules freely changes, while the macro molecules' receptor protein remains static, except for the rotatable amino acids. The approach is found to be suitable to analyse the interaction between proteins and ligand molecules [53, 54]. Another side, in flexible docking, the simulated conformation of both the ligand and receptor are free to change. This provides more detailed information by consuming more computing resources and the approach is ameliorated with better accuracy [55]. A few of the examples of widely accepted docking software with the successful application in target identifications are FlexX: Pneumococcal peptidoglycan deacetylase inhibitors [56]; SEED: Plasmepsin receptor [57]; Glide: Cytochrome P450 inhibitors [58], Falciparum inhibitors [59]; Surflex DOCK: Oncogene in B cell lymphomas, *etc* [60].

4.2. ADMET Predictions

Drug development and discovery process have more and more relied on the pharmacokinetics profile of a molecule with respect to its movement and time span within the biological system and include key parameters like absorption, distribution, metabolism, and excretion (ADME), and toxicity (T) [61]. Enormous shreds of evidence indicate 80% of drug failure at the drug development phase is due to poor pharmacokinetics and toxicity along with the consumption of time and money [62]. This traditional system of drug discovery is summarized in Fig. (4).

This trend and information from failure influenced and made a strong bend of researchers and medicinal chemists towards early predictions of these parameters with the possibilities to reduce the late-stage attrition rate, thereby influencing the efficacy and potency of the molecule. Physicochemical descriptors of a molecule have a great influence on its pharmacokinetic profile [63, 64].

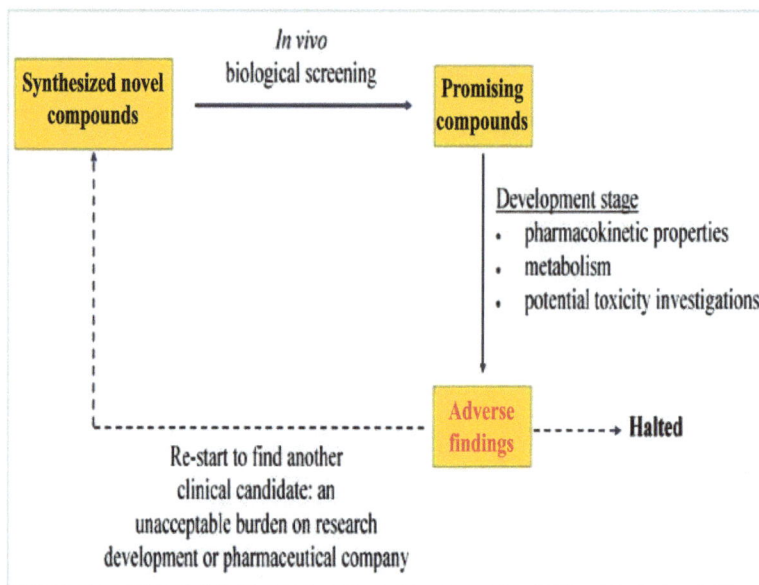

Fig. (4). The traditional system of drug discovery.

A deep understanding of relationships between descriptors or pharmacophores and their impact on the pharmacokinetic fate inside the body, coupled with *in silico*-based algorithms forms the basis for the prediction of the ADMET model. Successful adoption of *in silico*-based ADMET tools, with early-stage prediction, has allowed the scientists to focus on the compounds, considering the high probability to meet the requirement of pharmacokinetic profile and safety. These advanced technologies are contributing to accelerating the drug discovery process, and a few of the key players in ADMET software are summarized in Fig. (5) [65 - 67].

Schrodinger, being the most widely adopted software, integrates predictive physics-based simulation for various modeling, analysis, and computational tasks in drug design. It provides a variety of suits that facilitate the study of the chemical and biological structure, ligand-receptor interactions, molecular dynamics, and properties and reactivity of chemical systems (Figs. **6** and **7**).

Advanced features to accelerate the evaluation and optimization of drug molecules at lower cost are believed to have a higher success rate as compared to traditional methods [68, 69].

Company/ Institute	Product	URL
Accelrys	Cerius2, C2.ADME, Topkat	www.accelrys.com
Iconix	DrugMatrix	www.iconixpharm.com
Molecular Discovery	VolSurf	www.moldiscovery.com
Schrödinger	QikProp	www.schrodinger.com
Spotfire	Spotfire	www.spotfire.com
ZyxBio	OraSpotter	www.zyxbio.com
Cerep	Bioprint	www.cerep.fr
Cyprotex	Cloe PK	www.cyprotex.com
ZyxBio	OraSpotter	www.zyxbio.com

Fig. (5). Commercially available ADMET software.

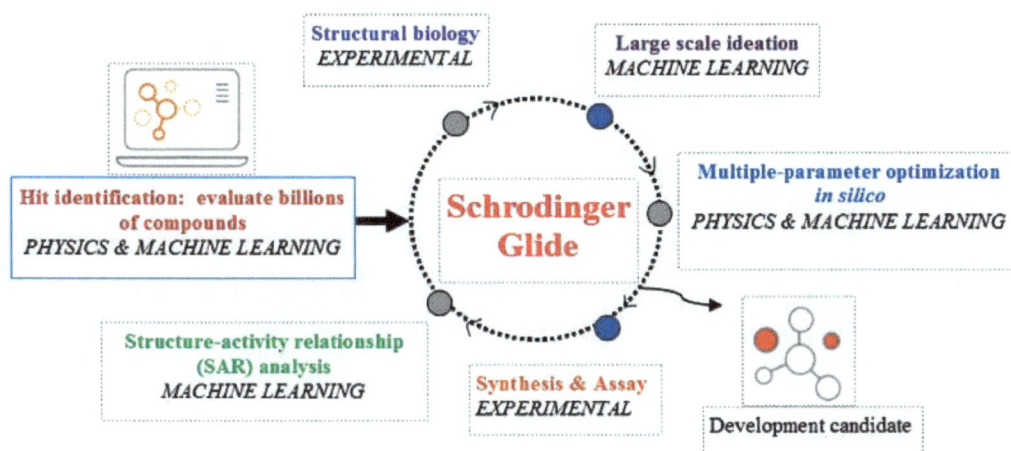

Fig. (6). Schrodinger platform for drug discovery and development.

Drug Discovery	Quantum Chemistry	Binding Site & Structure Analysis
- Faster lead discovery - Accurate property prediction - Large-scale molecule exploration- - Large-scale molecule evaluation -Integrated data management and visualization	- QM/MM simulation - Chemical isomer calculaion- - Chemical shift calculation - Chemical reaction & mechanism calculation	- Identify binding pockets - Perform hydration site analysis - Discover cryptic binding sites - Automate antibody modeling - Engineer and design proteins - Evaluate structures rapidly
Hit Identification & Virtual Screening	**Structure Enablement**	**Hit Discovery**
- Perform ligand preparation - Screen prepared compound libraries - Predict pKa using a knowledge-based system - Predict pKa using quantum mechanics	- Predict novel protein-ligand complexes. -Correct & assess structure to create reliable models - Create homology models for structure-based design - Refine ligand poses in Cryo-EM structures	-Screen using pharmacophores screen using 3D ligand shape - Dock novel ligands - Virtually screen billion+ libraries with docking and machine learning - Enrich hits from screens
Property Predictions		**Lead Optimization**
- Predict affinity, selectivity, and solubility - Create and apply expert-level QSAR & QSPA models - Predict membrane permeability - Predict ADME properties		- Interactive design and enumeration - Hit-to-lead library design - Lead optimization library design

Fig. (7). Applications of schrodinger platform.

CONCLUSION

Recent advancements in computational tools have revolutionized the use of *in silico* methods in the drug discovery and development process. In this book chapter, we have discussed about the various *in silico*-based software that are playing a major role in drug designing and drug discovery. Coupling sophisticated *in silico* approaches and high throughput technologies with a rational drug design process has become an essential tool for the development of new therapies. It is expected that the power of computer-aided drug design will grow as the technology continues to evolve.

REFERENCES

[1] Seyhan AA. Lost in translation: The valley of death across preclinical and clinical divide – identification of problems and overcoming obstacles. Transl Med Commun 2019; 4(1): 18.
[http://dx.doi.org/10.1186/s41231-019-0050-7]

[2] Mohs RC, Greig NH. Drug discovery and development: Role of basic biological research. Alzheimer's & Dementia: Translational Research & Clinical Interventions. 2017; 3(4): 651-7.
[http://dx.doi.org/10.1016/j.trci.2017.10.005]

[3] Mohs RC, Greig NH. Drug discovery and development: Role of basic biological research. Alzheimer's & Dementia: Translational Research & Clinical Interventions. 2017; 3(4): 651-7.
[http://dx.doi.org/10.1016/j.trci.2017.10.005]

[4] Paul D, Sanap G, Shenoy S, Kalyane D, Kalia K, Tekade RK. Artificial intelligence in drug discovery and development. Drug Discov Today 2021; 26(1): 80-93.
[http://dx.doi.org/10.1016/j.drudis.2020.10.010] [PMID: 33099022]

[5] Deore AB, Dhumane JR, Wagh R, Sonawane R. The stages of drug discovery and development process. Asian Journal of Pharmaceutical Research and Development 2019; 7(6): 62-7.
[http://dx.doi.org/10.22270/ajprd.v7i6.616]

[6] Fleming N. How artificial intelligence is changing drug discovery. Nature 2018; 557(7707): S55-7.
[http://dx.doi.org/10.1038/d41586-018-05267-x] [PMID: 29849160]

[7] Kiriiri GK, Njogu PM, Mwangi AN. Exploring different approaches to improve the success of drug discovery and development projects: A review. Future Journal of Pharmaceutical Sciences 2020; 6(1): 27.
[http://dx.doi.org/10.1186/s43094-020-00047-9]

[8] Vijayan RSK, Kihlberg J, Cross JB, Poongavanam V. Enhancing preclinical drug discovery with artificial intelligence. Drug Discov Today 2022; 27(4): 967-84.
[http://dx.doi.org/10.1016/j.drudis.2021.11.023] [PMID: 34838731]

[9] Drago T, Cahill K, Grealy A, Lucey K, Mahmoud M. Pharmaceutical company in-house research and licensing transaction review. Advanced Journal of Social Science 2021; 8(1): 77-85.
[http://dx.doi.org/10.21467/ajss.8.1.77-85]

[10] Kim S, Thiessen PA, Bolton EE, *et al.* PubChem substance and compound databases. Nucleic Acids Res 2016; 44(D1): D1202-13.
[http://dx.doi.org/10.1093/nar/gkv951] [PMID: 26400175]

[11] Irwin JJ, Sterling T, Mysinger MM, Bolstad ES, Coleman RG. ZINC: A free tool to discover chemistry for biology. J Chem Inf Model 2012; 52(7): 1757-68.
[http://dx.doi.org/10.1021/ci3001277] [PMID: 22587354]

[12] Hristozov DP, Oprea TI, Gasteiger J. Virtual screening applications: A study of ligand-based methods and different structure representations in four different scenarios. J Comput Aided Mol Des 2007; 21(10-11): 617-40.
[http://dx.doi.org/10.1007/s10822-007-9145-8] [PMID: 18008169]

[13] Ekins S, Mestres J, Testa B. *In silico* pharmacology for drug discovery: methods for virtual ligand screening and profiling. Br J Pharmacol 2007; 152(1): 9-20.
[http://dx.doi.org/10.1038/sj.bjp.0707305] [PMID: 17549047]

[14] Sliwoski G, Kothiwale S, Meiler J, Lowe EW. Computational methods in drug discovery. Pharmacological reviews. 2014; 66(1): 334-95.

[http://dx.doi.org/10.1124/pr.112.007336]

[15]   Mandal S, Moudgil M, Mandal SK. Rational drug design. Eur J Pharmacol 2009; 625(1-3): 90-100.
[http://dx.doi.org/10.1016/j.ejphar.2009.06.065] [PMID: 19835861]

[16]   Kolb P, Ferreira RS, Irwin JJ, Shoichet BK. Docking and chemoinformatic screens for new ligands and targets. Curr Opin Biotechnol 2009; 20(4): 429-36.
[http://dx.doi.org/10.1016/j.copbio.2009.08.003] [PMID: 19733475]

[17]   Garman EF. Developments in x-ray crystallographic structure determination of biological macromolecules. Science 2014; 343(6175): 1102-8.
[http://dx.doi.org/10.1126/science.1247829] [PMID: 24604194]

[18]   de Chadarevian S. John Kendrew and myoglobin: Protein structure determination in the 1950s. Protein Sci 2018; 27(6): 1136-43.
[http://dx.doi.org/10.1002/pro.3417] [PMID: 29607556]

[19]   Protein Data Bank. Available from: https://pdb101.rcsb.org/

[20]   Shekhar C. *In silico* pharmacology: Computer-aided methods could transform drug development. Chem Biol 2008; 15(5): 413-4.
[http://dx.doi.org/10.1016/j.chembiol.2008.05.001] [PMID: 18482690]

[21]   Anthony CS, Masuyer G, Sturrock ED, Acharya KR. Structure based drug design of angiotensin-I converting enzyme inhibitors. Curr Med Chem 2012; 19(6): 845-55.
[http://dx.doi.org/10.2174/092986712799034950] [PMID: 22214449]

[22]   Kaldor SW, Kalish VJ, Davies JF II, *et al.* Viracept (nelfinavir mesylate, AG1343): A potent, orally bioavailable inhibitor of HIV-1 protease. J Med Chem 1997; 40(24): 3979-85.
[http://dx.doi.org/10.1021/jm9704098] [PMID: 9397180]

[23]   Cui W, Aouidate A, Wang S, Yu Q, Li Y, Yuan S. Discovering anti-cancer drugs *via* computational methods. Front Pharmacol 2020; 11: 733-9.
[http://dx.doi.org/10.3389/fphar.2020.00733] [PMID: 32508653]

[24]   Chen X, Yan CC, Zhang X, *et al.* Drug–target interaction prediction: Databases, web servers and computational models. Brief Bioinform 2016; 17(4): 696-712.
[http://dx.doi.org/10.1093/bib/bbv066] [PMID: 26283676]

[25]   Martin YC, Kofron JL, Traphagen LM. Do structurally similar molecules have similar biological activity? J Med Chem 2002; 45(19): 4350-8.
[http://dx.doi.org/10.1021/jm020155c] [PMID: 12213076]

[26]   Baum ZJ, Yu X, Ayala PY, Zhao Y, Watkins SP, Zhou Q. Artificial intelligence in chemistry: Current trends and future directions. Journal of Chemical Information and Modeling. 2021; 61(7): 3197-212.
[http://dx.doi.org/10.1021/acs.jcim.1c00619]

[27]   Neves BJ, Braga RC, Melo-Filho CC, Moreira-Filho JT, Muratov EN, Andrade CH. QSAR-based virtual screening: Advances and applications in drug discovery. Front Pharmacol 2018; 9: 1275-81.
[http://dx.doi.org/10.3389/fphar.2018.01275] [PMID: 30524275]

[28]   Lin X, Li X, Lin X. A review on applications of computational methods in drug screening and design. Molecules 2020; 25(6): 1375-89.
[http://dx.doi.org/10.3390/molecules25061375] [PMID: 32197324]

[29]   Muratov EN, Bajorath J, Sheridan RP, *et al.* QSAR without borders. Chem Soc Rev 2020; 49(11): 3525-64.
[http://dx.doi.org/10.1039/D0CS00098A] [PMID: 32356548]

[30]   Wermuth CG, Ganellin CR, Lindberg P, Mitscher LA. Glossary of terms used in medicinal chemistry (IUPAC Recommendations 1998). Pure Appl Chem 1998; 70(5): 1129-43.
[http://dx.doi.org/10.1351/pac199870051129]

[31]   Seidel T, Bryant SD, Ibis G, Poli G, Langer T. 3D pharmacophore modeling techniques in computer-

aided molecular design using ligandScout. Tutorials Chemoinformatics 2017; 28(2): 279-309.
[http://dx.doi.org/10.1002/9781119161110.ch20]

[32]   Chang C, Ekins S, Bahadduri P, Swaan PW. Pharmacophore-based discovery of ligands for drug transporters. Adv Drug Deliv Rev 2006; 58(12-13): 1431-50.
[http://dx.doi.org/10.1016/j.addr.2006.09.006] [PMID: 17097188]

[33]   Ehrlich P. Über den jetzigen Stand der Chemotherapie. Ber Dtsch Chem Ges 1909; 42(1): 17-47.
[http://dx.doi.org/10.1002/cber.19090420105]

[34]   Gund P. Three-dimensional pharmacophoric pattern searching. Prog Mol Subcell Biol 1977; 5: 117-43.
[http://dx.doi.org/10.1007/978-3-642-66626-1_4]

[35]   Pirhadi S, Shiri F, Ghasemi JB. Methods and applications of structure based pharmacophores in drug discovery. Curr Top Med Chem 2013; 13(9): 1036-47.
[http://dx.doi.org/10.2174/1568026611313090006] [PMID: 23651482]

[36]   Valasani KR, Vangavaragu JR, Day VW, Yan SS. Structure based design, synthesis, pharmacophore modeling, virtual screening, and molecular docking studies for identification of novel cyclophilin D inhibitors. J Chem Inf Model 2014; 54(3): 902-12.
[http://dx.doi.org/10.1021/ci5000196] [PMID: 24555519]

[37]   Eswar N, Webb B, Marti-Renom MA, et al. Current protocols in protein science. Curr Protoc Protein Sci. 2007.

[38]   Schwede T, Kopp J, Guex N, Peitsch MC. SWISS-MODEL: An automated protein homology-modeling server. Nucleic Acids Res 2003; 31(13): 3381-5.
[http://dx.doi.org/10.1093/nar/gkg520] [PMID: 12824332]

[39]   Lu X, Yang H, Chen Y, et al. The development of pharmacophore modeling: Generation and recent applications in drug discovery. Curr Pharm Des 2018; 24(29): 3424-39.
[http://dx.doi.org/10.2174/1381612824666180810162944] [PMID: 30101699]

[40]   Chiang YK, Kuo CC, Wu YS, et al. Generation of ligand-based pharmacophore model and virtual screening for identification of novel tubulin inhibitors with potent anticancer activity. J Med Chem 2009; 52(14): 4221-33.
[http://dx.doi.org/10.1021/jm801649y] [PMID: 19507860]

[41]   Katsila T, Spyroulias GA, Patrinos GP, Matsoukas MT. Computational approaches in target identification and drug discovery. Comput Struct Biotechnol J 2016; 14: 177-84.
[http://dx.doi.org/10.1016/j.csbj.2016.04.004] [PMID: 27293534]

[42]   Schneider G. Virtual screening: An endless staircase? Nat Rev Drug Discov 2010; 9(4): 273-6.
[http://dx.doi.org/10.1038/nrd3139] [PMID: 20357802]

[43]   Leelananda SP, Lindert S. Computational methods in drug discovery. Beilstein J Org Chem 2016; 12: 2694-718.
[http://dx.doi.org/10.3762/bjoc.12.267] [PMID: 28144341]

[44]   Prada-Gracia D, Huerta-Yepez S, Moreno-Vargas LM. Application of computational methods for anticancer drug discovery, design, and optimization. Bol Med Hosp Infant Mex Engl Ed 2016; 73(6): 411-23.

[45]   Brogi S, Ramalho TC, Kuca K, Medina-Franco JL, Valko M. In silico methods for drug design and discovery. Front Chem 2020; 8: 612.
[http://dx.doi.org/10.3389/fchem.2020.00612] [PMID: 32850641]

[46]   Pinzi L, Rastelli G. Molecular docking: Shifting paradigms in drug discovery. Int J Mol Sci 2019; 20(18): 4331-9.
[http://dx.doi.org/10.3390/ijms20184331] [PMID: 31487867]

[47]   Ferreira L, dos Santos R, Oliva G, Andricopulo A. Molecular docking and structure-based drug design

strategies. Molecules 2015; 20(7): 13384-421.
[http://dx.doi.org/10.3390/molecules200713384] [PMID: 26205061]

[48]   Sherman W, Day T, Jacobson MP, Friesner RA, Farid R. Novel procedure for modeling ligand/receptor induced fit effects. J Med Chem 2006; 49(2): 534-53.
[http://dx.doi.org/10.1021/jm050540c] [PMID: 16420040]

[49]   Coleman RG, Carchia M, Sterling T, Irwin JJ, Shoichet BK. Ligand pose and orientational sampling in molecular docking. PLoS One 2013; 8(10): e75992.
[http://dx.doi.org/10.1371/journal.pone.0075992] [PMID: 24098414]

[50]   Kitchen DB, Decornez H, Furr JR, Bajorath J. Docking and scoring in virtual screening for drug discovery: Methods and applications. Nat Rev Drug Discov 2004; 3(11): 935-49.
[http://dx.doi.org/10.1038/nrd1549] [PMID: 15520816]

[51]   Collins JG, Shields TP, Barton JK. 1H-NMR of Rh (NH3) 4phi^{3+} bound to d (TGGCCA) 2: Classical intercalation by a nonclassical octahedral metallointercalator. J Am Chem Soc 1994; 116(22): 9840-6.
[http://dx.doi.org/10.1021/ja00101a004]

[52]   Matsuzaki Y, Uchikoga N, Ohue M, Akiyama Y. Rigid-docking approaches to explore protein–protein interaction space. Netw Biol 2017; pp. 33-55.

[53]   Sauton N, Lagorce D, Villoutreix BO, Miteva MA. MS-DOCK: Accurate multiple conformation generator and rigid docking protocol for multi-step virtual ligand screening. BMC Bioinformatics 2008; 9(1): 184.
[http://dx.doi.org/10.1186/1471-2105-9-184] [PMID: 18402678]

[54]   Salmaso V, Moro S. Bridging molecular docking to molecular dynamics in exploring ligand-protein recognition process: An overview. Front Pharmacol 2018; 9: 923-7.
[http://dx.doi.org/10.3389/fphar.2018.00923] [PMID: 30186166]

[55]   Wong CF. Flexible receptor docking for drug discovery. Expert Opin Drug Discov 2015; 10(11): 1189-200.
[http://dx.doi.org/10.1517/17460441.2015.1078308] [PMID: 26313123]

[56]   Rarey M, Kramer B, Lengauer T, Klebe G. A fast flexible docking method using an incremental construction algorithm. J Mol Biol 1996; 261(3): 470-89.
[http://dx.doi.org/10.1006/jmbi.1996.0477] [PMID: 8780787]

[57]   Bui NK, Turk S, Buckenmaier S, *et al.* Development of screening assays and discovery of initial inhibitors of pneumococcal peptidoglycan deacetylase PgdA. Biochem Pharmacol 2011; 82(1): 43-52.
[http://dx.doi.org/10.1016/j.bcp.2011.03.028] [PMID: 21501597]

[58]   Friedman R, Caflisch A. Discovery of plasmepsin inhibitors by fragment-based docking and consensus scoring. ChemMedChem 2009; 4(8): 1317-26.
[http://dx.doi.org/10.1002/cmdc.200900078] [PMID: 19472268]

[59]   Caporuscio F, Rastelli G, Imbriano C, Del Rio A. Structure-based design of potent aromatase inhibitors by high-throughput docking. J Med Chem 2011; 54(12): 4006-17.
[http://dx.doi.org/10.1021/jm2000689] [PMID: 21604760]

[60]   Shah F, Wu Y, Gut J, *et al.* Design, synthesis and biological evaluation of novel benzothiazole and triazole analogs as falcipain inhibitors. MedChemComm 2011; 2(12): 1201-7.
[http://dx.doi.org/10.1039/c1md00129a]

[61]   Cerchietti LC, Ghetu AF, Zhu X, *et al.* A small-molecule inhibitor of BCL6 kills DLBCL cells *in vitro* and *in vivo*. Cancer Cell 2010; 17(4): 400-11.
[http://dx.doi.org/10.1016/j.ccr.2009.12.050] [PMID: 20385364]

[62]   Dunnington K, Benrimoh N, Brandquist C, Cardillo-Marricco N, Di Spirito M, Grenier J. Application of pharmacokinetics in early drug development. Faqi AS, Ed Pharmacokinetics and Adverse Effects of Drugs-Mechanisms and Risk Factors. Intech Open 2018.
[http://dx.doi.org/10.5772/intechopen.74189]

[63]   Faqi AS, Ed. A comprehensive guide to toxicology in preclinical drug development. Academic Press 2012.

[64]   van de Waterbeemd H, Gifford E. ADMET *in silico* modelling: Towards prediction paradise? Nat Rev Drug Discov 2003; 2(3): 192-204.
[http://dx.doi.org/10.1038/nrd1032] [PMID: 12612645]

[65]   van de Waterbeemd H. High-throughput and *in silico* techniques in drug metabolism and pharmacokinetics. Curr Opin Drug Discov Devel 2002; 5(1): 33-43.
[PMID: 11865671]

[66]   Han Y, Zhang J, Hu CQ, Zhang X, Ma B, Zhang P. In silico ADME and toxicity prediction of ceftazidime and its impurities. Front Pharmacol 2019; 10: 434-9.
[http://dx.doi.org/10.3389/fphar.2019.00434] [PMID: 31068821]

[67]   Göller AH, Kuhnke L, Montanari F, *et al.* Bayer's *in silico* ADMET platform: A journey of machine learning over the past two decades. Drug Discov Today 2020; 25(9): 1702-9.
[http://dx.doi.org/10.1016/j.drudis.2020.07.001] [PMID: 32652309]

[68]   Shou WZ. Current status and future directions of high-throughput ADME screening in drug discovery. J Pharm Anal 2020; 10(3): 201-8.
[http://dx.doi.org/10.1016/j.jpha.2020.05.004] [PMID: 32612866]

[69]   Cox PB, Gupta R. Contemporary computational applications and tools in drug discovery. ACS Med Chem Lett 2022; 13(7): 1016-29.
[http://dx.doi.org/10.1021/acsmedchemlett.1c00662] [PMID: 35859884]

<h1>CHAPTER 13</h1>

Data Interpretation and Management Tools for Application in Pharmaceutical Research

Arvinder Kaur[1,*], Avichal Kumar[1], Kavya Manjunath[2], Deepa Bagur Paramesh[1], Shilpa Murthy[3] and Anjali Sinha[1]

[1] *Department of Pharmaceutics, KLE College of Pharmacy, Constituent Unit of KLE Academy of Higher Education and Research (Deemed to be University), Rajajinagar, Bengaluru-560010, Karnataka, India*

[2] *Department of Pharmacology, KLE College of Pharmacy, Constituent Unit of KLE Academy of Higher Education and Research (Deemed to be University), Rajajinagar, Bengaluru-560010, Karnataka, India*

[3] *Department of Pharmaceutical Chemistry, KLE College of Pharmacy, Constituent Unit of KLE Academy of Higher Education and Research (Deemed to be University), Rajajinagar, Bengaluru-560010, Karnataka, India*

Abstract: The information flow in pharmaceutical research before data interpretation and management was largely manual and simple, with limited application of technology. Establishing the research objective, designing the study, collecting data, analyzing data, and interpreting the result were laborious, tedious, and time-consuming processes. Manually entering and sorting a large amount of data made researchers more prone to human errors, leading to incorrect and invalid results. The chapter draws on data mining, data abstracting, and intelligent data analysis to collectively improve the quality of drug discovery and delivery methods. To develop new drugs and improve existing treatments, software can be used to analyze large datasets and identify patterns that help understand how drugs interact with the body. Virtual models of organs and cells are employed to study the effects of drugs, automate drug testing, and predict adverse drug reactions. Pharmaceutical management tools, such as pharmacy management software, electronic prescription software, inventory management software, and automated dispensing systems, are highly valuable for managing inventory, tracking patient prescriptions, monitoring drug interactions, maintaining patient information and history, and providing up-to-date drug information. The main objective of this chapter is to highlight the various tools and software solutions available and how they can facilitate the research process to ensure compliance with relevant regulations and laws regarding human healthcare safety.

*** Corresponding author Arvinder Kaur:** Department of Pharmaceutics, KLE College of Pharmacy, Constituent Unit of KLE Academy of Higher Education and Research (Deemed to be University), Rajajinagar, Bengaluru-560010, Karnataka, India; E-mail: kaurarvinder@outlook.com

Dilpreet Singh and Prashant Tiwari (Eds.)

Keywords: Automated dispensing system, Electronic prescription software, Intelligent data analysis, Inventory management software, Pharmaceutical management tools, Virtual model.

1. INTRODUCTION

Pharmaceutical research is a crucial field of study because it involves the constant development of new medicines and treatments for a wide range of medical conditions. This process demands extensive experimentation and analysis to identify novel potential treatments and assess their safety and efficacy [1].

These tools are intended to boost research productivity and success, producing more trustworthy and accurate results. This has led to an increased reliance on cutting-edge technology and software in pharmaceutical research, allowing scientists to study complex biological systems and create new treatments and pharmaceuticals more swiftly than before. This essay will discuss some of the relevance.

Interpreting data is an essential part of pharmaceutical research and is necessary for drawing valid conclusions about the development, efficacy, and safety of medications. The increasing volume of data generated by contemporary pharmacological research has made data interpretation more challenging [2]. It involves combining data from various sources to make informed judgments that can guide the creation of novel therapies and drugs. Effective data interpretation in the pharmacy sector requires a combination of specialized expertise, scientific acumen, and subject knowledge. Additionally, examining complex data sets and drawing conclusions necessitate the use of sophisticated software and statistical techniques. The importance of accurate data comprehension in this context cannot be overstated since it enables scientists to make decisions regarding patient care, public policy, and the course of action based on readily available facts. The highest level of patient care and the financial sustainability of healthcare organizations depend on the efficient management of pharmaceutical resources, including medications, equipment, and personnel. Pharmacies employ various management technologies, such as staff scheduling software, performance monitoring tools, and inventory management systems to achieve these objectives [3]. These tools assist pharmacists in streamlining their processes, reducing waste, and improving patient outcomes. This article outlines the frequently used management tools in pharmacies and describes their applications in diverse pharmacy contexts. It also examines the advantages and challenges associated with implementing these innovations, highlighting the importance of thorough planning and research when introducing new administrative systems. This study underscores the significance of management tools in optimizing pharmaceutical operations and enhancing patient care.

2. HISTORY

Data management and storage require specialized equipment. Various methodologies have been employed throughout the history of medical research, and contemporary technology has also simplified the management of complex data. Edward H. Shortliffe examines the importance of striking a successful balance between modern thinking methods and advancements in information infrastructure, adherence to international standards, and education [4].

2.1. Conventional Approach

In the past, research in medical science relied solely on manual data entry and data screening methods. Since 1993, research data and information processing have undergone significant advancements. In fact, the data collected and exchanged across various research laboratories and healthcare facilities are evolving rapidly. What used to be isolated databases or laboratory information systems are now integrated into departmental, hospital, community, and research-based medical information systems. With the increasing volume of data, extracting relevant information for decision support has become increasingly challenging. Traditional manual data analysis is no longer sufficient, necessitating the development of efficient computer-based analysis techniques. Two notable examples of these techniques are data abstraction and data mining [5].

2.2. Computer and Software Resolutions

Artificial intelligence serves as a reminder that computers in medicine are by no means a recent development. Numerous industries use computers in similar ways, and few technical advances have been explicitly sparked by what can be called "business computing" in the medical field. It doesn't appear realistic that these computerized medical applications will "reshape" medicine, nevertheless. Computers haven't significantly changed the kind of decisions made or how they are made during the previous 20 years.

We recognize that the lack of valid perspectives on enhancing executives' critical thinking abilities can significantly exacerbate this issue [6]. Similarly, during that time, most medical business computers had a minimal impact on the work of doctors. A secondary, less common use of computers in medicine was focused on the content rather than the format of healthcare. Computers were primarily tasked with maintaining medical records, laboratory data, clinical trial information, and other data, provided they could effectively manage billing records. Furthermore, if beneficial for data storage, a computer should also assist in data analysis, organization, and retrieval. To date, this second category of medical computing has been primarily addressed through the use of decision theory, comparison of new cases with extensive databases of past cases, and the utilization of flowcharts or clinical algorithms as the three primary approaches.

3. TYPES OF RESEARCH DATA

There are two main types of research data: qualitative data and quantitative data.

3.1. Qualitative Data

Qualitative data are non-numerical and descriptive in nature. This type of data is often gathered through methods such as interviews, focus groups, observations, and other approaches that involve direct interaction with participants [7]. Qualitative data are used to gain a comprehensive understanding of participants' experiences, attitudes, and behaviors. Examples of qualitative data include interview transcripts, observational notes, and field notes.

3.2. Quantitative Data

Quantitative data are numerical and can be measured and statistically analyzed. This type of data is often collected through methods such as surveys, experiments, and other approaches that involve quantifying variables of interest [8]. Quantitative data can be utilized to identify patterns and relationships among variables and to test hypotheses. Examples of quantitative data include survey responses, test scores, and measurements such as percentage and weight.

Another kind of data that can be used in research is mixed-methods data, which is data from both qualitative and quantitative sources. These are the most typical types of research data. Two additional categories of data exist: secondary data, which can be used in research, and meta-data, which is information about data that can be used to understand how research studies were conducted and how data were gathered and analyzed. Information that has been proactively gathered for one more purpose and can be used in research is considered optional information. For a better understanding of research data management, the five major types of research data are as follows:

3.3. Observational Data

This type of data is collected through the observation of people, objects, or phenomena without any form of manipulation or intervention. Various techniques, including direct observation, video recordings, and other methods, can be employed to gather observational data [9].

3.4. Experimental Data

Experiments are conducted to gather this type of data, involving the manipulation of one or more variables and the observation of how these changes impact other variables. Experimental data are usually collected in laboratory settings, but they can also be conducted outdoors [10].

3.5. Survey Data

Surveys are employed to collect this type of data by presenting individuals with a series of questions concerning their beliefs, attitudes, behaviors, or experiences. Survey data can be obtained through various means, including online surveys, in-person interviews, and telephone surveys [11].

3.6. Secondary Data

This type of data is collected by third parties for purposes unrelated to the researcher's current study. Secondary data can be obtained from a range of sources, including academic institutions, non-profit organizations, and government agencies [12].

3.7. Meta-Data

This type of data offers information about other types of data, including details on how the data was collected, analyzed, and the conclusions drawn from it. Metadata can be utilized to assist in the interpretation of research findings and assess the quality of research studies [13].

4. DATA ANALYSIS

Recent papers on data analysis, knowledge discovery, and machine learning, including conference proceedings and journals, frequently include statements of this nature. They all share a common concern: how to "make sense" of the vast volumes of data that have been generated, particularly over the past several years [14]. The outcomes of computer-based analysis need to be effectively communicated to users in a comprehensible format. In this context, analysis tools must produce clear and unambiguous results and, in most cases, enable human involvement in the analysis process. Real-world examples of such approaches include representative AI computations that establish an iconic model, such as a decision tree or a set of rules, with a preference for minimal complexity while maintaining high simplicity and accuracy.

4.1. Intelligent Data Analysis

We must offer a clear definition of what we mean by the term "intelligent data analysis" and clarify its connection to "knowledge discovery in databases" and "data mining." The stages in knowledge discovery in databases (KDD) are often depicted as a systematic process [15]:

• Comprehending the subject matter.

• Creating the dataset and cleaning the data.

• Uncovering data-hidden regularities to create patterns, rules, and other forms of knowledge.

• Data mining (DM), postprocessing of discovered knowledge.

• The utilization of the results is a term used to describe this stage within the broader KDD process.

Intelligent data analysis (IDA) is typically associated with KDD. The relationship between IDA, knowledge discovery in databases, and data mining is illustrated in Fig. (1).

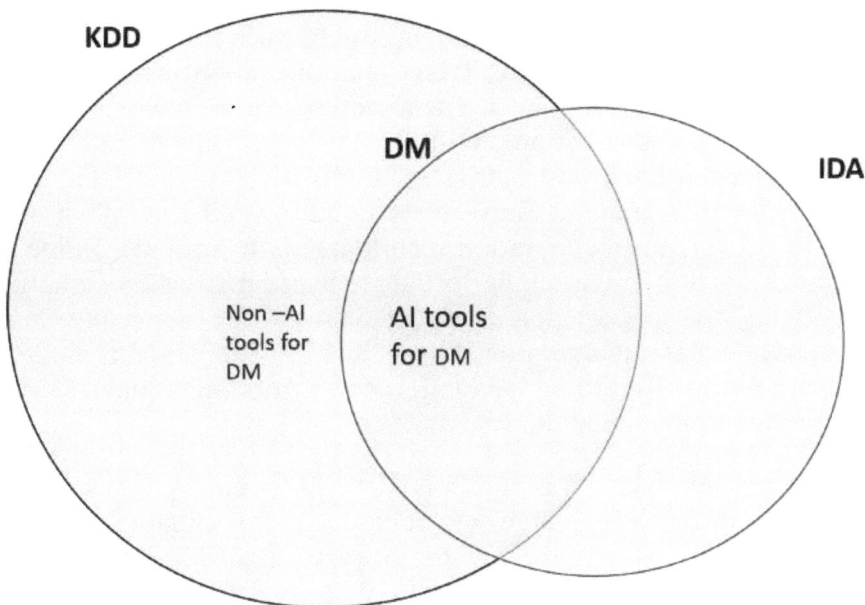

Fig. (1). Relationship of IDA, KDD and DM in data analysis.

KDD and IDA are intertwined in a significant way. The topic, data analysis, is shared by both fields, as are many of their methods. The main difference between IDA and KDD, as previously stated, is that IDA uses AI tools and methods, whereas KDD uses both AI and non-AI techniques. Another perspective involves the scale of data: KDD typically focuses on extracting information from extremely large datasets, whereas in IDA, datasets can be either large or moderately sized. This also influences the types of data mining tools used: in KDD, data mining tools are predominantly executed in the cluster mode (despite the entire KDD process being intelligent), whereas in IDA, the tools can be used either in a clustered fashion or as intelligent assistants [16].

Two primary perspectives define the significance and need for intelligent data analysis in medicine and pharmacology:

The first significant perspective revolves around the utilization of explicit knowledge-based critical thinking processes (such as inference, prediction, observation, *etc.*) through intelligent analysis of raw data from a specific patient, such as a time series of monitoring data. Much of this data consists of numerical values and is often noisy and incomplete. The objective is to dynamically extract meaningful abstractions, such as summaries of the patient's historical, current, and anticipated future conditions, which can be compared to relevant information (*e.g.*, diagnostic, prognostic, monitoring data, *etc.*). These data analysis techniques are referred to as information deliberation strategies, a term initially coined by Clancey in his now-classic proposal on heuristic classification [17]. In this proposal, these strategies constitute a crucial component of the reasoning process. Recent focus on data abstraction methods has been on the interpretation of temporal data (temporal data abstraction). The primary forms of such abstractions include temporal trends and more intricate temporal patterns. As the primary goal of (temporal) information deliberation techniques is to provide online decision support, their evaluation hinges on performance-based metrics. For example, does a method effectively support diagnostic and prognostic reasoning, and can it accurately predict a trend or value in the near future? In this context, data visualization is pivotal for aiding navigation and surprisingly plays a crucial role in effectively executing critical thinking tasks.

The second aspect involves the generation of novel medical and pharmacological knowledge by mining representative sets of example cases characterized by either numerical or symbolic descriptors. A majority of available datasets are either noisy (containing errors) or incomplete (with missing data). Data mining techniques are employed to extract symbolic knowledge that is both meaningful and comprehensible. The evaluation of these methods includes assessing their performance, the quality of the discovered information, and the significance and clarity of the extracted knowledge [18].

4.2. Data Mining

The objective of data mining techniques is to obtain information, preferably in a symbolic form that is meaningful and easy to comprehend. In this context, supervised symbolic learning techniques are commonly employed. For instance, there are powerful tools for inductive learning that can be used to generate meaningful diagnostic and prognostic rules. This framework also encompasses the learning of probabilistic causal networks and symbolic clustering.

Subsymbolic learning and case-based reasoning strategies can also be categorized within the field of data mining [19]. However, the contributions to data mining in this book are limited to symbolic and subsymbolic learning methods. By generalizing from multiple cases, their common objective is to discover the known 1-edge in symbolic or sub-symbolic forms. The representative learning approaches remembered from this book are isolated into propositional (characteristic-based) approaches, inductive rationale programming, and a deterioration way to deal with finding an idea order. The most extensive subset of symbolic learning approaches is the propositional learning approach. Within this subgroup, methods for learning rules from example cases and eliminating noise from example cases are further subdivided. Rules were initially introduced as a primary formalism for symbolically representing knowledge in knowledge-based systems, particularly expert systems. Despite their simplicity and lack of variables, propositional rules are regarded as sufficiently expressive for many real-world applications [20]. The quality of the processed case examples has a significant impact on the viability of rule-learning methods. Commotion in information is an undeniable peculiarity in numerous genuine spaces, in both medicine and pharmacology, noise is a factor that must be acknowledged and managed through computational means. There are two approaches to noise management. either eliminating the noise as a preprocessing step before learning the rules. The majority of the data mining contributions come from symbolic learning methods. Anyway, for a more adjusted image of the significant cutting edge, the incorporation of certain commitments from the sub-symbolic learning region was considered significant.

4.3. Data Abstracting

The identification of temporal abstractions is frequently carried out in a subtle or potentially distributed manner, and it is applicable to both discrete and continuous data streams. Trends, periodic events, and various types of temporal patterns represent valuable forms of temporal abstractions. Visualization techniques can also aid in revealing these temporal abstractions [15, 19]. The abstraction can be carried out on a single case, such as a single patient, or a collection of cases. The interpretation of patient data is influenced by relevant contexts in numerous medical fields. For example, the results of certain laboratory assessments can be interpreted differently in two scenarios: "the patient is not receiving any treatment" and "the patient is receiving treatment X" [21]. Regularly occurring events are crucial in medical practice, as symptoms can manifest in repetitive patterns. Periodic occurrences, by definition, involve composite events where successive components are presumed to remain dormant for a specified duration. In general, periodic events constitute a significant type of temporal abstraction. The benefits of distributed processing are well-recognized, and the advent of the Internet has facilitated the rapid growth of telematics tools and applications in handling data. Consequently, the approach to distributed temporal data abstraction aligns with these advancements [22].

An abstract case comprises a set of temporal abstractions derived from a specific case, such as a patient. By synthesizing numerous unique cases, new insights within the specific domain can be generated. From a temporal perspective, visualizing and clustering multiple concrete cases offer an efficient, user-friendly, and computationally economical method for identifying relevant temporal patterns in recent time intervals. These patterns can then be correlated across longer time windows to reveal more generalized temporal patterns. Fig. (**2**) illustrates the types and categories of intelligent data analysis.

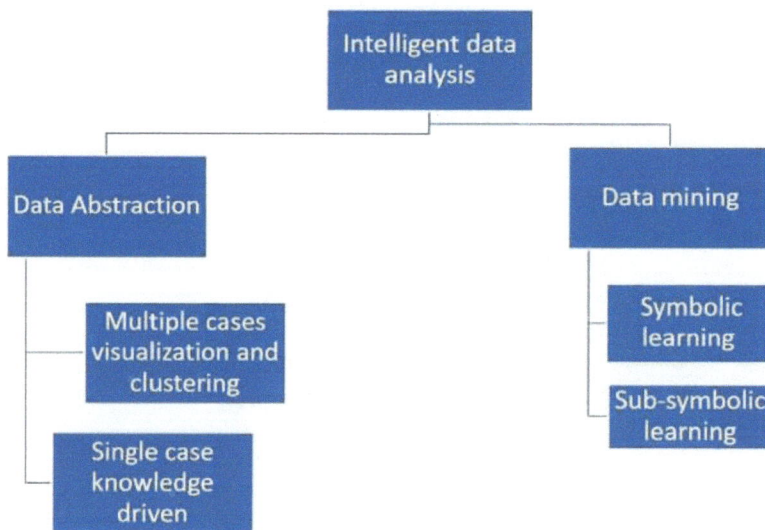

Fig. (2). Types of intelligent data analysis (IDA).

5. DATA REPRESENTATION

Vast amounts of data are available in nearly all fields of the sciences, yet the analysis of massive, integrated datasets remains challenging. Moreover, because the field of life sciences is highly fragmented, so too are its data and principles. Consequently, integrating data analysis and knowledge gathering across subfields becomes a formidable task. The simultaneous integration of various research perspectives and data types is crucial for achieving a comprehensive understanding of organismal complexity and biological processes [23].

Since the advent of artificial intelligence and the utilization of semantic structures for data representation, diagrams have been employed to formalize how human knowledge is represented. An increasing number of graph-based representation frameworks, such as labeled property graph databases like Neo4j and Resource Description Framework (RDF) triple stores, are known as "knowledge graphs" (KG). General-purpose KGs include Wikidata and the Google Knowledge Graph. In the biomedical sciences, examples include SemMedDB, Hetionet, Implicitome, Ruler Drive, the biological subset of Wikidata, Talked, and KGCOVID19 [24]. A graph with real-world items as nodes and known relationships between those entities as edges may be the most natural definition of KGs, although KGs have been defined in a variety of ways. A KG models the information or "realities" as proclamations, with every assertion being made out of two hubs associated by an edge that portrays their relationship. The statements may include additional features, metadata, and qualifying qualities that further define the properties of nodes and edges and the statement's meaning [25].

Information stored within a Knowledge Graph (KG) can be heterogeneous and diverse, yet it can still be represented in a structured manner within the graph due to the KG's fundamental architecture. The iterative and rapid development of KGs is made possible by representing data as clear connections between fundamental elements. Furthermore, using the graph data structure and a range of inference methods, it is possible to infer additional edges or connections between nodes within the network.

5.1. Software Applications

5.1.1. SPSS (Statistical Package for the Social Sciences)

The statistical software known as SPSS (Statistical Package for the Social Sciences) finds extensive usage in the social sciences, including pharmacy [26]. SPSS can serve various purposes in pharmacy, including but not limited to:

5.1.1.1. Statistical Descriptions

In pharmacy research, SPSS can be employed for data summarization and description. SPSS is capable of generating descriptive statistics, such as the mean, standard deviation, and frequency.

5.1.1.2. Statistical Inference

In pharmacy research, hypotheses can be tested with SPSS. SPSS can be used to carry out inferential statistics like t-tests, ANOVAs, and regression analyses.

5.1.1.3. Information Perception

Using SPSS, you can create graphs and charts that make it easier to comprehend the pharmacy research data. SPSS can generate histograms, scatter plots, and box plots.

5.1.1.4. Factor Examination

SPSS can be utilized to identify fundamental factors contributing to variations observed in pharmacy data. Factor analysis can assist in reducing data dimensionality and identifying significant variables.

5.1.1.5. Analysis of Clusters

With SPSS, it is possible to group similar observations based on their characteristics. For example, patient subgroups can be identified using cluster analysis, which considers their medication usage patterns.

5.1.1.6. Analyses of Survival

In pharmacy research, SPSS can be used to analyze time-to-event data. For instance, the time it takes for a patient to discontinue medication can be influenced by various factors, and survival analysis within SPSS can aid in identifying and understanding these factors.

In general, SPSS is a versatile tool for interpreting and analyzing data in pharmacy research.

5.1.2. GraphPad Prism

The statistical software package known as GraphPad Prism is commonly used in scientific research, particularly in the fields of biology, medicine, and pharmacy [27]. GraphPad Prism offers a range of methods, including but not limited to:

5.1.2.1. Analyses of Data

In pharmacy research, data can be analyzed using GraphPad Prism. This includes conducting descriptive statistics, such as calculating the mean and standard deviation, as well as inferential statistics, including t-tests, ANOVA, and regression analyses.

5.1.2.2. Information Perception

GraphPad Prism can be employed to create charts and graphs that aid in visualizing the data collected in pharmacy research. It offers the capability to generate various types of graphs, including scatterplots, bar graphs, and line graphs.

5.1.2.3. A Fitting Curve

Utilizing GraphPad Prism to fit curves to data can assist in identifying trends or patterns within the data. This encompasses fitting various types of curves, including linear, exponential, and sigmoidal curves, among others.

5.1.2.4. Portion Reaction Examination

GraphPad Prism can be used to analyze dose-response data, which is commonly collected in pharmacology research. It facilitates the generation of dose-response curves and the calculation of parameters like EC50 and IC50.

5.1.2.5. Modelling Statistical Data

GraphPad Prism can be utilized to construct statistical models that are employed for making predictions or testing hypotheses. This encompasses survival analysis, meta-analysis, and the development of both linear and nonlinear regression models.

Overall, GraphPad Prism stands out as a robust tool for analyzing and interpreting data in pharmacy research. Its user-friendly interface and comprehensive collection of statistical and graphical capabilities make it a valuable asset. Table **1** provides a list of other state-of-the-art software tools [28 - 34].

5.2. Implementation Application

Applications can play a significant role in enhancing patient care, optimizing processes, and increasing efficiency within the pharmacy industry, especially in drug management and distribution. Some examples of implementation applications in the field of pharmacy include:

Table 1. Latest software used for data handling of pharmaceutical research.

Sr. No.	Tool/Software	Description	Features
1	Microsoft Excel	A widely used spreadsheet program for organizing and analyzing data.	Data manipulation, formulae, charts/graphs, pivot tables, data validation.
2	Tableau	A business intelligence and data visualization tool for creating interactive dashboards and reports.	Data exploration, visualization, drag-and-drop interface, data blending, geospatial mapping.
3	R	A programming language and environment for statistical computing and graphics.	Data manipulation, statistical analysis, data visualization, machine learning, reproducible research.
4	Python	A general-purpose programming language with a wide range of libraries and frameworks for data analysis.	Data manipulation, statistical analysis, machine learning, data visualization, web scraping.
5	SAS	A suite of software tools for data management, analysis, and visualization.	Data cleaning, statistical analysis, machine learning, data visualization, business intelligence.
6	SPSS	A statistical software package for data analysis and visualization.	Descriptive statistics, inferential statistics, data visualization, predictive analytics, survey research.
7	Power BI	A business analytics service by Microsoft for creating interactive visualizations and reports.	Data exploration, visualization, data modelling, drag-and-drop interface, cloud-based.

5.2.1. EHRs, or Electronic Health Records

Electronic Health Records (EHRs) are digital repositories of patient health information that healthcare providers can access and update. EHRs contribute to improving patient safety by providing accurate and up-to-date medication records, as well as alerts for drug interactions and allergies [35].

5.2.2. Systems for Dispensing Medications

Automated medication dispensing systems can significantly enhance the efficiency of pharmacy operations while reducing medication errors. These systems have the capability to store and dispense medications in accordance with patient orders, in addition to monitoring inventory levels and expiration dates [36].

5.2.3. Information Systems for Pharmacies

Pharmacy information systems can support the management of pharmacy workflows, encompassing tasks like medication dispensing, inventory management, and billing. These systems can also generate reports and analyses to help identify trends and optimize pharmacy operations [37].

5.2.4. Apps for Mobile Devices

Patients can leverage mobile applications to access medication information, request prescription refills, and receive medication reminders. Furthermore, healthcare providers can use these applications to communicate with other providers and access patient health records [38].

5.2.5. Tele Pharmacy

Technology plays a vital role in tele-pharmacy, enabling services such as medication management and counseling to be provided remotely. This is particularly beneficial for patients residing in rural or underserved areas who may lack access to a nearby pharmacy [39].

In summary, the implementation of applications in the pharmacy setting can contribute to enhancing patient care, increasing efficiency, and reducing medication errors. These applications can improve various aspects of pharmacy operations, including medication dispensing, inventory management, and patient education.

5.3. Examples of Implementation

Several instances of implementations adopted in pharmacies have been widely embraced to improve patient outcomes and streamline pharmacy operations. Some examples include:

5.3.1. Electronic Recommending

Electronic prescribing, or e-prescribing, enables healthcare providers to electronically transmit prescriptions directly to the pharmacy [35]. This can help reduce medication errors and enhance patient safety by providing accurate and up-to-date medication information.

5.3.2. Automating Pharmacies

Automated medication dispensing cabinets and dispensing robots, for instance, can aid pharmacies in reducing medication errors, enhancing productivity, and conserving staff time [36].

5.3.3. The Clinical Choice was Emotionally Supportive Networks

Clinical decision support systems offer healthcare providers alerts, reminders, and real-time clinical information to enhance patient care. These systems may also provide drug information, dosage recommendations, and clinical guidelines, all aimed at improving patient outcomes [37].

5.3.4. Management of Medical Treatment

Medication therapy management (MTM) is a comprehensive approach to managing medication use and enhancing patient outcomes. It includes services such as medication reviews, drug therapy monitoring, patient education, and support for medication adherence, all aimed at optimizing patient care [38].

5.3.5. Purpose of Care Testing

Point-of-care testing is performed at the patient's location, such as in a pharmacy, and enables healthcare providers to conduct diagnostic tests. This approach, which leads to quicker treatment and faster results, can contribute to enhanced patient outcomes [39].

CONCLUSION

Since the turn of the century, pharmaceutical research has made substantial progress and has greatly contributed to society. However, the effective utilization of high-quality research data remains a significant challenge in both research and its application in the medical field. This article has explored both traditional and contemporary software developments in the field of medical data management. Such software can be valuable not only for managing research data but also for routine hospital operations. The historical section delved into the early manual data management practices and the subsequent evolution of software solutions. Additionally, various categories and subcategories of data types were explained. We aimed to encompass its implementation in both research and medical practices, emphasizing the proper utilization of data management software.

REFERENCES

[1]   Carracedo-Reboredo P, Liñares-Blanco J, Rodríguez-Fernández N, *et al*. A review on machine learning approaches and trends in drug discovery. Comput Struct Biotechnol J 2021; 19: 4538-58.

[http://dx.doi.org/10.1016/j.csbj.2021.08.011] [PMID: 34471498]

[2]   Okeh UM. Statistical problems in medical research. East Afr J Public Health 2009; 6(1) (Suppl.): 1-7.
       [PMID: 20088069]

[3]   Sohrabi B, Raeesi Vanani I, Nikaein N, Kakavand S. A predictive analytics of physicians prescription
       and pharmacies sales correlation using data mining. Int J Pharm Healthc Mark 2021; 13(3): 346-63.
       [http://dx.doi.org/10.1108/IJPHM-11-2017-0066]

[4]   Shortliffe EH. The adolescence of AI in Medicine: Will the field come of age in the '90s? Artif Intell
       Med 1993; 5(2): 93-106.
       [http://dx.doi.org/10.1016/0933-3657(93)90011-Q] [PMID: 8358494]

[5]   Khodke HE, Bhalerao M, Gunjal SN, Nirmal S, Gore S, Dange BJ. An Intelligent Approach to
       Empowering the Research of Biomedical Machine Learning in Medical Data Analysis using PALM.
       Int J Intell Syst Appl Eng 2023; 11(10s): 429-36.

[6]   Ngiam KY, Khor IW. Big data and machine learning algorithms for health-care delivery. Lancet
       Oncol 2019; 20(5): e262-73.
       [http://dx.doi.org/10.1016/S1470-2045(19)30149-4] [PMID: 31044724]

[7]   Bailey J. First steps in qualitative data analysis: Transcribing. Fam Pract 2008; 25(2): 127-31.
       [http://dx.doi.org/10.1093/fampra/cmn003] [PMID: 18304975]

[8]   Blaikie N. Analyzing quantitative data: From description to explanation. Sage 2003.

[9]   Kalincik T, Butzkueven H. Observational data: Understanding the real MS world. Mult Scler 2016;
       22(13): 1642-8.
       [http://dx.doi.org/10.1177/1352458516653667] [PMID: 27270498]

[10]   Mandel J. The statistical analysis of experimental data. Courier Corporation 2012.

[11]   Fink A. How to manage, analyze, and interpret survey data. Sage 2003.
       [http://dx.doi.org/10.4135/9781412984454]

[12]   Vartanian TP. Secondary data analysis. Oxford University Press 2010.
       [http://dx.doi.org/10.1093/acprof:oso/9780195388817.001.0001]

[13]   Castiello C, Castellano G, Fanelli AM. Second International Conference, MDAI Tsukuba, Japan.
       2005; pp. 457-68.

[14]   Patel V, Shah M. Artificial intelligence and machine learning in drug discovery and development.
       Intelligent Medicine 2022; 2(3): 134-40.
       [http://dx.doi.org/10.1016/j.imed.2021.10.001]

[15]   Fayyad U, Piatetsky-Shapiro G, Smyth P. The KDD process for extracting useful knowledge from
       volumes of data. Commun ACM 1996; 39(11): 27-34.
       [http://dx.doi.org/10.1145/240455.240464]

[16]   Greenhill AT, Edmunds BR. A primer of artificial intelligence in medicine. Techniques and
       Innovations in Gastrointestinal Endoscopy 2020; 22(2): 85-9.
       [http://dx.doi.org/10.1016/j.tgie.2019.150642]

[17]   Clancey WJ. Heuristic classification. Artif Intell 1985; 27(3): 289-350.
       [http://dx.doi.org/10.1016/0004-3702(85)90016-5]

[18]   Lavrač N, Keravnou ET, Zupan B. Intelligent data analysis in medicine and pharmacology: An
       overview. Springer 1997.
       [http://dx.doi.org/10.1007/978-1-4615-6059-3]

[19]   Piateski G, Frawley W. Knowledge discovery in databases. MIT press 1991.

[20]   Chang AC. Big data in medicine: The upcoming artificial intelligence. Prog Pediatr Cardiol 2016; 43:
       91-4.
       [http://dx.doi.org/10.1016/j.ppedcard.2016.08.021]

[21]   Wang L, Ding J, Pan L, Cao D, Jiang H, Ding X. Artificial intelligence facilitates drug design in the big data era. Chemom Intell Lab Syst 2019; 194: 103850.
[http://dx.doi.org/10.1016/j.chemolab.2019.103850]

[22]   Elbadawi M, Gaisford S, Basit AW. Advanced machine-learning techniques in drug discovery. Drug Discov Today 2021; 26(3): 769-77.
[http://dx.doi.org/10.1016/j.drudis.2020.12.003] [PMID: 33290820]

[23]   Pavel A, Saarimäki LA, Möbus L, Federico A, Serra A, Greco D. The potential of a data centred approach & knowledge graph data representation in chemical safety and drug design. Comput Struct Biotechnol J 2022; 20: 4837-49.
[http://dx.doi.org/10.1016/j.csbj.2022.08.061] [PMID: 36147662]

[24]   Hutchins W. The first decades of machine translation. Early years. Mach Transl 2000; 1: 1-15.

[25]   Hogan A, Blomqvist E, Cochez M, *et al.* Knowledge graphs. ACM Comput Surv 2022; 54(4): 1-37.
[http://dx.doi.org/10.1145/3447772]

[26]   Lu XN, Ma QG. Risk analysis in software development project with owners and contractors. In: IEEE International Engineering Management Conference 2004; 789-93.
[http://dx.doi.org/10.1109/IEMC.2004.1407488]

[27]   Swift ML. GraphPad prism, data analysis, and scientific graphing. J Chem Inf Comput Sci 1997; 37(2): 411-2.
[http://dx.doi.org/10.1021/ci960402j]

[28]   Carlberg C. Statistical analysis: Microsoft excel 2013. Que Publishing 2014.

[29]   Baader F, Sattler U. An overview of tableau algorithms for description logics. Stud Log 2001; 69(1): 5-40.
[http://dx.doi.org/10.1023/A:1013882326814]

[30]   Chambers JM. Software for data analysis: Programming with R. Springer 2008; Vol. 2.
[http://dx.doi.org/10.1007/978-0-387-75936-4]

[31]   Vallat R. Pingouin: Statistics in Python. J Open Source Softw 2018; 3(31): 1026.
[http://dx.doi.org/10.21105/joss.01026]

[32]   Rodriguez RN. Sas. Wiley Interdiscip Rev Comput Stat 2011; 3(1): 1-11.
[http://dx.doi.org/10.1002/wics.131]

[33]   Okagbue HI, Oguntunde PE, Obasi ECM, Akhmetshin EM. Trends and usage pattern of SPSS and Minitab Software in Scientific research. Journal of Physics: Conference Series. IOP Publishing 2021; p. 12017.
[http://dx.doi.org/10.1088/1742-6596/1734/1/012017]

[34]   Shaulska L, Yurchyshena L, Popovskyi Y. Using MS Power BI tools in the university management system to deepen the value proposition. In: 11th International Conference on Advanced Computer Information Technologies 2021; 294-8.
[http://dx.doi.org/10.1109/ACIT52158.2021.9548447]

[35]   Ammenwerth E, Hoerbst A. Electronic health records. A systematic review on quality requirements. Methods Inf Med 2010; 49(4): 320-36.
[http://dx.doi.org/10.3414/ME10-01-0038] [PMID: 20603687]

[36]   Anacleto TA, Perini E, Rosa MB, César CC. Medication errors and drug-dispensing systems in a hospital pharmacy. Clinics 2005; 60(4): 325-32.
[http://dx.doi.org/10.1590/S1807-59322005000400011] [PMID: 16138240]

[37]   Lin SJ. Access to community pharmacies by the elderly in Illinois: A geographic information systems analysis. J Med Syst 2004; 28(3): 301-9.
[http://dx.doi.org/10.1023/B:JOMS.0000032846.20676.94] [PMID: 15446619]

[38]   Ventola CL. Mobile devices and apps for health care professionals: Uses and benefits. P&T 2014; 39(5): 356-64.
[PMID: 24883008]

[39]   Baldoni S, Amenta F, Ricci G. Telepharmacy services: Present status and future perspectives: A review. Medicina 2019; 55(7): 327.
[http://dx.doi.org/10.3390/medicina55070327] [PMID: 31266263]

SUBJECT INDEX

A

Acid 89, 213
  amino 213
  hydrochloride 89
Activity 2, 5, 7, 8, 9, 110, 112, 113, 115, 116, 117, 120, 125, 126
  antitumor 115
  antiviral 126
Adoption, regulatory 198, 199
Adsorption, adherence 228
Alcohol 35, 236, 239
  polyvinyl 239
  vinyl 236
Algorithms 6, 8, 9, 12, 14, 15, 115, 118, 119, 120, 122, 125, 259
  genetic 12, 119
  geometric matching 12
Alzheimer's disease 2
Amphiphilic copolymer 242
Analysis 92, 93, 144, 182, 187, 280, 301
  bioinformatic 280
  chemical 93
  chromatographic 92
  computational 144
  intelligent 301
  network-based 187
  neurological 182
Analytical 83, 84, 85, 87, 98, 100, 101
  quality by design (AQbD) 83, 84, 85, 87, 101
  target profile (ATP) 84, 98, 100
ANOVA test 95
Anticancer 116, 118, 180
  agents 116, 118
  response 180
Antifungal agents 115
Applications 13, 15, 88, 178
  of QSAR modelling in drug discovery 13
  of virtual screening in drug discovery 15
  therapeutic 88, 178
AQbD method 86

Artificial intelligence 7, 14, 15, 124
  algorithms 124
  approaches 7, 14, 15
Atomistic simulation methods 229
ATP hydrolysis 137

B

Bayes' theorem 160
Bayesian 160, 161
  framework 160
  method 160
  pharmacokinetics 161
Behaviors 1, 155, 182, 227, 235, 237, 238, 240, 243, 264, 266, 298, 299
  clustering 240
  defining drug-induced patient 155
  drug's diffusion 266
  system's 227
  time-dependent growth 238
Bernstein polynomial basis functions 56
Bidirectional encoder representations 220
  from transformers (BERT) 220
Bimolecular molecule dynamics 62
Binary systems 242
Binding affinity 11, 12, 14, 15, 108, 116, 117, 119, 120, 123, 124, 126, 188, 285
  compound's 14
  computing 188
  protein-ligand 120
  putative energetic 285
Binding thermodynamics 11
Bioactive agents 182
Bioactivity measurements aids 212
Bioartificial organs 195
Bioavailability, oral 135
Biological 2, 7, 13, 14, 15, 32, 110, 111, 112, 114, 115, 117, 122, 126, 156, 197, 199, 304
  activity 2, 7, 13, 14, 15, 110, 111, 112, 114, 115, 117, 122, 126
  processes 156, 197, 199, 304

reactions 32
Box-Behnken design (BBD) 88, 89, 90, 95, 96, 97
Brain disorders 194
Breakdown process 38
Brunner film theory 34
Brunner's theory 34

C

Camellia sinensis 217
Cancer 89, 97, 116
  cell lung 116
  therapy 89
Carbamazepine 164
Chemical fragility 227
Chemometric 79, 144
  methodologies 79
  techniques 144
Chemotherapy 180
Chromatographic techniques 84
Chronic myeloid leukemia 119
Committee for control and supervision of experiments on animals (CPCSEA) 190
Comparative molecular similarity indices analysis (CoMSIA) 114, 115
Compositions, nanoparticle 94
Computational 1, 4, 6, 7, 9, 13, 14, 16, 17, 62, 109, 110, 116, 119, 120, 121, 126, 127, 137, 147, 161, 162, 185, 217, 250
  approaches 4, 6
  chemists 62
  data 9
  drug design 162
  hardware 6
  medicine 185
  methodologies 16, 161, 250
  methods 1, 4, 6, 7, 9, 14, 17, 116, 119, 120, 121, 126, 127
  modelling software 110
  techniques 13, 14, 109, 121, 137, 147, 217, 250
Computer(s) 8, 39, 46, 48, 120, 144, 190, 233, 280, 281, 297, 298
  aided drug design techniques 8
  based applications 281
  electronic 48
  modelling systems 190
  simulation techniques 233
  vision 46

Computing 3, 5, 298
  medical 298
  resources, high-performance 3, 5
Concentrations 193, 236, 239
  metabolite 193
  polymer 236, 239
Conformation 120, 113, 187, 284, 286
  bioactive 113
  stable 284
Consumer efforts 220
Covariance 48, 49, 52, 53, 58, 159
  analysis 159
  matrix 48, 58
  measures 52
Covariate dependencies 159
COVID-19 pandemic 186
Crank-Nicholson technique 232
Critical 68, 71, 72, 74, 76, 77, 78, 79, 80, 84, 85, 87, 91, 93, 98, 100
  material attributes (CMAs) 74, 76, 77, 78, 79, 85, 87, 93, 98
  method parameters (CMPs) 85, 87
  process variables (CPVs) 91
  quality attributes (CQAs) 68, 71, 72, 74, 76, 77, 78, 79, 80, 84, 98, 100
  quality parameters (CQPs) 91
  sample attributes (CSAs) 91
Cynomolgus monkeys 162
CYP 217
  inhibition 217
  isozyme 217

D

Data 46, 60, 214, 295, 297, 300, 301, 302, 307
  mining 60, 295, 297, 300, 302
  mining tools 301
  repository 214
  screening methods 297
  transformation 46
  visualization tool 307
Database(s) 121, 126, 207, 208, 210, 211, 212, 213, 214, 215, 216, 218, 219, 282, 284, 300
  chemical compound 121
  commercial 208
  massive 211
  of therapeutic targets 211
  rugged 216
  semantic scholar 214

Designed HIV protease inhibitor 282
Designing software 108
Devices, translation 61
Digital 188, 300
    breast tomosynthesis (DBT) 188
    mammography (DM) 188, 300
Diseases 61, 62, 180, 181, 182, 184, 185, 187,
        189, 192, 211, 213, 280, 281, 282
    cardiovascular 180
    heart 62
    ischemic heart 61
Dissipative particle dynamics (DPD) 228,
        229, 233, 238, 239
Docking techniques 11
Drug 27, 39, 68, 89, 153, 154, 155, 178, 180,
        195, 227, 228, 230, 233, 249
    diffusion coefficient 27, 249
    delivery systems (DDSs) 39, 68, 153, 154,
        155, 227, 228, 230, 233
    loading capacity 89
    metabolism 178, 195
    molecular signalling 180
Drug development 2, 9, 11, 12, 13, 14, 15, 87,
        117, 146, 147, 153, 154, 155, 164, 165,
        183, 184, 186, 189, 193, 250, 281, 286
    and discovery process 286
    process 2, 9, 11, 14, 117, 154, 155, 165,
        250, 281
    research 87
Drug discovery 1, 4, 5, 6, 10, 11, 12, 13, 15,
        16, 17, 55, 60, 108, 109, 110, 115, 116,
        118, 119, 120, 126, 127, 138, 139, 140,
        146, 177, 281, 285, 288, 289
    and biomedical data 55
    and development 6, 16, 108, 127, 138, 139,
        146, 177, 288
    campaign 285
    effective 10
    industry 140
    process 6, 11, 13, 15, 17, 108, 109, 110,
        116, 118, 120, 126, 281
Drug dissolution 39, 258, 259, 260
    data 39
    profile analysis 260
    profiles 258, 259
Dysfunction, intestinal 179, 183

E

Electronegativity 13

Electronic 62, 295, 296, 307
    cardiac pacemakers 62
    health records (EHRs) 307
    prescription software 295, 296
EMA's recommendation 32
Embryonic stem cells (ESCs) 179, 180, 194
Environmental stress 179
Enzymes 118, 123, 124
    histone deacetylase 124
    integrase 118
    protein kinase 123
Epidermal growth factor receptor (EGFR) 116
European 32, 70, 99, 198
    chemicals agency (ECHA) 198
    medicines agency (EMA) 32, 70, 99
Expression 154, 194
    transporter 154
    tyrosine hydroxylase 194

F

Factors 48, 71, 73, 74, 84, 85, 86, 88, 89, 90,
        94, 136, 146, 182, 192, 198, 305
    environmental 192
    impacting drug permeability 146
    microenvironmental 182
Failure mode effect analysis (FMEA) 70, 71,
        78, 82, 98
Fault tree analysis (FTA) 82
Food-and-mouth disease 98
Force 5, 9, 27, 120, 187
    fields 5, 9, 120, 187
    movement 27
Freeze 75, 76, 77, 78, 79
    dried products 76, 77, 78, 79
    drying process 75, 76
Frequency, contaminant transport 28
Functional data analysis (FDA) 55, 69, 70, 73,
        99, 158, 163, 164, 180, 189

G

GastroPlus software 190
Gaussian chain density function 232
Glycosidases 187
Green 94, 95, 96
    nanotechnology 94
    synthesis process 95
    synthesis reaction 96

Growth-factor signals 180

H

Healthcare 72, 208, 297, 309
   facilities 297
   providers alerts 309
   sector 72
   system, contemporary 208
Herb-drug interactions stocking 217
Herbal 91, 92, 215, 216, 220
   chemical marker ranking system 92
   components 216
   drug interactions 215
   medicine products 91, 92
   medicines 220
High-performance 3, 54, 84, 93
   computer resources 3
   liquid chromatography (HPLC) 54, 84, 93
HPLC method 85
HPTLC method 86
Human 118, 119, 194
   disease tissues 194
   immunodeficiency virus (HIV) 118, 119
Hybrid approaches 112
Hydrogen 114, 124, 139, 141
   bond donors 114, 124, 141
   bonding 139
Hydrophilic blocks 243
Hydrophobic blocks 243
Hydrophobicity 114, 115, 139, 233
Hygroscopicity 81

I

IAM chromatography 139
ILC 140
   column 140
   technique 140
Immobilized 137, 138, 139, 140
   artificial membrane (IAM) 137, 138, 139
   liposome chromatography (ILC) 137, 138, 140
Immunoassays 98
Immunotherapy 180
Intelligent data analysis (IDA) 295, 296, 300, 301, 303
Interactions 7, 114, 139, 163, 164, 179, 211, 236, 238

chemical-protein 211
crucial 7
drug-disease 163, 164
drug-membrane 139
electrostatic 114
hydrophobic 238
oxygen-sensitive bacterial 179
protein-substance 211
thermodynamic 236
International 178, 284
   union of pure and applied chemistry (IUPAC) 284
   workgroup of genetic toxicology (IWGT) 178
Inventory management systems 296
Iron precursors 95, 96

K

Key performance indicators (KPIs) 71, 72
Korsmeyer-Peppas 39, 40
   equations 40
   relationship 39

L

Laboratory information management system (LIMS) 272
Laser-induced breakdown spectroscopy (LIBS) 46
Lexicomp drug interactions (LDI) 218
Ligand 5, 7, 8, 9, 281, 283, 287
   based drug design (LBDD) 5, 7, 8, 9, 283
   receptor interaction 281, 287
Lipoprotein 61
Liposome(s) 88, 89, 140
   properties 89
   thermosensitive 89
Liquid chromatography 54, 84, 93, 141
   high-performance 54, 84, 93
Lyomesophase 242
Lyotropic mesophase 242

M

Machine learning 3, 7, 14, 15, 51, 58, 108, 115, 120, 124, 125, 299, 307
   algorithms 3, 115, 120, 125
   approaches 120

methods 51
Management 116, 309
  inflammation 116
  medication therapy 309
Managing assay transfers 98
Mass 46, 156, 172
  balance principles 156
  poisoning 172
  spectrometry 46
MATLAB software 40
Mean 39, 142, 267
  absolute error (MAE) 142
  dissolution time (MDT) 39, 267
Medication(s) 15, 26, 28, 38, 91, 92, 135, 136,
    137, 155, 162, 163, 173, 192, 193, 194,
    213, 217, 219, 296, 308, 309
  antihypertensive 162
  botanical 92
  combination database 213
  delivery systems 155
  herbal 91
  licensed 15
  ligand 217
  permeability 136
  therapy management (MTM) 309
  transportation 137
Methods 17, 75, 86, 93, 189, 303
  chromatographic 86
  chromatography 93
  computational-based 189
  computer-based 17
  economical 303
  freeze-drying 75
Microbiological systems 194
Microcrystalline cellulose 93
Microfluidic devices 171, 195, 196, 197
Mobile devices 308
Molecular 120, 144, 235
  access system (MACCS) 144
  dynamic simulation (MDS) 120, 235
  fingerprinting 144
Monoclonal antibodies 163, 181
Monte Carlo simulations 87, 93

N

Nanoscale drug delivery systems 241
Natural language processing (NLP) 214, 219
Near-infrared spectroscopy 25, 46
NMR technology 281

Noyes-Whitney 27, 34, 40
  equation 27, 34, 40
  Rule 34
Nuclear magnetic resonance (NMR) 46

P

Parallel artificial membrane permeability
    (PAMP) 136, 138, 178
Parkinson's disease 2
Process analytical technologies (PAT) 72, 74,
    79, 100
Production processes, contemporary 100
Products, biotechnological 97
Prostate cancer 217
Protein(s) 187, 282
  data bank (PDB) 282
  polymer-binding 187

Q

Quantitative structure 1, 2, 9, 13, 17, 110, 114,
    115, 116, 127, 138, 140, 141, 143, 144,
    145, 146, 147, 177, 184, 185, 283
  activity relationship (QSAR) 1, 2, 9, 13, 17,
    110, 114, 115, 116, 127, 145, 146, 147,
    177, 184, 185, 283
  permeability relationship (QSPR) 138, 140,
    141, 143, 144, 145, 146, 147, 283
Quantum mechanical methods 144

R

Relative risk matrix analysis (RRMA) 70, 72
Renal disease 217
Renewable energy sources 175
Resource description framework (RDF) 304
Risk 81, 82, 84, 207
  herbal product drug interaction 207
  management strategy 84
  management tool 81, 82

S

Scoring techniques 15, 187
Screening, machine-learning-based 127
Sensory assessment tool 61
Signal 45, 155
  processing 45

transmission 155
Signaling cascades 197, 280
Singular value decomposition (SVD) 47
Software 12, 39, 110, 118, 119, 120, 157, 273,
      284, 307
   programs 110, 118, 119, 120
   theory-based 284
   tools 12, 39, 118, 157, 273, 307
Solid lipid nanoparticles 90
Solubility, heterogeneity 228
Spectrophotometric methods 87
Structure 1, 2, 13, 60, 109, 110, 114, 115, 145,
      146, 147, 184, 185, 186
   activity relationship (SAR) 1, 2, 13, 60,
      109, 110, 114, 115, 145, 146, 147, 184,
      185, 186
   based virtual screening (SBVS) 186

T

Techniques 6, 12, 15, 16, 45, 60, 82, 87, 98,
      126, 127, 138, 140, 141, 146, 147, 173,
      188, 199, 284, 301, 302
   data mining 301, 302
   imaging 188
   implemented systematic 98
   machine learning 12, 141, 284
   multidisciplinary 6
Technologies 188, 195, 196
   imaging 188
   microfluidic 195, 196
Therapeutic 69, 94, 136, 249, 281
   agents 136, 249
   effects 69, 94, 281
Tumour growth inhibition (TGI) 163, 164

U

UV-Vis spectroscopy 141

V

Vaccines 179, 186
   effective antiviral 179
Virtual 12, 186
   physiological human (VPH) 186
   screening, high-throughput 12

W

Web services 210

X

X-ray 10, 46, 281
   absorption spectroscopy 46
   crystallography 10
   diffraction 281

www.ingramcontent.com/pod-product-compliance
Lightning Source LLC
Chambersburg PA
CBHW050808220326
41598CB00006B/149